Genetics, Speciation,
and the Founder Principle

Hampton Lawrence Carson in the rainforest at Pauahi, Island of Hawaii, June 1980. *(Photo by L.V. Giddings)*

Genetics, Speciation, and the Founder Principle

Edited by

LUTHER VAL GIDDINGS
Office of Technology Assessment
Congress of the United States

KENNETH Y. KANESHIRO
University of Hawaii

WYATT W. ANDERSON
University of Georgia

New York   Oxford
OXFORD UNIVERSITY PRESS
1989

Oxford University Press

Oxford   New York   Toronto
Delhi   Bombay   Calcutta   Madras   Karachi
Petaling Jaya   Singapore   Hong Kong   Tokyo
Nairobi   Dar es Salaam   Cape Town
Melbourne   Auckland

and associated companies in
Berlin   Ibadan

Published by Oxford University Press, Inc.,
200 Madison Avenue, New York, New York   10016

Oxford is a registered trademark of Oxford University Press

Library of Congress Cataloging-in-Publication Data
Genetics, speciation, and the Founder principle.
Bibliography: p.   Includes index.
1. Population genetics. 2. Population biology.
3. Species. I. Giddings, Luther Val. II. Kaneshiro,
Kenneth Y. III. Anderson, Wyatt W. IV. Title:
Founder principle.
QH455.G467   1989   575.1′5   88-15249
ISBN 0-19-504315-4

1   2   3   4   5   6   7   8   9

Printed in the United States of America
on acid-free paper

Dedicated to Hampton L. Carson:
"Maybe Mother Nature is trying to tell us something . . ."

in the spirit of William Bateson:
"Treasure your exceptions."

Preface

This book explores the nature and importance of a genetic phenomenon—the founder effect—that is increasingly perceived to be of importance to speciation in a wide variety of plants and animals. Founder effects and their genetic concomitants are the focus of a growing body of research and a number of scientific controversies. The chapters in this book explore this research and some of the controversies.

Intellectual antecedents have a strong effect on the work of a scientist. Understanding the scientist's lifework depends on understanding the context of thought during his or her formative years, and in tracing the development of the scientist's work as thinking in the field changes. Interactions with other scientists—the influences of collaborators, students, and critics—all exert major effects on the evolution of a research program. The book begins with an interview with Hampton Carson who discusses the origins and development of his work in evolutionary biology and his contributions to speciation theory. In this interview, Carson responded to the editors questions about what we felt were the most important aspects of his professional work. We have not attempted any detailed historical analysis, but instead present the interview as a conversation with an eminent scientist about his life's work. Carson's account of the other researchers influencing his thinking presents an interesting international array including Theodosius Dobzhansky, Nikolai P. Dubinin, Nikolai Vladimirovich Timofeef-Ressovsky, and S.S. Tschetverikov from the early Russian school of evolutionary genetics; Kenneth Mather and C.D. Darlington from the British school; and Ernst Mayr. Surprisingly, the impact of Sewall Wright came later, stimulated by Dobzhansky's book. From the beginning, Carson makes it clear that his research has been largely self-directed and his interest in evolutionary genetics and speciation based, like Darwin's, in his love of natural history. This was also true of Mayr, Dobzhansky, and many other contributors to the Modern Synthesis. The range of Carson's interests in natural history is apparent in the list of his publications that follows the interview.

Will Provine (Chapter 2) provides a historian's critical perspective on the antecedents of Carson's work. He examines the currents of thought that led to the early stages of Carson's work on founder effects and speciation, and shows that the ideas of founder effect and speciation theory can be traced back much further than is generally assumed. Provine provides a clear analysis of the different lines of thought that have influenced recent studies of speciation and founder effects as the research strategies themselves evolved along complex and sometimes unclear paths.

While the founder principle has primarily been used to explain instances of speciation in invertebrates and birds, it is an important mechanism in other organisms as well. Carr et al. (Chapter 3) show the striking parallelism of speciation and adaptive radiation between silverswords and the Hawaiian *Drosophila*. They raise the intriguing possibility that founder events may have played a significant role in this group of plants. The silverswords' radiation, coupled with the lack of genetic differentiation, is compatible with the transilience model Templeton describes in Chapter 16. The Hawaiian beggarticks have radiated into at least 27 taxa that differ more in morphology and ecology than in allozymes or chromosomes. Ganders (Chapter 4) concludes that the number of genes involved in speciation in this group may be quite small, and speciation may have followed the colonization of new islands and habitats by founder populations. Stebbins (Chapter 5) surveys plant speciation with particular attention to the role of population bottlenecks. He concludes that dramatic reductions in population size accompanied by founder effects are involved in the formation of many recent plant species. These founder effects are sometimes the most important factor in speciation, while divergent selection seems to predominate in other cases.

Perhaps the best-studied examples of speciation in animals come from the multidisciplinary studies of Hawaiian *Drosophila*. Genetic changes at the level of chromosome organization have provided some of the most reliable data on speciation events in this large complex of *Drosophila*, including Carson's best-known paper on speciation and the founder principle, "Chromosomal tracers of the origin of species." The patterns of overlapping inversions among the Hawaiian picture-winged *Drosophila* established by Carson and Harrison Stalker permitted them to construct an elegant phylogenetic tree that subsequently provided the foundation for understanding morphological, behavioral, ecological, and molecular data. Sturtevant and Dobzhansky used this method of chromosomal inversion study nearly 50 years ago to establish what Sturtevant called "honest to God phylogenies." In Hawaii, as in few other places, these phylogenies can be anchored by the geological history of the archipelago and the isolation of most species on one island (or, in the case of the Maui complex, one biological unit). Yoon (Chapter 6) reviews the large body of literature on chromosome evolution in Hawaiian *Drosophila*, including his own contributions to karyology, the possible role of pseudochromo-centers in generating rearrangements, and chromosomal models of speciation and the cytogenetic phylogeny of Hawaiian species. Lyttle (Chapter 7) directs his attention to a new mechanism that may influence the evolution of karyotypes. He raises the intriguing possibility that meiotic drive may fix chromosomal rearrangements in populations, leading to the production of new karyotypes. While having no necessary connection with founder events, this phenomenon could lead to a pattern of local differentiation like that produced by varying selection in a heterogeneous environment, or that produced by drift and selection following founder events.

Until 20 years ago evolution was studied with the classical methods of morphology, physiology, zoogeography, and genetics. The use of gel electrophoresis opened questions of genetic variability to direct experimental measurement at the level of proteins, the end products of gene action. Questions of gene activity and gene regulation are now testable. Recombinant DNA techniques have moved us another step so that we may directly assay genetic changes at the DNA level.

Hunt et al. (Chapter 8) have applied three different techniques to the problem of DNA variation among Hawaiian *Drosophila*. They have studied nuclear DNA by DNA/DNA hybridization, by restriction mapping, and by nucleotide sequencing in the ADH region. They used their data to construct a phylogenetic tree of five species from the planitibia subgroup, a phylogeny based on relationships at the level of DNA itself.

Dickinson (Chapter 9) explores the effects of regulatory differences on evolutionary change. He has detailed the effects of regulatory elements positioned *cis* and *trans* to the structural gene for alcohol dehydrogenase. He has mapped the patterns of distribution of these regulatory elements in related species, constructed phylogenies, and tried to abstract evolutionary lessons from these patterns. He provides an analysis of the advantages and difficulties in using such patterns to make evolutionary inferences. He also considers the different roles structural and regulatory genes may play in evolution, and shows how developmental biology may contribute to evolutionary thinking.

The founder–flush theory of speciation is consistent with data on speciation in Hawaiian *Drosophila* as well as in some other organisms. It differs in many ways from the classical model of speciation (i.e., the gradual accumulation of genetic differences in allopatric populations followed by the establishment of barriers to gene flow). This classical model of allopatric speciation, strongly endorsed by Mayr, applies to a broad array of organisms. Nevo (Chapter 10) provides an example of speciation that appears to fit this model of allopatric speciation. He concludes that founder effects and genetic revolutions have not been an important element in the formation of *Spalax,* a fossorial rodent. Nevo's example serves to remind that founder events are not always involved in speciation.

Differences in reproductive behavior are often seen among species that appear to have formed by founder events. Thus, it is no surprise that sexual selection and ethological isolation are often among the consequences of founder events. Powell (Chapter 11) reports on an on-going experiment with *Drosophila*. His data reveal a stochastic pattern in the ethological differences that have arisen in lines derived from a common ancestral population via a series of founder–flush events. He also provides a synthesis of the controversial interpretations of asymmetrical mating preferences. Heed (Chapter 12) analyses data from Sonoran desert species of *Drosophila* in terms of Carson's theory of founder events, and reports on an extended effort to discover the results of founder effects in continental populations. His results suggest that founder effects and resulting asymmetries in courtship behavior could also play a role in the evolution of continental populations, although this is clearly not as important as it is with insular populations. Heed builds upon Kaneshiro's theory to test for the effects of founder events in speciation among populations and emphasizes transilience effects of the sort Templeton has described, rather than invoking simple bottlenecks. Kaneshiro (Chapter 13) interprets asymmetries in mate preference in terms of Carson's model. His theory of the evolution of differences in courtship behavior ties together in a new way a number of disparate observations in population and behaviour genetics extending back to R.A. Fisher and H.J. Muller.

The importance of founder effects in speciation must be reflected in genetic changes that occur in natural populations. Yet, how generally applicable founder models are in speciation is a subject of debate and discussion. Neel (Chapter 14) shows how founder–flush events may have played a role in the differentiation of human populations. The demographic studies he and his group have carried out provide a powerful demonstration of the biological and cultural events that can be involved in the fission and fusion of small tribal populations. Ohta (Chapter 15) has interpreted experimental data on oviposition behavior from the grimshawi group of Hawaiian *Drosophila* in terms of Carson's founder–flush theory and Templeton's closely related theory of genetic transilience. Templeton (Chapter 16) presents an experiment on founder effects and speciation that exploits the unique possibilities of examining founder events through parthenogenetic *D. mercatorum*. Templeton confirms that behavioral isolation may evolve rapidly, and he interprets these results in terms of Carson's theory of founder effects as well as his own theory of genetic transilience. He uses genetic analysis to dissect genetic differences involved in premating isolation.

In the concluding chapter, Carson provides a summary of his ideas on the role of founder effects in the formation of species. He organizes his synthesis as a series of succinct comments on the major principles associated with founder events and speciation. The broad reach of these principles and Carson's commentary on them exemplify his long interest in the genetic basis of speciation.

The chapters in this volume were presented at a symposium honoring the work of Hampton Carson. The International Symposium on Genetics, Speciation, and the Founder Principle was held in Honolulu, Hawaii, from June 5 to 7, 1985. The focus on speciation and the Founder Principle is in keeping with the dominant themes of Hamp's long and productive career as a scientist. The broad range of topics related to his major conceptual contributions is testimony to their importance in modern evolutionary biology.

This volume would not have been possible without the cooperation and encouragement of many people. We are indebted to those who made possible the symposium at which these chapters were originally delivered. These include James R. Barnes, Geoffrey Ashton, Joyce Sato Kurihara, Kelvin Kanegawa, Glen Shiraki, Al Ohta, Betty Kaneshiro, Daphne Rallita, and Joanne Chow Nakamoto. Generous financial support for the symposium was extended by Dean Terence W. Rogers of the John A. Burns School of Medicine at the University of Hawaii, the Offices of the President and the Manoa Chancellor of the University of Hawaii, John and Maile Kjargaard, and especially by Alvin Yoshinaga and the Yoshinaga Foundation.

An additional measure of gratitude is due those who made the symposium a success by attending. This applies especially to the invited speakers, who cooperated by amending their papers in light of the fruitful exchanges and cross fertilizations that took place during the symposium and the field trip that followed.

A final measure of gratitude is due to those who reviewed and critiqued the papers herein contained. More than 100 reviewers provided critical comments and suggestions for improving these papers, notably strengthening this book.

Washington, D.C.	L.V.G.
Honolulu, HI	K.Y.K.
Athens, GA	W.W.A.
January, 1989	

Contents

Contributors

Wyatt W. Anderson   Department of Genetics, University of Georgia, Athens, GA   30602.

John G. Bishop, III   Department of Genetics, University of Hawaii, Honolulu, HI   96822.

Laura M. Brezinsky   Department of Genetics, University of Hawaii, Honolulu, HI   96822.

Gerald D. Carr   Department of Botany, University of Hawaii, Honolulu, HI   96822.

Hampton L. Carson   Department of Genetics, University of Hawaii, Honolulu, HI   96822.

W. Joseph Dickinson   Department of Biology, University of Utah, Salt Lake City, UT   84112.

Fred R. Ganders   Department of Botany, University of British Columbia, Vancouver, B.C., Canada   V6T 2B1.

Luther Val Giddings   Agriculture and Rural Development, The World Bank, 1818 M Street, N.W., Washington, D.C.   20433

William B. Heed   Department of Ecology and Evolutionary Biology, University of Arizona, Tucson, AZ   85721.

Kathleen A. Houtchens   Department of Genetics, University of Hawaii, Honolulu, HI   96822.

John A. Hunt   Department of Genetics, University of Hawaii, Honolulu, HI   96822.

Kenneth Y. Kaneshiro   Hawaiian Evolutionary Biology Program, University of Hawaii at Manoa, Honolulu, HI   96822.

Donald W. Kyhos   Department of Botany, University of California, Davis, CA   95616.

Terence W. Lyttle   Department of Genetics, University of Hawaii, Honolulu, HI   96822.

James V. Neel   Department of Human Genetics, University of Michigan, Ann Arbor, MI   48109.

Eviatar Nevo   Institute of Evolution, University of Haifa, Haifa, 31999, Israel.

Alan T. Ohta   Hawaiian Evolutionary Biology Program, University of Hawaii at Manoa, Honolulu, HI   96822.

Jeffrey R. Powell   Department of Biology, Yale University, New Haven, CT   06511.

William B. Provine   Section of Ecology and Systematics, and Department of History, Cornell University, Ithaca, NY   14853.

Robert H. Robichaux   Department of Ecology and Evolutionary Biology, University of Arizona, Tucson, AZ   85721.

Farideh Shadravan   Department of Genetics, University of Hawaii, Honolulu, HI   96822.

G. Ledyard Stebbins   Department of Genetics, University of California, Davis, CA   95616.

Alan R. Templeton   Department of Biology, Washington University, Saint Louis, MO   63130.

Martha S. Witter   Department of Botany, University of Hawaii, Honolulu, HI   96822.

Jong Sik Yoon   Department of Biological Sciences, Bowling Green State University, Bowling Green, OH   43403.

I

HISTORICAL PERSPECTIVE

1

Hampton Lawrence Carson: Interviews Toward an Intellectual History; List of Publications from 1934 to 1989

WYATT W. ANDERSON, KENNETH Y. KANESHIRO,
AND LUTHER VAL GIDDINGS

Hampton Carson's scientific work spans five decades of active research on genetics and evolution. The genetic processes involved in speciation have been a continuing theme over the course of his work. The development of his ideas and the influences of earlier workers on his thinking are difficult to assess from the published record alone, since the space devoted to such matters in scientific papers is generally small. This symposium honoring Hamp's contributions to biology seemed a good occasion to explore his growth as a scientist, and to this end we arranged to interview him. Hamp responded to questions about his development as a scientist in a long session with Wyatt Anderson on April 27, 1986 and in a shorter telephone conversation with Anderson on September 26, 1986. The questions were drawn from a review of his research by the editors. The interviews were recorded on tape and transcribed. The material as presented below has been edited, selected, and arranged by the editors, Hamp having made changes to correct minor errors of fact in the transcript of the long interview. We present the interviews in the form of a conversation with Hamp. "A" indicates a question from Anderson, and "C" a response from Carson.

EARLY INTERESTS AND GRADUATE STUDY

A: In looking over your early work, I noticed you had a paper [Carson, 1935] on medicinal herbs, one [Carson, 1940] on the red crossbill, and one [Carson, 1945a] on an interesting point about late fertilization in a snake, which sounds very much like the modern interest in sperm storage. Your interest in biology has been strongly influenced by an interest in natural history, I take it?

C: Yes, that certainly is the case, and one of the most important influences was the Delaware Valley Ornithological Club which was at the Academy of Natural Sciences in Philadelphia. As a freshman student I met John Hess, a young fellow in my class at college who suggested that we go birding together. I became a bird watcher with him, and we went to the meetings of the DVOC. It was a very interesting group of people, most of them amateurs who shared an interest in birds. That was the way I really got interested in natural history. In fact, when I started graduate work, I had the idea that I wanted to work on birds, and I went down to Washington to visit Alexander Wetmore. I had concocted an idea that I would work on the salivary glands of birds because very little was known about them. I don't know who suggested that to me, but Wetmore let me into his office and was extremely gracious. We discussed all he knew about the salivary glands of birds, and salivary glands of woodpeckers in particular. I remember that his office was extremely long and was built like a set of cases to keep specimens in. He had a card file—rather a file-folder file—extending maybe a hundred yards down a long corridor in the Smithsonian. I told him that I was interested particularly in the salivary glands of the flicker, because it picks up ants by a mucoid substance that gets on the tongue. So he went to look up woodpeckers, and he pulled out this drawer and here was everything known to man on woodpeckers, including a whole file on Woody Woodpecker.

A: Did your interest in natural history begin in high school or earlier?

C: Oh, a little bit earlier. There was a microscope. My great-grandfather, Joseph Carson, was a biologist at the University of Pennsylvania. He was a professor of Materia Medica there, a botanist, and he had a tiny brass microscope which was really an antique. Somehow or another, it came down to my father, and he gave it to me. I decided I wanted to look at things under the microscope, and there was a set of slides that went with it. I had not had any biology in school at all, but I did look at things through the microscope. I captured a wasp and looked at the sting under the microscope and things like that.

A: Hamp, perhaps we could discuss your graduate study at the University of Pennsylvania. In looking at your publication list and your early papers, I have noticed that you worked with McClung, Metz, and Whiting, who constituted a distinguished group in early genetics.

C: Whiting gave me my formal genetics training in his two courses. He gave two semesters of it but was almost completely unintelligible as a genetics teacher. He was so clever and so erudite in genetics that he immediately jumped far ahead of all of the students in the class, and it was completely mystifying. He would say, "Today I'm going to discuss zero ratios," and then he would launch into three-quarters of an hour on something, while we didn't know what he meant by zero ratios. He never stopped to clarify that sort of thing. Anna Rachel Whiting was quite aware of that and also of the fact that Phineas was not communicating too well with the students. Once in a while, when he was sick, she would come in and give lectures, and it was absolutely marvelous. She was absolutely clear as a bell. In fact, she overcompensated, I think, so that when she gave a lecture you understood every single point.

A: I take it that you began cytological work with McClung, and then, when he retired, changed to work with Metz.

C: Yes, that's right. McClung gave a course also but he was very aloof and did not really work with students very closely; what he did was to give you a problem in the Louis Agassiz system of throwing you a fish. He gave you a problem to work out, a very detailed, long problem. You were supposed to spend weeks working the problem out and then come back and write a paper for him about what you found. It was quite interesting. I started working on the apical cell of the insect testis, which is a cell near the spermatogonia, and McClung had the idea that there was some sort of influence of the environment, that the cellular environment influenced the germ cells in some manner. He was kind of a Lamarckian, and he thought that this was the Lamarckian influence. The experience of the individual was being recorded in the apical cell and that was being transmitted to the spermatogonia. Therefore, he said, "Carson, this is a very interesting problem. I want you to study this." So I did almost two years on the apical cell as a beginning graduate student. Then McClung announced he was going to retire and that I'd better look for somebody else to do my thesis with. He never even looked at what I had done on the apical cell. He was a very aloof sort of person, and that is one of the reasons I spent seven years in graduate school—the first two years really produced nothing.

A: But you did get a paper [Carson, 1945b] on the apical cell?

C: Well, I put together the observations, yes, and published after I had done my other work. I still had the things there, and I enjoyed that old-fashioned kind of cytology where you get the testis out and section it and do the different kinds of stains and look for chromosomes and cellular relationships and so on.

A: But your doctoral work was on *Sciara*, because I know your paper [Carson, 1946] on inversion heterozygosity and dicentrics. As far as I know, there is only one other paper [Hinton and Lucchesi, 1960], in *Drosophila*, which actually shows the cytological consequences of crossing over in inversion heterozygotes.

C: Well, actually that paper on the inversion crossovers was done after I went to Washington University. My thesis was on the same species, *Sciara impatiens*, but it was mostly on the geographical distribution of inversions. This is the one that was published in the *Journal of Morphology* [Carson, 1944]. But the paper on dicentrics came along later.

A: Did you have a close relationship with Metz as a mentor?

C: Not particularly. But he was very sympathetic. He was much more approachable than McClung, for example. I would frequently go in and tell him how things were going, and he kept track of what I was doing. This pleased me very much because I had felt a little bit as if I were entirely on my own under McClung's sponsorship.

A: Coming back to your experience as a graduate student, is there any person who you feel served as a real mentor in the sense that's talked about now, as someone who guided you into your scientific work?

C: Yes, I would say Bob Stabler was very important in that. He was my instructor in Elementary Zoology at the university.

A: At Pennsylvania?

C: Yes, and he was very interested in natural history. He was interested in falconry and gave a very dramatic section of Introductory Zoo. He was always bringing in all kinds of things that he had found in his back yard. He was so dynamic about his approach that he did a lot to spark my interest in it. I remember that at the time I took Elementary Zoology I was already a junior and was on the track to go into the law profession. My father was a lawyer and I had the idea I was going to be a lawyer also, but that course caused me to begin to change my mind, although I was a junior and I had to change my major. The fact that I was a junior meant that I had a lot of courses to take in order to get enough zoology to graduate with a major in zoology. When I finished, I was really still uncertain whether I wanted to be a professional biologist, but I went to Woods Hole that summer and took a course. That helped settle it for me.

A: And your association with Stabler continued after you became a graduate student?

C: Yes. He was a parasitologist at the University of Pennsylvania who worked on protozoan parasites. He eventually went to Colorado College and was a professor there for many years. He became a personal friend, and when I was a graduate student, why, I lived in the same area outside of the city in an old, rented farmhouse.

A: Who influenced you to be interested in evolutionary biology? In your paper [Carson, 1980], "Hypotheses That Blur and Grow," you mention a visit by Dobzhansky that spurred your interest in the evolutionary side of the modern synthesis. Would you elaborate?

C: Well, actually it was before that. Dobzhansky did make a very exciting visit to the lab.

A: When was that?

C: It was a year or so after he gave the Jesup Lectures—a year or so after the publication of *Genetics and the Origin of Species* [Dobzhansky, 1937], which I think was 1937. I think it was in 1939 that he came to the Penn lab, and, oh, it was marvelous! He walked into my lab and I had an inversion configuration from *Sciara* under the microscope, which I thought was a transposition. In other words a section of bands was not in the right place. It was moved down the chromosome, but the rest of the chromosome was in the expected order, and he said, "Oh, Carson, that is two inversions, but they're overlapping one another." He made a diagram to show me how that worked.

But actually it was Metz who first interested me in evolutionary biology. I was studying some birds and bird distributions and animals and plants and nature, so that the idea of applying genetics to natural populations was already in my mind. It may have come from reading Dobzhansky's book. I guess that really is where it did come from.

A: Were there other authors writing about evolution or the modern synthesis, the early population geneticists like Fisher and Haldane, who influenced you?

C: Well, I did not read Fisher. For some reason I didn't get to the Fisher [1930] book very early. I think I read Fisher only after the courses with Whiting, but I did read Haldane's [1932] book at a very early stage.

A: This is *Causes of Evolution?*

C: Causes of Evolution, but I think mainly it was *Genetics and the Origin of Species* that was most exciting to me. The fact that you could put genetics together with evolutionary study, and especially natural populations, and get somewhere with it was what interested me the most.

A: Was your interest stimulated by personal contact with Dobzhansky, or by his book coupled with just an occasional visit?

C: I think it must have been the book. When he came and I met him, that was a tremendous stimulus to me because I was already familiar with the book, and the other students in the department were talking about Dobzhansky's work, and these polytene chromosomes in particular. And I had already, at Metz's suggestion, taken up the polytene chromosomes in *Sciara* as my thesis. When Metz arrived—you see, Metz came to replace McClung as chairman of the Department of Zoology at Penn—he looked around to see what the students were doing. McClung had said he was going to retire, and so I was without a sponsor; so that after a rather short while, we decided that I would study this one species. It was not exactly a geographical study because I had to get the species from greenhouses. Therefore, I did travel around the eastern part of the United States, visiting greenhouses and catching these flies.

A: I am interested in exploring whether Sewall Wright's work had a strong influence on your early work, or whether it came through reading Dobzhansky's treatment of Wright, or what?

C: Yes, it really was Dobzhansky's treatment of Wright.

A: Did you ever come to know Sewall Wright better?

C: Not ever well, but he and I used to sit down and have conversations at various meetings, brief conversations. I remember several times he came to hear short papers that I was giving. Once, he said to me: "That's a very promising approach." It was very exciting to hear him make a remark which indicated that he understood what I was trying to do, although he never really opened up. I never really tried to explore ideas with him.

HARRY STALKER AND *DROSOPHILA ROBUSTA*

A: You took a job at Washington University after graduating from Pennsylvania, and your list of publications indicates that you must have quickly begun work with Harry Stalker on *Drosophila robusta.* Could you expand on this?

C: It was just delightful to find Stalker installed at Washington University. He had been there, I think, six months when I arrived. He was already looking around for another species of *Drosophila* to work on, because he had worked on *americana* for his thesis [Stalker, 1942].

A: Whose student was he?

C: He was a student of Curt Stern at Rochester and also, in some respects, a product of Warren Spencer at Wooster College. The problem he did for his thesis was inspired more by Spencer than by Stern, but Stern was so generous in the way he treated people, that he let Stalker do just what Stalker wanted to do, and he did a beautiful thesis. But our cooperative work was begun very soon. I did work first on this *Sciara* problem, which you mentioned earlier. Then Stalker was in correspondence with Sturtevant, and Sturtevant said, "Why don't you work on *Drosophila robusta*—because it has inversions, I know it has inversions in it." Stalker at that time was interested in morphometrics. So he said to me, "Why don't you do the inversions?" And, because I had done inversions in *Sciara*, that appealed to me very much. So we set to work on *Dorosphila robusta* from that point on.

A: Did Stalker himself know chromosome techniques at that time, or did you teach him?

C: I taught Stalker chromosome techniques, and for a long time I did all the cytology and he did all the morphometrics. Then he slowly changed. He became more and more interested in some of the genetic aspects of comparing salivary gland chromosomes, and then he took over that role, such as the photographic techniques, which he developed entirely.

A: You mentioned earlier that a part of your interest in *robusta* had been stimulated by Sturtevant. Would you say a little more about that?

C: Yes. Sturtevant had corresponded with Stalker. I think Stalker had asked him what species he ought to be working on, and he suggested *Drosophila robusta*. But Stalker did not particularly want to do the cytology and he said, "Why don't you do the cytology?" I agreed to try the cytology, so Sturtevant really helped us locate this particular problem. But after I had been working on it for a while, Sturtevant sent an unpublished manuscript which he had prepared on the salivary gland chromosomes of *Drosophila robusta*. It was not the correct name—I think it was *sulcata* he was calling it. But anyway, he had depicted some of the inversions, and I thought it was an extremely generous gesture. I never forgot that, but unfortunately the letter and remains of the manuscript have long since disappeared. It was to me a gesture of generosity by a great scientist, which Stalker seemingly took for granted. He thought, too, that of course Sturtevant would do that, because Stalker was that kind of person.

A: Would you comment on Harry Stalker's personality? I met him only toward the end of his life and unfortunately never came to know him well.

C: He was a very modest man. I feel that he was extremely modest but very hard-boiled about other people's ideas. Every time I would go to Stalker with an idea, he would find ten reasons why it was useless or worthless, and then I would say, "Well, maybe we could do this kind of a problem, but you'd already have a series of reasons why that wouldn't work." He loved to be very hard-boiled in science, and he was a very severe critic of himself also, of his own methods, his own way of doing things. He was able to self-criticize as he went along to an extraordinary degree, and was a delightful person. He and I got along extremely well. I think anybody would have to be somewhat insane to have not gotten along with Stalker. He was such a nice fellow, and we had a very close, personal relationship. My wife and I and Harry and Marian Stalker used to go out on various projects from the lab which were not at all critical to science.

A: What do you think you learned most from Stalker?

C: I think the quantitative methods—really the way to get enough data to say something. Stalker was one of those people who would do not just a hundred measurements; he would make three hundred measurements or five hundred measurements before he would begin to examine whether the data meant anything. He was not the kind of person to do pilot experiments to get an idea. He would often decide to do something, and we would decide together to do it. We would spend six months at it, and then we would sit down and look at it and there wouldn't be anything there. Stalker would put it all away in the file, and actually we would have lots of information in the files, some of which might be really quite valuable. But if it did not fit the particular notion we had in conceiving the experiment, then for Stalker it simply was not worth publishing. I think that a lot of the cytological data that we obtained is still unpublished. Of course, I think that is true of everybody's career, but he was a tremendous datagetter. He was always down there getting that data. He loved to work with flies. Sturtevant was his idol; Sturtevant was his model of perfect science. He also smoked. There were large clouds of smoke around him always when he worked, the same as with Sturtevant.

A: You mentioned that Stalker had influenced you in terms of experimentation and particularly the quantitative side of doing many replicates and extremely careful experiments. What kind of conceptual influence did Harry Stalker have on your thinking?

C: He did not like to speculate very much. He was not particularly interested in writing a paper that would deal with theoretical topics. He was not a theoretician. He was a data-gatherer, and he let the data speak to him. Then he wrote in his papers what the data said to him, and he did not generally try to tie that into the broader picture. There were several opportunities that came along for talks—symposium talks—and things of that sort. Stalker always avoided that. I think he felt that he just was not comfortable with theoretical notions because he feared that he might be ridiculed by somebody who knew more about it than he did. In other words, he was not willing to—or maybe just did not want to—operate in that area, so that when I started to do that he said, "Oh well, okay, I guess."

A: Did you and he discuss these matters of your early conceptual work or was it mostly your own?

C: I guess it was mostly my own. Of course we discussed these things, but he was always a little condescending with regard to my ideas. He did not think they were really all that well-founded, and he always—as I said earlier—he always had a dozen reasons why an idea or research plan or a theoretical notion was no good, and he would always trot these out for each occasion.

A: I take it Sturtevant had a great influence on your early work, through Stalker, then?

C: Through Stalker, yes. After Stalker wrote Sturtevant and said that Carson is going to do cytology, I was simply appalled that Sturtevant sent me an unpublished manuscript which he had prepared on salivary gland chromosomes of *Drosophila robusta,* including inversion breakpoints. He said, "You take it and do with it what you please. I'm not going to ever publish this and you and Stalker take it and use it in your work on *robusta.*" It was really a very generous suggestion.

A: So you utilized a bit of this in your early work?

C: Actually, at the time I could recognize what we were seeing as some of the same inversions that Sturtevant had seen, and so we did not actually use it in that way. We had our own terminology and worked it up independently. But he had made a start and discovered a number—I do not remember how many—of inversions. I tried to find that letter and manuscript from Sturtevant recently in my files but was unable to. I think it might be at Washington University still. I left many of the *robusta* things there.

A: Working with any person for a long period of time must create some difficulties. I know that when I first began studying population genetics, your name and Stalker's were closely linked in the *robusta* work, and then of course your work took off from there on more of a conceptual vein involving gene pools and genetic organization and speciation. I do not know if it is fair for me to ask, but was there any difficulty between you and Harry in terms of two good scientists who worked together for a long time?

C: Well, I think it was an extremely pleasant association with Stalker. He and I were very close personal friends as well as colleagues. In the work that we shared, I did the cytology in the early days and he did the morphometrics, on *Drosophila robusta,* and we worked together on the ecology. It was a lovely experience with Stalker. As a matter of fact, you know, we discovered the breeding site [Carson and Stalker, 1951] of *Drosophila robusta* on the sap exudations of the American elm, and the way it happened was quite characteristic. We were out on one of our field trips, and there was a large elm tree. We were scraping around looking for what we might find in the way of breeding sites. We had brought in a lot of humus and such, which had never produced anything. He went to one side of the tree and he scraped a little "goo" from the base of the trunk, and I went to the other side of the same tree and scraped a little goo from the trunk. After we got back to the lab, we found that there were *Drosophila robusta* larvae in both of these samples, so that actually we made a simultaneous discovery of the breeding site of *robusta*.

PROFESSIONAL INFLUENCES

A: I am curious as to what influence the early Russian work in population genetics had on you. Were you aware, for instance, of Chetverikov's [1926] paper, or of Dubinin's work?

C: Dubinin certainly, because he did the work on different populations in the neighborhood of Moscow—with *Drosophila funebris*, I think it was.

A: This would be with inversions?

C: It was with inversions, yes, and I was quite familiar with that. Dubinin and Tiniakov had a paper in the *American Naturalist* [1946], which I think was probably solicited by Dobzhansky; Dobzhansky had something to do with getting that paper out of Russia and getting it where we could see it. It involved urban versus rural differences in chromosome frequencies, and there was some information on inversion frequency changes during overwintering. That was the main effect that the Russian work had on me, through Dubinin and Tiniakov's paper. In fact, Chetverikov was really Dobzhansky's mentor, in a way.

A: I am interested in your comment that Chetverikov was Dobzhansky's mentor. Would you expand a little on that?

C: I really don't know about that, but Dobzhansky would always say . . . of course, I knew him fairly well . . . "You know, Chetverikov was making studies of mutant frequencies and so on, long ago in Russia." As to just how much influence, I rather imagine that he had quite a lot of influence on Dobzhansky, and that seems to be borne out by the articles in the Mayr–Provine volume [Mayr and Provine, 1980]. But I never read Chetverikov's papers for myself. A very important paper was one by Dubinin [1948] on *network*, on the *network* gene. It was a very large piece of work in which he selected different strains of *Drosophila melanogaster* for the expression of *network* and came to the conclusions that there was a large polygenic influence on such characters and that a great deal of variability must be for the quantitative, small genetic effects. That had a great effect on me at the time, in my thinking anyway, because it was quite obvious that these inversions just did not directly adapt the organism to particular facets of the environment. They had some much broader significance that may lie in these genes of minor effect.

A: I take it that Dobzhansky is the one who brought Dubinin's and the other Russian scholars' work to your attention, and perhaps to that of others as well?

C: I think so, and I guess it was Michael Lerner, who, through his association with Dobzhansky, translated the large paper by Dubinin

[1948] on *network*. I think it was Lerner who sent me a copy of the paper with a penciled remark: "It's lovely!"

A: Did you learn about inversions through Dobzhansky and his book, or his papers?

C: It is very hard to remember exactly how I got started on the inversions, but I think that was largely via Dobzhansky, and through Metz also. When Metz suggested the problem for my thesis work, he was already aware that there were inversions in some of the *Sciara* species. And although I believe *Sciara ocellaris* had no inversions, *Sciara impatiens* did. I guess it was jointly through Metz and Dobzhansky and the fact that I needed a problem that was different from the apical cell.

A: Metz must have been aware of Dobzhansky's work on the population genetics of inversions?

C: Yes. He was not really very interested in population genetics, however. He was more interested in chromosome structure, and he studied the polytene chromosomes with the idea that these giant chromosomes would finally give us a clue to the nature of the chromsome. It was rather a new idea that we should study geographical distribution within one species, *Sciara impatiens*. I do not know quite where that idea came from, whether it was my idea or whether I had borrowed it from reading Dobzhansky. I think it was probably the latter—reading Dobzhansky's book and seeing what he had done with *pseudoobscura*.

A: I would like to go back just a little bit to talk about your early work on recombination in populations. Were you influenced by the work of Kenneth Mather at all?

C: I certainly was, and I think that the Mather and Harrison [1949] papers (I guess that's 1949) were to me—and still are—very important papers. They treat the whole idea of balance and how genetic variability can be held by balance in both experimental and laboratory populations. His [Mather's] paper on what is called "relational and internal balance" had a very great effect on me. I mentioned it to Dobzhansky several times, but he was a bit scathing about it. I never could quite understand why Dobzhansky did not think that Mather's ideas were so good. Maybe it was that some of Mather's ideas did not last, like the idea that oligogenes and polygenes were utterly different.

A: Did Michael Lerner's [1954] concept of genetic homeostasis play a role in your thinking?

C: Yes indeed, it did, and it was important in setting up one [Carson, 1959] of the three Cold Spring Harbor lectures that I gave. I talked about certain heterotic properties and what I called homoselection and heteroselection.

A: Yes, I remember that.

C: I played around with that, of course, all very strongly influenced by Lerner.

A: Let me ask a question about another person who might have influenced you. Edgar Anderson, the botanist, was prominent at Washington

University when you were a faculty member there. Did he and his work influence you?

C: He certainly did. He taught a course which he called Genetics and Natural History, and he was giving that when I was hired at Washington University, in January of 1943, I think. He gave this course the next year and said, "Why don't you come in and take my course?" and so both Stalker and I took Anderson's course. Stalker did not like Anderson's very disorganized way of going about things, but I got a great deal more out of the association with Anderson than Stalker did. Anderson was great on looking at the natural population situation in nature, especially in plants, and seeing the variation there and trying to interpret to what degree that was due to hybridization, or so on. It was his method of looking at natural populations which made a strong impression, and that was part of the reason I used that same approach with *robusta,* because again it was a natural population situation we were looking at. He was a very exciting person to take a course with.

A: Did you talk with him often about professional matters?

C: Yes, and as a matter of fact we had planned a paper together which never quite materialized. It is really too bad, because I thought it was quite a good idea. He was particularly interested in the way in which one went about making observations. His favorite advice to give his graduate students was, "When you go out to your study area, I want you not to look at one of those plants which you are studying for the first half hour. I want you just to lie on your back in the grass and look up at the clouds and watch the clouds go by, for half an hour. Time it for half an hour. And then slowly allow your mind to come around to your plant." His whole idea was that somehow you had to clear your mind of all kinds of preconceptions. Several people have told me that this was Anderson's approach, and when they went out and tried it they thought, "Well, it's really pretty bad; I'm not learning anything this way." But according to Anderson, it was just at that time when he noticed something about the plants which he had not noticed before.

A: Let me ask about another person. We have talked about the influence many people might have had on your ideas. What about Darwin's *Origin of Species?* When did you read that, and what effect did it have on your thinking?

C: I guess I read that as a graduate student. There was a reading course for graduate students at Penn, and the *Origin* was one of the books we were supposed to read. We were supposed to read some works of Von Frisch, and this and that and the other thing, and one of the books we were supposed to read was the *Origin of Species.* I think it was a very good idea which we have not continued in many graduate schools. We also had to write book reports. I wrote one on the *Origin of Species,* I remember, and so I went through it at quite an early time. And, of course, Darwin's naturalistic approach was the kind of thing I was trying to develop.

A: We were talking about the influence of Darwin on your work.

C: As I said, he took a naturalistic approach to observation of populations, which of course Mayr [1942] also strongly advocated in *Systematics and the Origin of Species*. The Mayr–Darwin approach was very strong, and I had the feeling all along that we should put genetics to work to get to the bottom of some of these problems. That was my main thrust in those early days.

A: And, Dobzhansky's approach was more recognizably genetic to you, and experimental, I would guess.

C: Oh, yes indeed. It was definitely genetic, and that is what was unique in that regard.

A: Did E. B. Ford have much influence on your ideas?

C: Well, I read Ford's papers, but I think I was more impressed with the elegance of what you could do with inversion frequencies. You could use selection coefficients, and you could ask whether they were obeying the Hardy–Weinberg law and such. You could not do that with Ford's work because it dealt with quantitative variation. Actually, I think the approach to quantitative inheritance is really an extremely important one which we should be making, even today. However, there was a certain elegance about the homozygotes and heterozygotes for inversions which just transported all young workers at that stage.

SPECIATION AND FOUNDER EFFECT

A: I would like to turn now to the development of your ideas about speciation. Where did you learn first about speciation in the sense of the modern synthesis?

C: Oh, the way I look at it now, I am not sure just where that came from. I guess it is with Mayr. Really, I was probably more influenced by Mayr than anyone else, and by reading *Systematics and the Origin of Species* [Mayr, 1942]. The origin of species through marginal populations, and marginal subspecies which became species, and superspecies and such was mostly due to Mayr, because Dobzhansky did not operate in that particular area as much as Mayr did. I think my debt to Mayr is very great, and of course I have always been a very strong "allopatriot." I could never quite understand the arguments about sympatric speciation. Stalker did not either, but we didn't discuss that very much. As I said, he was not particularly interested in long harangues on philosophical topics of what might be or could be and so on. Stalker liked to know what was really there.

A: When did you first meet Mayr and hear him speak about his evolutionary concepts?

C: I think it was in Philadelphia. He came to give a talk at the Philosophical Society, or was it Chapin who came to Philadelphia and gave a talk at the Philosophical Society and talked about Mayr? I think that was it. I did not meet Mayr until much later, and meeting him in person was somehow relatively unimportant to me because I had dug through his writings so much. His writings were clear enough, so it was pretty clear what he was saying in the books. It was just a matter of meeting him at a lecture and saying, "Glad to know you." We never really sat down and talked about science very much.

A: When did you read Mayr's 1954 article about founder effect and genetic revolutions?

C: That was just prior to one of the Cold Spring Harbor talks which I gave. I gave three talks at Cold Spring Harbor in the period around 1955, and it was about that time when I became interested in the geographical idea of the marginal populations, the ebb and flow of populations at the margin that might be related to speciation. I read Timofeef-Ressovsky's [1940] paper in *The New Systematics,* which was very important to me. And as for Mayr's 1954 paper, I read it about that time, so that when I gave a talk at Cold Spring Harbor [Carson, 1955], I had read it then.

A: Did it have a big influence at the time or was it like any other work on speciation or genetics?

C: It seemed that the revolution idea was a novel one, and one that I really tried to build on. I guess Mayr thought I was in a sense stealing the idea from him. One time he said something about that, and I said, "Look, when you have a good idea, it's an honor to have somebody try to build on it." I told him that what I was trying to do was just to work out the genetic details. Actually, I have gotten along pretty well with Mayr over the years. He occasionally feels that I am just repeating what he has always said, and that may be true. I remember one little tangle. I do not remember what meeting it was—maybe it was one of the Cold Spring Harbor meetings—but Mayr was just a little annoyed that my ideas were so close to what he was saying.

A: You mentioned to me that you had written a fairly long letter to Mayr detailing some of your concerns about the way he had represented your theory in his book, *Animal Species and Evolution* [Mayr, 1963].

C: Yes.

A: Would you say something about that?

C: He cited me in a number of places, and, as we all do, I looked at the citations and was rather dissatisfied with some of the ways in which he had reported my work. I did not think the emphasis was of the kind that I had given it originally, particularly concerning the genetic constitution of the marginal populations, that is, the inversion frequencies and how they were interpreted. I had interpreted them rather differently from the way Dobzhansky had, and I found that he had reported them in a different fashion.

A: I want to pursue this idea of speciation in association with Mayr because it is a major part of your work, a part of your work that clearly is going to have an influence for many decades in evolutionary biology. How do you think your ideas about founder–flush speciation differ from the original idea of Mayr in his 1954 paper?

C: I think they differ in the kind of genetic changes that one might expect to occur during the founder effect. Mayr really did not spell out in any kind of detail what he meant by the genetic revolution. That is, there was a rather vague idea that the genes might be in some new genetic context due to recombination, but beyond that there was no real attempt to outline the details of what might happen under these circumstances. That is what I tried to do, and in a recent paper [Carson and Templeton, 1984], Alan Templeton and I tried to make a distinction between Mayr's original founder idea, Templeton's transilience idea [Templeton, 1980], and my old idea of the founder–flush [Carson, 1968]. They really have quite different characteristics. Each of them has a slightly different expectation from the genetic point of view, and that is again an area in which there are lots of polemics and lots of argumentation, for a point of view without quite enough data. I think that what we desperately need is to have more really good data on these things. I have always felt that it is amenable to experimental study, and that is the direction that it should go.

A: You mentioned a letter from Mayr that stimulated you to begin some thinking about speciation in relation to your *robusta* work. Would you say something about that?

C: Yes, it concerned the idea that the peripheral population was one that was, in a sense, cut off from the main population. Using natural history observations, he had noticed that many of the birds on the mainland of New Guinea had separate subspecies on the peripheral islands. He felt that a genetic study of the peripheral populations would be very important for species origins, to see how those changes became incorporated into the populations. He wrote me a letter in which he said, "Young guys like you who are studying inversions and morphometrics ought to look at the populations at the extreme periphery." So he specifically suggested an avenue that my work took for ten years or so, I guess, or maybe even more; I am still to some degree concerned with those things. I and some of my students, especially P. S. Nair and Satya Prakash, made excursions to extreme edges of the range in Florida or in North Dakota or in Nebraska, I guess it was Nebraska, to try to get the extreme edge of the population—to try to get data on the conditions there. However, in those days we were studying only inversions, and it was only later that this work was followed up by studying the electrophoretic variability.

A: What did Mayr's genetic revolution mean to you when you first read his paper?

C: It was not very clear just what it meant. That is what attracted me to it, because in *Drosophila robusta,* we had these nice marginal populations with particular genetic characteristics, and I thought maybe we might find out whether they were becoming homozygous or what was happening to them. I think my feeling was that there was just not a very good genetic basis in Mayr's original formulation of the idea, and that the contribution that a geneticist could make would be to try to spell this out in some manner. Of course, I decided that there was a lot of genetic variability in the marginal populations. Later on, Prakash's [1973] work showed that they are full of electrophoretic variability, just as much variability in the marginal populations as in the central, so that it was not just a matter of the attrition of genetic variability in the marginal populations.

A: What effect did Prakash's work have on your hypotheses about marginal and central populations—the fact that he found essentially the same level of variability at the periphery as at the center?

C: It was pretty much what it was expected to be, in the case of the electrophoretic loci, which did not seem to show geographical patterns in *pseudoobscura* nor in *robusta*. I never really had the idea—or did not cling to the idea—of homozygosity in marginal populations, which I think I had proposed in one of the Cold Spring Harbor papers [Carson, 1959]. But it is quite clear that these marginal populations were not being homo-selected, and it was, I think, a very important discovery that Prakash made. I think it helped to make us understand what might be happening in the marginal populations better than we had previously understood.

A: Did it cause you to feel more strongly about the importance of recombination and perhaps the role of inversions as regulators of recombination?

C: Yes. Actually I had done an experiment [Carson, 1958] on that very topic. I had selected for motility of adults toward light in *Drosophila robusta* and got very strong responses in the marginal populations—much stronger, actually, and more sustained in the marginal populations than in the central populations. I had the idea that there was genetic variability in the marginal populations, and therefore this seemed to fit with that idea. I had the idea that the central populations were tied up in genetic balances which could not respond to directional selection, whereas the marginal populations were free of inversions and could respond to directional selection. I had that idea before Praskash found his electrophoretic information.

A: Did you connect Ernst Mayr's paper on founder effect and genetic revolutions with Sewall Wright's genetic drift?

C: Yes, but at the time I think I was not at all clear as to just what the distinctions were between the two, if any, and it was only later that I came to look at this in a somewhat different light. I do not know whether Mayr drew anything from Wright on that or not. That would be an interesting problem, maybe peripheral to this discussion, but he certainly did not ascribe anything to Wright. And no, I really did not make a very strong connection between those two things. I certainly understood random drift and selective drift and the interplay between the two as Wright studied them, but I did not relate those to the founder effect in marginal populations, in speciation events, and so on.

A: What about the role of random drift in your theory? Does it fit in, except as founder effect?

C: I think at one time I had the idea that following the founder event one might wind up with a more homozygous genotype through drift, because of the necessary inbreeding following a founder effect. If you are going to have a single founder, then you are going to have inbreeding, and I had originally thought there was a connection there. But as I became more convinced of the enormous variability that is carried through even a single pair, then I gave up the idea that simple gene-frequency drift could have more than just a very, very small part in the idea. You would lose a few alleles—in more modern terms, lose electrophoretic alleles—but you would not, I think, necessarily lose a great deal of genetic variability.

A: Did the concept of effective population number from Wright enter into your thinking?

C: Yes, I think it did, in the sense that populations are always much smaller than they appear to be from the census. But I never utilized that in any kind of sophisticated way at all, and my arguments are almost always of the natural history sort.

A: Sewall Wright had a 1940 paper in the *American Naturalist* which was the report of a symposium on speciation. It is entitled "Breeding Structure of Populations in Relation to Speciation." Did this paper play any major role in your thinking about speciation?

C: I am not sure that I was aware of that paper. I do not remember any particular role that it played.

A: Which papers of Wright did you read that influenced you?

C: "Evolution in Mendelian Populations," in *Genetics* [Wright, 1931], and his paper in the *Proceedings of the Sixth International Congress of Genetics* in 1932. Of course, I did not follow Wright's mathematical formulations with any skill. Like many other people who have gotten into the game, I think I got my Wright by reading Dobzhansky's idea of what Wright said. I must admit that I was not someone who pored over Wright's papers with the idea of extracting from them the essence, so I would not say they had very much effect.

FOUNDER–FLUSH SPECIATION

A: When did the idea of founder–flush speciation first come to you?

C: That I do not quite remember, but I think that I wrote a paper [Carson, 1968] for a little book which Dick Lewontin edited. I have forgotten just when that was.

A: The Syracuse Symposium of 1967?

C: 1967, yes. It was about then that I was thinking in terms of what would happen if a population increased greatly in size and then was suddenly to crash down. Perhaps one might get a kind of attenuation or shift or change in the genetic structure of the population. That would do something to it that might necessitate a change in genetic adjustment, just due to the flush and crash. Of course that is slightly different from the founder effect itself. Probably there is a flush connected with the founder effect at some stage or other, but we are never quite sure whether that comes before or after. And we have not really ever observed such flushes in nature. That is one of those rather theoretical notions that we do not have too many observations on.

A: Could you say something about the relationship between population flush and speciation? Was the idea something of your own creation, or did the idea come from Mayr?

C: The flush idea was rather independent of anything I got from Mayr. It was something that I cooked up. As a matter of fact, I remember asking Dick Lewontin whether he liked that designation, and he said yes, it has a nice juicy sound. So I used the title "Population Flush" in the article that I put in the Syracuse Symposium. That article was, I think, formulated pretty much separately from the idea depending on Mayr. The way I got into that was through the fact that when we establish isofemale lines, we obviously do not cause a big revolution. In other words, a single isolate from a natural population does not seem to give the kind of shift that you might expect in a marginal population. Therefore, I felt that some other element was necessary to set the stage for genetic change to shift selection, perhaps to something else.

A: I would like to ask you about the relationship between your founder–flush ideas and founder speciation, because founder speciation rests on one or a few individuals, whereas founder–flush seems to be occurring in a fairly large population. How did one lead to the other?

C: I think it came about through the isofemale method of handling wild *Drosophila*. You bring a wild *Drosophila* into the laboratory and establish an isofemale line, and it seems to have a great deal of variability in it. A large amount of the genetic variance of the population seems to be in that isofemale line, and my idea was that things are so organized that they even retain that organization within the isofemale line. You are not getting a new species or a founder effect each time you establish a line, and thus I felt that it was necessary to invoke some other idea that would further disorganize or further disturb the status quo in a species before you could get a big break with the past. In other words, you have the large population existing, and if you are going to get something novel and new, I think you have to have some disorganization effect. That was where the flush idea came in—that if the population increases greatly, then you will have the lid taken off selection temporarily and you will have various kinds of odd recombinants that might survive under those circumstances. I developed that idea in addition to the founder effect because I did not think that just the simple founder effect was good enough, because we do that all the time in all kinds of artificial populations without getting new species each time. Actually, I think Dick Lewontin made the point that a great deal of the genetic variance of a population is existing in two individuals.

A: Were you familiar with the Ford and Ford paper of 1930 on variability released by population growth?

C: Yes. I dug that up later—only later, because I am afraid we were rather provincial in our approach in America. We did not really study Ford. Looking back on it, we could have learned a lot by reading some of his earlier work.

A: Where does selection come into the founder–flush speciation?

C: Following the founder–flush, selection resumes and finds itself with slightly new genetic elements to work upon. In other words, the old organization—the old balanced condition—is lost following the founder event and the flush, so that you have mostly a recombinational situation. I am not thinking of mass mutation or anything like that. I am thinking of unusual recombinations that occur following the founder–flush, and I feel that the new species or the new gene system can be carved out of one of these combinations. If you are going to get a new species, it has to be an integrated system. It has to be a population system that is capable of self-perpetuation, having its own variability system, and so on. You can not possibly get that without the intervention of selection to equilibrate things.

A: The selection that you picture is fairly intense, isn't it?

C: Yes, I am inclined to think in terms of quantum periods of change, the "quantum changes"—that is a term that has been used by Simpson [1944] and also by Grant [1977]. It is due to the periods of very rapid selective change followed by the new equilibrium. I am sticking to that formulation fairly well in my most recent formulas.

A: Where did the idea of quantum selection come to you from? Anybody's work?

C: I used a system proposed by Buzzati-Traverso. He had established certain base populations and then had introduced genes into them and gotten rather rapid change due to selection. I set up a model of that with *Drosophila melanogaster,* a vial population modeled on Buzzati's system in which I had a stock of *D. melanogaster* with three mutant genes. I have a paper [Carson, 1961] on it.

A: Did the ideas of Simpson play a role in your founder speciation theory?

C: No, I don't think so. I don't think he had any such real notions. I guess it was Verne Grant who used this quantum phase. I had difficulty finding a good name for periods of rapid change under selection followed by equilibrium. It could be "punctuation," but I avoided that for obvious reasons. I don't want to draw from a paleontological idea to explain an idea in population genetics.

A: If I may, I would like to talk a little bit more about the origins of your founder speciation ideas. Can you place the time when you moved from the founder–flush population ideas to speciation by one or a few females? You mentioned the isofemale lines, and I remember that you coined the term "isofemale line," if I am correct.

C: Whether I did or not I am not sure.

A: And you mentioned that the variability in isofemale lines was a part of your thinking about it. When did the idea of speciation by one or a few founders first really come to you?

C: I think that must have come about just after we were into the Hawaiian *Drosophila* for about two or three years. By that time we were aware that each island had its own unique set of species. These islands are often within sight of one another, and yet they carry many of the chromosome arrangements that you find in the so-called older population. The population in the newer volcano is clearly a different species, with a different mating behavior and all of the attributes of a different species. Yet they can be related directly to ancestral forms on an adjacent island. Within an island and within a volcano you find ordinary geographical distribution of the kind that we see in the continents. I think that is what drove me to the idea that they have to get over there somehow, and why do they always change when they go across these channels or when the colonizations occur across the channels? So that is the idea of the founder effect, following the event of one individual or a small number going across and founding something. I think it was when I wrote the paper [Carson, 1971] for the Stadler Symposium that I first really started to think about that kind of thing, and I gave some examples from Hawaiian *Drosophila* to back up that idea.

A: Had you talked about these ideas with other people, and if so, whom?

C: Well, I think we all talked about them in the Hawaiian *Drosophila* project.

A: Whom would that encompass?

C: Well, that would be Stone and Stalker and Spieth and Throckmorton and Heed and all the other people. Let's see, there were some others. I am not sure whether or not I am leaving out some important people.

A: Was Stone still alive when the founder speciation ideas were developed?

C: No, and it was really a tragedy that Stone never quite understood what we had found in Hawaii—I mean that he died just at the crucial time. The very semester in which Stone died, I had finally decided that we could map a dozen or twenty species on the same chromosome map and see the different morphologies and different ecologies all mapping onto the same set of chromosome maps, the bands in the same place, and so on. I never got that point across to him, so it probably did emerge after Stone. But we talked about it somewhat with Throckmorton. We talked about it with Herman Spieth. He tried to see where the jetstreams flowed and tried to obtain meteorological data, which is very hard to come by, to explain how flies might get from one island to another. So maybe I articulated it, but it was something that I think many of us talked about. As soon as you see the facts, you can not avoid that idea, so it is not a very complicated idea.

A: When did you first present a major statement of the founder speciation?

C: I think that was in the Stadler Symposium paper [Carson, 1971].

A: What other paper would you consider the strongest statement of it?

C: I had one in *Science* which I think I called "Chromosome Tracers of the Origin of Species" [Carson, 1970], or something like that. Maybe I was a little bit extreme there. I had some ideas in that paper that I do not think I would subscribe to now.

A: Which ones are they?

C: I would have to remember just how it went. It seemed to me that Steve Gould quoted a few things out of that, taken partly out of context, which really sound awful, as if the formation of species were a chance formation of the species by drift. The drift sets the stage for them, and I think selection takes over after that and makes the new species.

A: Your work is often mentioned by those who adhere to macroevolutionary ideas. I assume that your background still puts you squarely in the classic evolutionary mold, or does it?

C: Yes, certainly. I do see a little bit of an analogous relationship between punctuated equilibrium and some of the things that I see in population genetics.

A: We had talked a little bit about the association of your work with macroevolution. Your work seems to stem from the modern synthesis, and isn't it viewed by you as something of an extension?

C: Of the modern synthesis, yes. We have good genetic evidence for some of the things we have found in population genetics, such as the existence of balance in populations and the fact that many populations run along an equilibrium state. Now, when Gould and Eldredge [1977; Eldredge and Gould, 1972] talk about equilibrium, they are talking about something quite different. But there is a kind of an equilibrium in populations. There are also phases of quantum change which I think you can see experimentally, and perhaps see as well in some natural situations where you have very rapid directional selection which then levels off and comes out into some sort of equilibrium stage. So there is a kind of analogous situation there, but it is not homologous in that Gould is dealing with the fossil record. His idea of the almost instantaneous formation of species suggests macromutation as a basis for macroevolution—all of that—and I am completely against that. I have, in a number of recent papers [Carson, 1982, 1987] anyway, taken cracks at the oversimplification of this macroevolution idea.

HAWAIIAN *DROSOPHILA* PROJECT

A: I would like to turn now to the matter of the Hawaiian *Drosophila*, which is an area of work that most of us associate with you. Let me begin with some general questions about the effect of going to Hawaii on your own work. Did you go to Hawaii in order to test ideas about speciation?

C: No, not at all. I went there because Stone asked me to go. I was at a meeting in Texas, and Stone said about six words. He said, "We have a grant to study Hawaiian *Drosophila*. You want to come next summer?" And I said yes in a kind of adventurous way. We really did not know anything about it at all; in fact, it worked out very differently from what we had conceived. We did not even know there would be endemic species on different islands and things of that sort.

A: Was there any plan for what kind of concepts might be explored in the Hawaiian *Drosophila* project when it was first proposed?

C: There was an elaborate proposal. It was written by Stone and Hardy, although I guess it was mostly written by Stone. I hope that Will Provine some day will look at the original proposal to see what they thought they were going to find in Hawaii. But I never saw that; I was just asked to go out and look at some chromosomes and see whether we could rear the species. At that stage it was just a very loosely organized kind of exploratory study, and for two or three years it was just thoroughly exploratory. We would decide, well, where will we go to look for some more flies this time? And then we would decide to go to a particular place in Maui. We would not get anything, and then we would go somewhere else to try to get something there.

A: Who were the principal workers at the very beginning of the project?

C: Well, there was Hardy. Hardy was the key one, of course, because he knew the areas, he had collected the flies, and he knew something about getting them. And there was Stone, who came along and actually went out in the field a few times and swept flies. And there was Stalker, who went the first year, and Frances Clayton, and Marshall Wheeler who eventually worked out the culture methods. All of these people went, and each one of them did something slightly different, took some different aspect. I soon took over the study of polytene chromosome variability, and the first thing I did with Hawaiian *Drosophila* was to do a sort of *Drosophila robusta* study of one species, in which I could get as much polymorphism and different geographical situations as possible. But later on, I pursued what you might call the phylogenetic clustering of the species by inversion studies, which of course was lifted directly from Wasserman's approach to the *repleta* group. Wasserman is totally responsible for that technique.

A: Let me return to the Hawaiian *Drosophila* project and ask a few questions about the people who started the project. Wilson Stone was the organizer of the project as I understand it, and I know that he was a very taciturn person. What kind of scientific influence did he have? Was his effect all organizational and inspirational, or what?

C: It was both. Stone just absolutely loved it from one end to the other, but he never showed it. He did not effuse about data or things of this sort, but the way he used to show his interest, when he was in Hawaii, was that he would come and sit in the lab, not hanging over anybody but separately, not reading or talking. He would just sit there. He would not say anything and there would be people running electrophoreses or doing chromosomes and raising flies and pounding this way and that. Stone would just sit there in the atmosphere of the lab. It was Heed who pointed out to me that Stone had been here for two weeks and was just sitting in the laboratory every day. Well, there was something about that: A few grunts and a few remarks were often enough to start you in a new direction. I mean, his mind was working on these things all the time, and his influence on people was impossible to discern. And yet it was very strongly there. He would just drop a remark that you ought to try this, or why don't you do that, and very frequently it was exactly the right thing to do next. Now, he would not do it himself. Somebody else would do it. I think that if he were the kind of person who did work at the bench, he would know just what the next step would be—which some people don't know. They either go on doing the same thing or else cannot follow a lead—they don't see the lead when it turns up.

A: Was his death a severe blow to the project?

C: Oh, it was really a terrible, terrible blow. The project never really recovered from it, because he gave the intellectual leadership to the University of Texas. I think Marshall Wheeler and Dick Richardson tried to inherit that, but you could not mimic Stone. Stone had a unique approach. I think that it was a great loss to the project. And it was always very sad to me that he never quite saw how beautiful it became, because I think he had some notion it was going to be, but in those early years we really did not have much notion what we were going to find.

A: What role did Elmo Hardy play in the project?

C: He gave the very important taxonomic base or systematic base that you have to have. Somebody has to describe these species, put them on record, and get the geographical distributions straight. He did that very well, and to his credit I will say that a lot of the things we found were against many of the systematic principles that he had been brought up with. For instance, he had a genus of flies with an extra crossvein. When we found that some of the chromosomes were homosequential with ones that did not have the crossvein and you could actually hybridize them, and we argued that they were really very close, only a few inversions away, he accepted that. I think a real old-line taxonomist would have felt that his morphological evidence was far better than any of this genetic stuff, but he did not take that attitude. He was always very, very willing to accept our judgments on these things, even when they impinged on the systematics. And, of course, Kaneshiro was just about ready to put down all genera except two, and Hardy did not like that at all. There was bit of a falling out there, but he did go quite a distance with Kaneshiro.

A: Is Hardy still involved in the project?

C: No, not very much. I think that he put in a rather elaborate proposal just a year or so before he retired, and the proposal was turned down by the NSF Systematics Section. His project to continue systematics on Hawaiian *Drosophila* was not funded largely because numerical taxonomy was the big thing then; they wanted everything to be numerical, and Hardy did not feel that was what we needed. He drifted off then to study other Diptera, because he is really a dipterologist of broad interests, and *Drosophila* are not his only area.

A: What about Herman Spieth? You've mentioned him earlier, and I note that he was one of the authors of the first review [Carson, Hardy, Spieth, and Stone, 1970] of the Hawaiian *Drosophila* work in *Evolutionary Biology*.

C: As you say, he was a very important cog in the early wheel. In fact, he was the one who recognized that the behavioral attributes of these species were quite unique, and he was the one who really made it clear that the secondary sexual characters which Hardy was using to describe the species were really connected with behavior and courtship. Bringing these flies into the laboratory and seeing them perform their gyrations using these secondary sexual characters was a very exciting period, and of course he wrote extensively about Hawaiian *Drosophila*, about their behavior. In general, I think it is very solid and very important work. In fact, it set up my present concern with sexual selection, which I think is probably the most important mechanism that fuels speciation in Hawaiian *Drosophila*.

A: Would you elaborate on that just a little bit?

C: That would be almost like giving a paper, but I think that each species seems to be a unique sexual selection system. They have unique use of secondary sexual characters, and even within *Drosophila silverstris* we have two subspecies or sets of strains that are very different from each other in their secondary sexual characters. I think that some sort of founder effect or flush effect or disorganization effect may then reset the sexual selection along a slightly different pattern, which then reinforces itself and becomes something uniquely different from the ancestral forms.

A: Is the current work on behavioral aspects of mating and sexual selection developed from Spieth's work, or have some of these ideas been stimulated by direct suggestions or discussions with him?

C: They were to some extent. However, Herman always felt that the analysis of the sexual behavior is too difficult—too hard—to do. He had his students do cleaning behavior or male-to-male jousting behavior because it is easier to handle experimentally. I would say that conceptually, certainly, all the ideas are there, but he has not been a part of the development of ethograms, and of the analysis of sexual selection in the manner that Kaneshiro has done—the asymmetrical system that results in some isolation, and so on. That is quite independent of Spieth's influence. He has always been very interested in phylogenetic relationships between these species and always has lots to say on that subject.

A: When you moved to Hawaii from Washington University, did your role in the project change?

C: Yes, it did change. I thought for a while—when I moved there—that we would continue to have joint grants with Hardy. Actually I had a joint grant with Hardy for a while. Then the NSF decided that they wanted to separate us and they insisted that we have separate grants and separate funding; after that, I was funded in the Genetics Department. He was funded in Entomology, and then we shared some of the work and some of the space. So, yes, the relationships there were very close for a while— between Hardy and myself—and there was again a division of labor as it had been with Stalker. I was doing the genetics and biology and he was doing the systematics. It was then necessary to establish a stock collection, a stockroom set of Hawaiian *Drosophila* stocks we could work with experimentally. I began to narrow down to one species again—to study a single species in great detail rather than extend myself over the hundred or so picture-winged species that I had been studying in the early stages of the project. We had a period when we just were doing chromosome mapping of all these different picture-wings. With each new picture-wing, the question was, where is that going to fit? We would have a little pool and write down in a sealed envelope where the species was going to fit. And then we would do the chromosome analysis and see where it fit. It was a very exciting time, but it was not really the dynamics of speciation. It was more or less patterns of the past.

A: Is it fair to say that the Hawaiian *Drosophila* project led you to your ideas of speciation?

C: Well, it was already implanted by the marginal population idea. When you examine the geographical distribution of some of these Hawaiian *Drosophila*, you are almost forced to look at the extreme attenuation of populations, and the founding of populations, and volcanoes that have been covered with lava and that have young forests growing on them. The species have to recolonize them continually. The natural history situation there directly supports the idea that some kind of founder effects have got to be involved.

LITERATURE CITED

Carson, H. L. 1935. Use of medicinal herbs among the Labrador Eskimo. Gen. Mag. 37:436–439.

Carson, H. L. 1940. Red crossbill in North Carolina in summer. Auk 57:421.

Carson, H. L. 1944. An analysis of natural chromosome variability in *Sciara impatiens* Johannsen. J. Morphol. 75:11–59.

Carson, H. L. 1945a. Delayed fertilization in a captive indigo snake, with notes on feeding and shedding. Copeia 4:222–225.

Carson, H. L. 1945b. A comparative study of the apical cell of the insect testis. J. Morphol. 77:141–161.

Carson, H. L. 1946. The selective elimination of inversion dicentric chromatids during meiosis in the eggs of *Sciara impatiens*. Genetics 31:95–113.

Carson, H. L. 1955. The genetic characteristics of marginal populations of *Drosophila*. Cold Spring Harbor Symp. Quant. Biol. 20:276–287.

Carson, H. L. 1958. Response to selection under different conditions of recombination in *Drosophila*. Cold Spring Harbor Symp. Quant. Biol. 23:291–306.

Carson, H. L. 1959. Genetic conditions which promote or retard the formation of species. Cold Spring Harbor Symp. Quant. Biol. 24:87–105.

Carson, H. L. 1961. Heterosis and fitness in experimental populations of *Drosophila melanogaster*. Evolution 15:496–509.

Carson, H. L. 1968. The population flush and its genetic consequences. In R. C. Lewontin (ed.), Population Biology and Evolution, pp. 123–137. Syracuse University Press, Syracuse, NY.

Carson, H. L. 1970. Chromosome tracers of the origin of species. Science 168:1414–1418.

Carson, H. L. 1971. Speciation and the founder principle. Stadler Genet. Symp. 3:51–70.

Carson, H. L. 1980. Hypotheses that blur and grow. In E. Mayr and W. B. Provine (eds.), The Evolutionary Synthesis, pp. 383–384. Harvard University Press, Cambridge, MA.

Carson, H. L. 1982. Speciation as a major reorganization of polygenic balances. In C. Barigozzi (ed.), Mechanisms of Speciation, pp. 411–433. Alan R. Liss, New York.

Carson, H. L. 1987. Population genetics, evolutionary rates and Neo-Darwinism. In K. S. W. Campbell and M. F. Day (eds.), Rates of Evolution, pp. 208–217. Allen & Unwin, London.

Carson, H. L., and H. D. Stalker. 1951. Natural breeding sites for some wild species of *Drosophila* in the eastern United States. Ecology 32:317–330.

Carson, H. L., and A. R. Templeton. 1984. Genetic revolutions in relation to speciation phenomena: the founding of new populations. Annu. Rev. Ecol. Systematics 15:97–131.

Carson, H. L., D. E. Hardy, H. T. Spieth, and W. S. Stone. 1970. The evolutionary biology of the Hawaiian Drosophilidae. In M. K. Hecht and W. C. Steere (eds.), Essays in Evolution and Genetics in Honor of Th. Dobzhansky, pp. 437–543. Appleton-Century Crofts, New York.

Chetverikov, S. S. 1926. On certain aspects of the evolutionary process from the standpoint of modern genetics. Zhur. Eksp. Biol. A2:3–54. [English translation by M. Barker in Proc. Am. Philos. Soc. 105:167–195 (1961).]

Dobzhansky, Th. 1937. Genetics and the Origin of Species. Columbia University Press, New York.

Dubinin, N. P. 1948. Experimental investigation of the integration of heredity systems in the processes of evolution of populations. Zhur. Obshch. Biol. 9:203–244. English translation by I. M. Lerner.

Dubinin, N. P., and G. G. Tiniakov. 1946. Structural chromosome variability in urban and rural populations of *Drosophila funebris*. Am. Natur. 80:393–396. [See also their article in J. Hered. 37:39–44 (1946).]

Eldredge, N., and S. J. Gould. 1972. Punctuated equilibria: an alternative to phyletic gradualism. In T. J. M. Scopf (ed.), Models in Paleobiology, pp. 82–115. Freeman, Cooper, San Francisco.

Fisher, R. A. 1930. The Genetical Theory of Natural Selection. Clarendon Press, Oxford.

Ford, H. D., and E. B. Ford. 1930. Fluctuation in numbers and its influence on variation in *Melitaea aurinia*. Trans. R. Entomol. Soc. Lond. 78:345–351.

Gould, S. J., and N. Eldredge. 1977. Punctuated equilibria: the tempo and mode of evolution reconsidered. Paleobiology 3:115–151.

Grant, V. 1977. Organismic Evolution. W. H. Freeman, San Francisco.

Haldane, J. B. S. 1932. The Causes of Evolution. Harper, New York.

Hinton, C. W., and J. C. Lucchesi. 1960. A cytogenetic study of crossing over in inversion heterozygotes of *Drosophila melanogaster*. Genetics 45:87–94.

Lerner, I. M. 1954. Genetic Homeostasis. Wiley, New York.

Mather, K., and B. J. Harrison. 1949. The manifold effect of selection. Heredity 3:1–52, 131–162.

Mayr, E. 1942. Systematics and the Origin of Species. Columbia University Press, New York.

Mayr, E. 1954. Change of genetic environment and evolution. In J. Huxley, A. C. Hardy, and E. B. Ford (eds.), Evolution as a Process, pp. 157–180. Allen & Unwin, London.

Mayr, E. 1963. Animal Species and Evolution. Belknap Press of Harvard University Press, Cambridge, MA.

Mayr, E., and W. B. Provine (eds.). 1980. The Evolutionary Synthesis. Harvard University Press, Cambridge, MA. [See especially the chapter by Th. Dobzhansky, pp. 229–242: "The birth of the genetic theory of evolution in the Soviet Union in the 1920's."]

Prakash, S. 1973. Patterns of gene variation in central and marginal populations of *Drosophila robusta*. Genetics 75:347–369.

Simpson, G. G. 1944. Tempo and Mode in Evolution. Columbia University Press, New York.

Stalker, H. D. 1942. Sexual isolation studies in the species complex *D. virilis*. Genetics 27:238–257.

Templeton, A. R. 1980. The theory of speciation via the founder principle. Genetics 94:1011–1038.

Timofeef-Ressovsky, N. W. 1940. Mutations and geographical variation. In J. Huxley (ed.), The New Systematics, pp. 73–136. Clarendon Press, Oxford.

Wright, S. 1931. Evolution in Mendelian populations. Genetics 16:97–159.

Wright, S. 1932. The roles of mutation, inbreeding, crossbreeding and selection in evolution. Proc. Sixth Int. Cong. Genet. 1:356–366.

Wright, S. 1940. Breeding structure of populations in relation to speciation. Am. Natur. 74:232–248.

HAMPTON LAWRENCE CARSON: LIST OF PUBLICATIONS FROM 1934 TO 1989

1. Carson, H. L. 1934. Labrador quarry. Gen. Mag. 37(1):97–104.

2. Carson, H. L. 1935. Use of medicinal herbs among the Labrador Eskimo. Gen. Mag. 37(4):436–439.

3. Carson, H. L. 1940. Red crossbill in North Carolina in summer. Auk 57:421.

4. Carson, H. L. 1941. Linkage, interference and semilethals in the white group of *Habrobracon*. Am. Natur. 75:608–614.

5. Carson, H. L. 1943. Cytological analysis of natural populations of *Sciara impatiens*. Genetics 28:71–72 [abstract].

6. Carson, H. L. 1944. An analysis of natural chromosome variability in *Sciara impatiens* Johannsen. J. Morphol. 75(1):11–59.

7. Carson, H. L. 1945a. A comparative study of the apical cell of the insect testis. J. Morphol. 77(2):141–161.

8. Carson, H. L. 1945b. Delayed fertilization in a captive indigo snake, with notes on feeding and shedding. Copeia 1945(4):222–225.

9. Carson, H. L. 1946a. The selective elimination of inversion dicentric chromatids during meiosis in the eggs of *Sciara impatiens*. Genetics 31:95–113.

10. Carson, H. L., and H. D. Stalker. 1946b. Chromosome Studies on *Drosophila robusta*. Genetics 31(2):213 [abstract].

11. Stalker, H. D., and H. L. Carson. 1946c. Geographical variation in the morphology of *Drosophila robusta*. Genetics 31(2):231 [abstract].

12. Carson, H. L., and H. D. Stalker. 1947a. A seasonal study of gene arrangement frequencies and morphology in *Drosophila robusta*. Genetics 32(1):81 [abstract].

13. Carson, H. L., and H. D. Stalker. 1947b. Gene arrangements in natural populations of *Drosophila robusta* Sturtevant. Evolution 1(3):113–133.

14. Stalker, H. D., and H. L. Carson. 1947c. Morphological variation in natural populations of *Drosophila robusta* Sturtevant. Evolution 1(4):237–248.

15. Carson, H. L., and H. D. Stalker. 1948a. Reproductive diapause in *Drosophila robusta*. Proc. Natl. Acad. Sci. USA 34:124–129.

16. Carson, H. L., and H. D. Stalker. 1948b. An altitudinal transect of gene arrangement frequencies in *Drosophila robusta*. Genetics 33:100 [abstract].

17. Stalker, H. D., and H. L. Carson. 1948c. Seasonal changes in gene arrangement frequencies and morphology of *Drosophila robusta*. Genetics 33:629–630 [abstract].

18. Stalker, H. D., and H. L. Carson. 1948d. An altitudinal transect of *Drosophila robusta* Sturtevant. Evolution 2:295–305.

19. Stalker, H. D., and H. L. Carson. 1949a. Seasonal variation in the morphology of *Drosophila robusta* Sturtevant. Evolution 3:330–343.

20. Carson, H. L., and H. D. Stalker. 1949b. Seasonal variation gene arrangement frequencies over a three-year period *Drosophila robusta* Sturtevant. Evolution 3:322–329.

21. Carson, H. L., and H. D. Stalker. 1950. Natural breeding sites for *Drosophila robusta*. Genetics 35:100 [abstract].

22. Carson, H. L., and H. D. Stalker. 1951a. Natural breeding sites for some wild species of *Drosophila* in the eastern United States. Ecology 32:317–330.

23. Carson, H. L. 1951b. Breeding sites of *Drosophila pseudoobscura* and *Drosophila persimilis* in the transition zone of the Sierra Nevada. Evolution 5:91–96.

24. Carson, H. L. 1951c. Interfertile sympatric sibling species within *D. bocainensis* Pavan and da Cunha 1947. Dros. Info. Serv. 25:103–104.

25. Carson, H. L., and W. C. Blight. 1952a. Sex chromosome polymorphism in a population of *Drosophila americana*. Genetics 37:572 [abstract].

26. Carson, H. L. 1952b. Contrasting types of population structure in *Drosophila*. Am. Natur. 86:239–248.

27. Carson, H. L. 1952c. A new case of cryptic species in *Drosophila*. Science 116:518 [abstract].

28. Carson, H. L. 1953a. The effects of inversions on crossing over in *Drosophila robusta*. Genetics 38:168–186.

29. Carson, H. L. 1953b. *Evolution in the Genus* Drosophila, by J. T. Patterson and W. S. Stone, 1952 [review]. Am. Natur. 87:271–272.

30. Levitan, M., H. L. Carson, and H. D. Stalker. 1954a. Triads of overlapping inversions in *Drosophila robusta*. Am. Natur. 88:113–114.

31. Carson, H. L. 1954b. Interfertile sibling species in the *willistoni* group of *Drosophila*. Evolution 8:148–165.

32. Carson, H. L. 1954c. Hybridization experiments with two sympatric sibling species of the willistoni group of *Drosophila*. Caryologia 6(Suppl.):653 [abstract].

33. Carson, H. L. 1955a. Variation in genetic recombination in natural populations. J. Cell. Comp. Physiol. 45 (Suppl. 2):221–236.

34. Carson, H. L. 1955b. The genetic characteristics of marginal populations of *Drosophila*. Cold Spring Harbor Symp. Quant. Biol. 20:276–287.

35. Dorsey, C. K., and H. L. Carson. 1956a. Selective response of wild Drosophilidae to natural and artificial attrahents. Ann. Entomol. Soc. Am. 49:177–181.

36. Carson, H. L. 1956b. Marginal homozygosity for gene arrangement in *Drosophila robusta*. Science 123:630–631.

37. Carson, H. L., E. P. Knapp, and H. J. Phaff. 1956c. Studies on the ecology of *Drosophila* in the Yosemite Region of California. III. The yeast flora of the natural breeding sites of some species of *Drosophila*. Ecology 37(3):538–544.

38. Dobzansky, T., D. M. Cooper, H. J. Phaff, E. P. Knapp, and H. L. Carson. 1956d. Studies on the ecology of *Drosophila* in the Yosemite Region of California. IV. Differential attraction of species of *Drosophila* to different species of yeasts. Ecology 37(3):544–550.

39. Carson, H. L. 1956e. Response of *Drosophila robusta* to selection for motility. Genetics 41:636–637 [abstract].

40. Carson, H. L. 1956f. A female-producing strain of *D. borealis* Patterson. Dros. Info. Serv. 30:109–110.

41. Carson, M. R. Wheeler, and W. B. Heed. 1957a. A parthenogenetic strain of *Drosophila mangabeirai* Malogolowkin. Univ. Texas Publ. 5721:115–122.

42. Carson, H. L. 1957b. The species as a field for gene recombination. In The Species Problem, Am. Assoc. Adv. Sci. Publ. No. 50:23–38.

43. Carson, H. L. 1957c. Parrot-watching. Bull. St. Louis Audubon Soc. 26:1–3.

44. Carson, H. L. 1957d. Production of biomass as a measure of fitness of experimental populations of *Drosophila*. Genetics 42:363–364.

45. Wolfson, M., H. D. Stalker, and H. L. Carson. 1957e. A serious parasite of laboratory *Drosophila*. Dros. Info. Serv. 31:170.

46. Carson, H. L. 1958a. The population genetics of *Drosophila robusta*. Adv. Genet. 9:1–40.

47. Carson, H. L. 1958b. Increase of fitness in experimental populations following introduction of one haploid set of autosomes. Proc. Xth Int. Cong. Genet. 2:44–45 [abstract].

48. Carson, H. L. 1958c. Response to selection under different conditions of recombination in *Drosophila*. Cold Spring Harbor Symp. Quant. Biol. 23:291–306.

49. Carson, H. L. 1958d. Increase in fitness in experimental populations resulting from heterosis. Proc. Natl. Acad. Sci. USA 44:1136–1141.

50. Susman, M., and H. L. Carson. 1958e. Development of balanced polymorphism in laboratory populations of *Drosophila melanogaster*. Am. Natur. 92:359–364.

51. Carson, H. L. 1958f. A study of experimental populations of *Drosophila*. Proc. Entomol. Soc. Am. No. Cent. Br. 13:22 [abstract].

52. Murdy, W. H., and H. L. Carson. 1959a. Parthenogenesis in *Drosophila mangabeirai*. Am. Natur. 93:355–363.

53. Carson, H. L. 1959b. Effect of irradiation on artificial populations under strong natural selection. Rec. Genet. Soc. Am. 28:63 [abstract].

54. Carson, H. L. 1959c. Genetic conditions which promote or retard the formation of species. Cold Spring Harbor Symp. Quant. Biol. 24:87–105.

55. Carson, H. L. 1960. Survival of newly-induced chromosome aberrations in experimental populations of *Drosophila melanogaster*. Genetics 45:980–981 [abstract].

56. Carson, H. L. 1961a. Variation. In The Encyclopedia of the Biological Sciences, pp. 1047–1049. Reinhold Publ., New York.

57. Carson, H. L. 1961b. Rare parthenogenesis in *Drosophila robusta*. Am. Natur. 95:81–86.

58. Carson, H. L. 1961c. Relative fitness of genetically open and closed experimental populations of *Drosophila robusta*. Genetics 46:553–567.

59. Carson, H. L. 1961d. Heterosis and fitness in experimental populations of *Drosophila melanogaster*. Evolution 15:496–509.

60. Heed, W. B., H. L. Carson, and M. S. Carson. 1961e. A list of flowers utilized by drosophilids in the Bogotá region of Colombia. Dros. Info. Serv. 34:84–85.

61. Carson, H. L. 1962a. Fixed heterozygosity in a parthenogenetic species of *Drosophila*. Univ. Texas Publ. 6205:55–62.

62. Carson, H. L. 1962b. Selection for parthenogenesis in *Drosophila mercatorum*. Genetics 47:946 [abstract].

63. Carson, H. L. 1963a. Silent Spring, by Rachel L. Carson, a review. College and University 38:294–296.

64. Carson, H. L. 1963b. Transitory increase in genetic load in irradiated laboratory populations of *Drosophila melanogaster*. Proc. XIth Int. Cong. Genet. 1:74.

65. Stalker, H. D., and H. L. Carson. 1963c. A very serious parasite of laboratory *Drosophila*. Second report. Dros. Info. Serv. 38:96.

66. Carson, H. L. 1963d. Heredity and Human Life. Columbia Univ. Press, New York.

67. Carson, H. L. 1963e. Humanism and the new biology. Wash. Univ. Mag. 34:7–9.

68. Carson, H. L. 1963f. Introduction to *The Origin of Species* by Charles Darwin, pp. xi–xviii. Washington Square Press, New York.

69. Carson, H. L. 1964a. Population size and genetic load in irradiated populations of *Drosophila melanogaster*. Genetics 49:521–528.

70. Carson, H. L., and W. B. Heed. 1964b. Structural homozygosity in marginal populations of Neoarctic and Neotropical species of *Drosophila* in Florida. Proc. Natl. Acad. Sci. USA 52:427–430.

71. White, M. J. D., H. L. Carson, and J. Cheney. 1964c. Chromosomal races in the Australian grasshopper *Moraba viatica* in a zone of geographic overlap. Evolution 18:417–429.

72. Carson, H. L. 1965a. Eugenics: Hereditarian Attitudes in American Thought, by Mark H. Haller; a review. Am. J. Sociol. 70:505.

73. Carson, H. L., and M. Wasserman. 1965b. A widespread chromosomal polymorphism in a widespread species, *Drosophila buzzatti*. Am. Natur. 99:111–115.

74. Carson, H. L. 1965c. Chromosomal morphism in geographically widespread species of *Drosophila*. In H. G. Baker and G. L. Stebbins (eds.), The Genetics of Colonizing Species, pp. 503–531. Academic Press, New York.

75. Carson, H. L. 1966. Chromosomal races of *Drosophila crucigera* from the islands of Oahu and Kauai, State of Hawaii. Univ. Texas Publ. 6615:405–412.

76. Carson, H. L. 1967a. Inbreeding and gene fixation in natural populations. In R. A. Brink (ed.), Heritage from Mendel, pp. 281–308. Univ. Wisconsin Press, Madison.

77. Carson, H. L. 1967b. Selection for parthenogenesis in *Drosophila mercatorum*. Genetics 55:157–171.

78. Carson, H. L. 1967c. Permanent heterozygosity. In Th. Dobshansky, M. K. Hecht, and W. C. Steere (eds.), Evolutionary Biology, Vol. 1, pp. 143–168. Appleton-Century-Crofts, New York.

79. Carson, H. L., F. E. Clayton and H. D. Stalker. 1967d. Karyotypic stability and speciation in Hawaiian *Drosophila*. Proc. Natl. Acad. Sci. USA 57:1280–1285.

80. Carson, H. L. 1967e. The association between *Drosophila carcinophila* Wheeler and its host, the land crab *Gecarcinus ruricola* (L.) Am. Midl. Nat. 78:324–343.

81. Carson, H. L. 1967f. Chromosomal polymorphism in altitudinal races of *Drosophila*. Proc. Jpn. Soc. Syst. Zool. 3:10–16.

82. Carson, H. L. 1967g. Genetics and Evolution of Hawaiian Drosophilidae. Preliminary report of collections made in Kipahulu Valley, Maui. In R. E. Warner (ed.), Scientific Report of the Kipahulu Valley Expedition, pp. 87–91. The Nature Conservancy, Arlington, Va.

83. Carson, H. L., and M. R. Wheeler. 1968a. *Drosophila endobranchia*, a new drosophilid associated with land crabs in the West Indies. Ann. Entomol. Soc. Am. 61:675–678.

84. Carson, H. L. 1968b. Parallel inversion polymorphisms in different species of Hawaiian *Drosophila*. Proc. XIIth Int. Cong. Genet. 1:321.

85. Carson, H. L., and H. D. Stalker. 1968c. Polytene chromosome relationships in Hawaiian species of *Drosophila*. I. The *D. grimshawi* subgroup. Univ. Texas Publ. 6818:335–354.

86. Carson, H. L., and H. D. Stalker. 1968d. Polytene chromosome relationships in Hawaiian species of *Drosophila*. II. The *D. planitibia* subgroup. Univ. Texas Publ. 6818:335–365.

87. Carson, H. L., and H. D. Stalker. 1968e. Polytene chromosome relationships in Hawaiian species of *Drosophila*. III. The *D. adiastola* and *D. punalua* subgroups. Univ. Texas Publ. 6818:367–380.

88. Carson, H. L. 1968f. The population flush and its genetic consequences. In R. C. Lewontin (ed.), Population Biology and Evolution, pp. 123–137. Syracuse Univ. Press, Syracuse, N. Y.

89. Carson, H. L. 1969a. Parallel polymorphisms in different species of Hawaiian *Drosophila*. Am. Natur. 103:323–329.

90. Carson, H. L., and J. E. Sato. 1969b. Microevolution within three species of Hawaiian *Drosophila*. Evolution 23:493–501.

91. Carson, H. L. 1969c. Maintenance of lethal and detrimental genes in natural populations. Introduction and synthesis by the Chairman. Symposium 15, the XII International Congress of Genetics. Jpn. J. Genet. 44 (Suppl. 1):225–227.

92. Carson, H. L., and H. D. Stalker. 1969d. Polytene chromosome relationships in Hawaiian species of Drosophila. IV. The *D. primaeva* subgroup. Univ. Texas Publ. 6918:85–94.

93. Carson, H. L., I. Y. Wei, and J. A. Niederkorn, Jr. 1969e. Isogenicity in parthenogenetic strains of *Drosophila mercatorum*. Genetics 63:619–628.

94. Carson, H. L. 1969f. Drosophilidae of Hawaii. Ann. Missouri Bot. Gard. 56:417–418.

95. Carson, H. L., D. E. Hardy, H. T. Spieth, and W. S. Stone. 1970a. The evolutionary biology of the Hawaiian Drosophilidae. In M. K. Hecht and W. C. Steere (eds.), Essays in Evolution and Genetics in Honor of Theodosius Dobzhansky, pp. 437–543. Appleton-Century-Crofts, New York.

96. Carson, H. L. 1970b. Chromosome tracers of the origin of species. Science 168:1414–1418.

97. Carson, H. L. 1970c. Chromosomal tracers of founder events. Biotropica 2:3–6.

98. Carson, H. L. 1971a. Polytene chromosome relationships in Hawaiian species of Drosophila. V. Additions to the chromosomal phylogeny of the picture-winged species. Univ. Texas Publ. 7103:183–191.

99. Carson, H. L. 1971b. The ecology of *Drosophila* breeding sites. University of Hawaii, Honolulu. Harold L. Lyon Arboretum Lecture No. 2:1–27.

100. Carson, H. L. 1971c. Speciation and the founder principle. Stadler Genet. Symp. 3:51–70.

101. Carson, H. L., and S. H. Snyder. 1972a. Screening by parthenogenesis for induced mutations in *Drosophila mercatorum*. Egypt. J. Genet. Cytol. 1:256–261.

102. Carson, H. L. 1972b. Evolutionary biology: its value to society. Bioscience 22:349–352.

103. Carson, H. L. 1972c. Ancient chromosomal polymorphism and its use in phylogeny. Abstracts 14th Int. Cong. Entomol., Canberra, Australia, p. 50.

104. Clayton, F. E., H. L. Carson, and J. E. Sato. 1972d. Polytene chromosome relationships in Hawaiian species of *Drosophila*. VI. Supplementary data on metaphases and gene sequences. Univ. Texas Publ. 7213:163–177.

105. Carson, H. L. 1972e. Microevolution. A review of Ecological Genetics and Evolution. Essays in Honour of E. B. Ford. Robert Creed, Ed., Blackwell, Oxford and Appleton-Century-Crofts, New York, 1971. Science 178:855–856.

106. Carson, H. L. 1973a. Ancient chromosomal polymorphism in Hawaiian *Drosophila*. Nature 241:200–202.

107. Carson, H. L. 1973b. The genetic system in parthenogenetic strains of *Drosophila mercatorum*. Proc. Natl. Acad. Sci. USA 70:1772–1774.

108. Yoon, J. S., and H. L. Carson. 1973c. Codification of polytene chromosome designations for Hawaiian Drosophilidae. Genetics 74:s303–s304 [abstract].

109. Carson, H. L., and M. R. Wheeler. 1973d. A new crab fly from Christmas Island, Indian Ocean (Diptera: Drosophilidae). Pacific Insects 15:199–208.

110. Kaneshiro, K. Y., H. L. Carson, F. E. Clayton, and W. B. Heed. 1973e. Niche separation in a pair of homosequential *Drosophila* species from the island of Hawaii. Am. Natur. 107:766–774.

111. Carson, H. L., D. E. Hardy, L. H. Throckmorton, M. Wasserman, and M. R. Wheeler. 1973f. *Drosophila carinata* Grimshaw, 1901 (Insecta, Diptera): proposed suppression under the plenary powers in order to preserve *Drosophila*

mercatorum Patterson and Wheeler, 1942 Z. N. (S) 2035. Bull. Zool. Nomencl. 30(2):112–117.

112. Carson, H. L. 1973g. Reorganization of the gene pool during speciation. In N. E. Morton (ed.), *Genetic Structure of Populations*. Population Genetics Monographs, Vol. 3, pp. 274–280. Honolulu, University of Hawaii Press.

113. Ikeda, H., and H. L. Carson. 1973h. Selection for mating reluctance in females of a diploid parthenogenetic strain of *Drosophila mercatorum*. Genetics 75:541–555.

114. Steiner, W. W. M., W. E. Johnson, and H. L. Carson. 1973i. Molecular differentiation in *D. grimshawi*. Dros. Info. Serv. 50:100–101.

115. Carson, H. L. 1974a. The biston affair. A view of *The Evolution of Melanism*, by B. Kettlewell. New York, Oxford University Press, 1973. Science 183:67.

116. Ahearn, J. N., H. L. Carson, Th. Dobzhansky, and K. Y. Kaneshiro. 1974b. Ethological isolation among three species of the planitibia subgroup of Hawaiian *Drosophila*. Proc. Natl. Acad. Sci. USA 71:901–903.

117. Carson, H. L. 1974c. Human genetics. In Encyclopaedia Britannica Macropaedia, Vol. 7, pp. 996–1010. Helen Hemingway Benton, Publ., Chicago.

118. Carson, H. L. 1974d. Patterns of speciation in Hawaiian *Drosophila* inferred from ancient chromosomal polymorphism. In M. J. D. White (ed.), Genetic Mechanisms of Speciation in Insects, pp. 81–93. Australia and New Zealand Book Co., Sydney.

119. Carson, H. L. 1974e. Natural History of Islands. A review of *Island Biology*, by S. Carlquist and M. J. Cole (Columbia University Press, New York 1974). Science 186:252–253.

120. Carson, H. L. 1974f. Three flies and three islands: parallel evolution in *Drosophila*. Proc. Natl. Acad. Sci. USA 71:3517–3521.

121. Steiner, W. W. M., and H. L. Carson. 1974g. Genetic structure and variability in two species of endemic Hawaiian *Drosophila*. U.S./I.B.P. Island Ecosystems IRP. Technical Report No. 50:iv and 66 pp.

122. Carson, H. L. 1975a. The genetics of speciation at the diploid level. Am. Natur. 109:82–92.

123. Carson, H. L., and W. E. Johnson. 1975b. Genetic variation in Hawaiian *Drosophila* I. Chromosome and allozyme polymorphism in *D. setosimentum* and *D. ochrobasis* from the island of Hawaii. Evolution 29:11–23.

124. Johnson, W. E., H. L. Carson, K. Y. Kaneshiro, W. W. M. Steiner, and M. M. Cooper. 1975c. Genetic variation in Hawaiian *Drosophila* II. Allozymic differentiation in the *D. planitibia* subgroup. In C. L. Markert (ed.), Isozymes IV. Genetics and Evolution, pp. 563–584. Academic Press, New York.

125. Carson, H. L., W. E. Johnson, P. S. Nair, and F. M. Sene. 1975d. Genetic similarities based on allozymic and chromosomal data. Genetics 80:s19 [abstract].

126. Craddock, E. M., and H. L. Carson. 1975e. Chromosome variability in an endemic Hawaiian Drosophila species. Genetics 80:s23 [abstract].

127. Johnson, W. E., and H. L. Carson. 1975f. Allozymic variation in *Drosophila silvestris*. Genetics 80:s46 [abstract].

128. Nair, P. S., F. M. Sene, and H. L. Carson. 1975g. Regulatory influence on isozyme expression in *Drosophila*. Genetics 80:s60 [abstract].

129. Carson, H. L., P. S. Nair, and F. M. Sene. 1975h. *Drosophila* hybrids in nature: proof of gene exchange between sympatric species. Science 189:806–807.

130. Carson, H. L., W. E. Johnson, P. S. Nair, and F. M. Sene. 1975i. Genetic variation in Hawaiian *Drosophila* III. Allozymic and chromosomal similarity in two *Drosophila* species. Proc. Natl. Acad. Sci. USA 12:4521–4525.

131. Carson, H. L. 1976a. Inference of the time of origin of some *Drosophila* species. Nature 259:395.

132. Templeton, A. R., H. L. Carson, and C. F. Sing. 1976b. The population genetics of parthenogenetic strains of *Drosophila mercatorum* II. The capacity for parthenogenesis in a natural, bisexual population. Genetics 82:527–542.

133. Sene, F. M., and H. L. Carson. 1976c. Close allozymic similarity within and between two sympatric species of Hawaiian *Drosophila*. Genetics 83:s69–70 [abstract].

134. Carson, H. L. 1976d. Genetic differences between newly formed species. Bioscience 26:700–701.

135. Carson, H. L. 1976e. The unit of genetic change in adaptation and speciation. Ann. Missouri Bot. Gard. 63:210–233.

136. Carson, H. L., and K. Y. Kaneshiro. 1976f. *Drosophila* of Hawaii: systematics and ecological genetics. Annu. Rev. Ecol. Systematics 7:311–345.

137. Carson, H. L. 1976g. Ecology of rare *Drosophila* species in Hawaii volcanoes national park. In Proc. First Conf. in Natural Sciences, Hawaii Volcanoes National Park, pp. 39–45. Cooperative National Park Resources Studies Unit. Department of Botany, University of Hawaii, Honolulu.

138. Carson, H. L., L. T. Teramoto, and A. R. Templeton. 1977a. Behavioral differences among isogenic strains of *Drosophila mercatorum*. Behav. Genet. 7:189–197.

139. Sene, F. M., and H. L. Carson. 1977b. Genetic variation in Hawaiian *Drosophila* IV. Allozymic similarity between *D. silvestris* and *D. heteroneura* from the island of Hawaii. Genetics 86:187–198.

140. Nair, P. S., H. L. Carson, and F. M. Sene. 1977c. Isozyme polymorphism due to regulatory influence. Am. Natur. 111:789–791.

141. Carson, H. L. 1977d. Introductions to a pivotal subject: review of *Evolution*, by Theodosius Dobzhansky et al., and *Organismic Evolution*, by Verne Grant. Science 197:1272–1273.

142. Fontdevila, A., and H. L. Carson. 1978a. Spatial distribution and dispersal in a population of Drosophila. Am. Natur. 112:365–380.

143. Carson, H. L. 1978b. Chromosomes and species formation. Evolution 32:925–927.

144. Carson, H. L. 1978c. Speciation and sexual selection in Hawaiian *Drosophila*. In P. F. Brussard (ed.), Ecological Genetics: The Interface, pp. 93–107. Springer-Verlag, New York.

145. Carson, H. L. 1978d. Hawaii IBP Synthesis: 6. Genetic variation and population structure in island species. In. Proc. Second Conf. in Natural Sciences, Hawaii Volcanoes National Park, p. 41 [abstract]. Cooperative National Park Resources Studies Unit, Department of Botany, University of Hawaii, Honolulu.

146. Carson, H. L. 1978e. Genetic distance, sexual selection and speciation. U.S.–Japan Cooperative Science Program. In H. I. Oka and O. Kitagawa (eds), Dynamics of Speciation in Plants and Animals, pp. 84–87. Japan Soc. Prom. Science, Tokyo.

147. Carson, H. L., and P. J. Bryant. 1979a. Genetic variation in Hawaiian *Drosophila* VI. Change in a secondary sexual character as evidence of incipient speciation in *Drosophila silvestris*. Proc. Natl. Acad. Sci. USA 76:1929–1932.

148. Bryant, P. J., and H. L. Carson. 1979b. Genetics of an interspecific difference in a secondary sexual character in Hawaiian *Drosophila*. Genetics 91:s15–s16 [abstract].

149. Carson, H. L. 1979c. Local variation in a secondary sexual character in *Drosophila*. Genetics 91:s18 [abstract].

150. Dickinson, W. J., and H. L. Carson. 1979d. Regulation of the tissue specificity of an enzyme by a cis-acting genetic element: evidence from interspecific *Drosophila* hybrids. Proc. Natl. Acad. Sci. USA 76:4559–4562.

151. Carson, H. L., and T. Okada. 1980a. Drosophilidae associated with flowers in Papua, New Guinea. I. *Colocasia esculenta*. Kontyû 48:15–29.

152. Carson, H. L., and A. T. Ohta. 1980b. Origin of the genetic basis of colonizing ability. In Abstracts Second Int. Cong. Syst. Evol. Biol., p. 39 [abstract]. The University of British Columbia, Vancouver.

153. Carson, H. L. 1980c. Evolution of studies on Hawaiian Drosophilidae. In Abstracts Second Int. Cong. Syst. Evol. Biol., p. 110. The University of British Columbia, Vancouver.

154. Carson, H. L., and T. Okada. 1980d. The ecology and evolution of some flower-breeding Drosophilidae of New Guinea. In Abstracts XVI Int. Cong. Entomol., Kyoto, Japan, p. 7.

155. Carson, H. L., and L. T. Teramoto. 1980e. Differences in copulatory success among laboratory males of *Drosophila silvestris*. Genetics 94:s14 [abstract].

156. Carson, H. L. 1980f. Homosequential species of Hawaiian *Drosophila*. In Proc. 7th Int. Chromosome Conf. Oxford, England, p. 18 [abstract].

157. Carson, H. L. 1980g. Cytogenetics and the Neo-Darwinian synthesis. In E. Mayr and W. B. Provine. (eds.), The Evolutionary Synthesis, pp. 86–95. Harvard Univ. Press, Cambridge, MA.

158. Carson, H. L. 1980h. Hypotheses that blur and grow. In E. Mayr and W. B. Provine (eds.), The Evolutionary Synthesis, pp. 383–384. Harvard Univ. Press, Cambridge, MA.

159. Carson, H. L. 1980i. Variation among males of *Drosophila silvestris* in the Olaa tract, Hawaii Volcanoes National Park. In Proc. Third Conf. in Natural Sciences, Hawaii Volcanoes National Park, p. 51 [abstract]. Cooperative National Park Resources Studies Unit, Department of Botany, University of Hawaii, Honolulu.

160. Carson, H. L. 1980j. A provocative view of the evolutionary process. In L. K. Piternick (ed.), Richard Goldschmidt, Controversial Geneticist and Creative Biologist. Birkhauser, Basel. Experientia [Suppl] 35:24–26.

161. Okada, T., and H. L. Carson. 1980k. Drosophilidae associated with flowers in Papua, New Guinea II. *Alocasia* (Araceae). Pacific Insects 22:217–236.

162. Carson, H. L. 1980l. Chromosomes and Evolution in some relatives of *Drosophila grimshawi* from Hawaii. Symp. R. Entomol. Soc. Lond. 10:195–205.

163. Carson, H. L. 1981a. Macroevolution conference: a letter to the editor. Science 211:773.

164. Carson, H. L. 1981b. Genetics in Russia. A review of *Animal Genetics and Evolution*, eds. N. N. Vorontsov and J. M. van Brink. The Hague: Junk, 384 pp., 1978. Science 211:932–933.

165. Carson, H. L. 1981c. Chromosomal tracing of evolution in a phylad of species related to *Drosophila hawaiiensis*. In W. R. Atchley and D. S. Woodruff (eds.), Evolution and Speciation, pp. 286–297. Cambridge Univ. Press, Cambridge.

166. Carson, H. L. 1981d. Age of chromosomal polymorphism in *Drosophila silvestris*. Genetics 97:s17 [abstract].

167. Sene, F. M., J. M. Amabis, H. L. Carson, and T. H. F. S. Cyrino. 1981e. Chromosome polymorphism in *Drosophila mercatorum pararepleta* in South America. Rev. Brasil. Genet. 4:1–10.

168. Carson, H. L. 1981f. Homosequential species of Hawaiian *Drosophila*. In M. D. Bennett, M. Bobrow, and G. Hewitt (eds.), Chromosomes Today, Vol. 7, pp. 150–164. Allen & Unwin, London.

169. Mueller-Dombois, D., K. W. Bridges, and H. L. Carson (eds.). 1981g. Island Ecosystems: Biological Organization in Selected Hawaiian Communities, 583 pp. Hutchinson-Ross Publ., Stroudsburg, PA.

170. Carson, H. L. 1981h. Genetic variation within Island Species: Introduction. In D. Mueller-Dombois, K. W. Bridges, and H. L. Carson (eds.), Island Ecosystems: Biological Organization in Selected Hawaiian Communities, pp. 431–437. Hutchinson-Ross Publ., Stroudsburg, PA.

171. Spiess, E. B., and H. L. Carson. 1981i. Sexual selection in *Drosophila silvestris* of Hawaii. Proc. Natl. Acad. Sci. USA 78:3088–3092.

172. Stuart, W. D., J. G. Bishop, H. L. Carson, and M. B. Frank. 1981j. Location of the 18/28S ribosomal RNA genes in two Hawaiian *Drosophila* species by monoclonal immunological identification of RNA–DNA hybrids in situ. Proc. Natl. Acad. Sci. USA 78:3751–3754, 1981.

173. Carson, H. L. 1981k. Genetic variation within and between endemic Hawaiian species of *Drosophila*. In D. Mueller-Dombois, K. W. Bridges, and H. L. Carson (eds.), Island Ecosystems: Biological Organization in Selected Hawaiian Communities, pp. 451–455. Hutchinson-Ross Publ., Stroudsburg, PA.

174. Carson, H. L. 1981l. Microevolution in insular ecosystems. In D. Mueller-Dombois, K. W. Bridges, and H. L. Carson (eds.), Island Ecosystems: Biological Organization in Selected Hawaiian Communities, pp. 471–482. Hutchinson-Ross, Publ., Stroudsburg, PA.

175. Ashburner, M., H. L., Carson, and J. N. Thompson, Jr. (eds.). 1981 m. The Genetics and Biology of *Drosophila*, Vol. 3a, 429 pp. Academic Press, London.

176. Carson, H. L., and A. T. Ohta. 1981n. Origin of the genetic basis of colonizing ability. In G. G. E. Scudder and J. L. Reveal (eds.), Evolution Today, Proc. Second Int. Cong. Syst. and Evol. Biol., pp. 365–370. Hunt Institute, Carnegie Mellon Univ. Pittsburgh, PA.

177. Carson, H. L., and T. Okada. 1982a. Drosophilidae of New Guinea. In J. L. Gressitt. (ed.), Biogeography of New Guinea. W. Junk, Amsterdam. Monographiae Biologicae 42(2):675–687.

178. Carson, H. L., F. C. Val, C. M. Simon, and J. W. Archie. 1982b. Morphometric evidence for incipient speciation in *Drosophila silvestris* from the island of Hawaii. Evolution 36:132–140.

179. Carson, H. L. 1982c. Evolution of *Drosophila* on the newer Hawaiian volcanoes. Heredity 48:3–25.

180. Kaneshiro, K. Y., and H. L. Carson. 1982d. Selection experiments on mating behavior in *Drosophila silvestris*. Genetics 100:s34 [abstract].

181. Giddings, L. V., and H. L. Carson. 1982e. Behavioral phylogeny of populations of *Drosophila crucigera*. Genetics 100:s26–27 [abstract].

182. Ashburner, M., H. L. Carson, and J. N. Thompson, Jr. (eds.). 1982f. The Genetics and Biology of *Drosophila*, Vol 3b, 428 pp. Academic Press, London.

183. Carson, H. L., and J. S. Yoon. 1982g. Genetics and evolution of Hawaiian *Drosophila*. In M. Ashburner, H. L. Carson, and J. N. Thompson, Jr. (eds., The

Genetics and Biology of *Drosophila*, Vol. 3b, pp. 297–344. Academic Press, London.

184. Okada, T. and H. L. Carson. 1982h. Drosophilidae associated with flowers in Papua New Guinea III. *Zingiberales*. Kontyû 50(3):396–410.

185. Carson, H. L., L. S. Chang, and T. W. Lyttle. 1982i. Decay of female sexual behavior under parthenogenesis. Science 218:68–70.

186. Carson, H. L. 1982j. Hawaii: showcase of evolution. An introduction. Nat. Hist. 91(12):16–18.

187. Carson, H. L. 1982k. Hawaii: showcase of evolution. A cloudy future. Nat. Hist. 91(12):72.

188. Carson, H. L., and L. T. Teramoto. 1982l. Bibliography of Hawaiian Drosophilidae. Dros. Info. Serv. 58:215–226.

189. Carson, H. L., C. B. Krimbas, and M. Loukas. 1982m. Slime fluxes a larval niche of *D. subobscura* Col. Droso. Info. Serv. 58:34–35.

190. Carson, H. L. 1982n. Speciation as a major reorganization of polygenic balances. In C. Barigozzi (ed.), Mechanisms of Speciation, pp. 411–433. Alan R. Liss, New York.

191. Carson, H. L. 1982o. Fluctuations in size of certain *Drosophila* populations in the Olaa Tract, Hawaii Volcanoes National Park. In Proc. Fourth Conf. in Natural Sciences, Hawaii Volcanoes National Park, p. 40 [abstract]. Cooperative National Park Resources Unit, Department of Botany, University of Hawaii, Honolulu.

192. Okada, T., and H. L. Carson. 1982p. Drosophilidae associated with flowers in Papua New Guinea IV. Araceae, Compositae, Convolvulaceae, Leguminosae, Malvaceae, Rubiaceae. Kontyû 50:511–526.

193. Carson, H. L., and T. Okada. 1982q. Drosophilidae associated with flowers in Papua New Guinea (Diptera: Drosophilidae). Entomol. Generalis 8:13–16.

194. Carson, H. L. 1983a. Genetical processes of speciation on high oceanic islands. In 15th Cong. Pacific Science Association, Vol. 1, p. 34 [abstract]. University of Otago, Dunedin, N.Z.

195. Carson, H. L. 1983b. Chromosomal sequences and interisland colonizations in Hawaiian *Drosophila*. Genetics 103:465–482.

196. Hunt, J. A., and H. L. Carson. 1983c. Evolutionary relationships of four species of Hawaiian *Drosophila* as measured by DNA reassociation. Genetics 104:353–364.

197. Carson, H. L., and R. Lande. 1983d. Inheritance of a secondary sexual character in *Drosophila silvestris*. Genetics 104:s12 [abstract].

198. Ashburner, M., H. L. Carson, and J. N. Thompson, Jr. (eds.), 1983e. The Genetics and Biology of *Drosophila* Vol. 3c, 428 pp. Academic Press, London.

199. Ashburner, M., and H. L. Carson. 1983f. A checklist of maps of polytene chromosomes of Drosophilids. Dros. Info. Serv. 39:148–151.

200. Carson, H. L. 1983g. The genetics of the founder effect. In C. M. Schonewald-Cox, S. M. Chambers, B. MacBryde and L. Thomas (eds.), Genetics and Conservation, Chapter 11, pp. 189–200. Benjamin/Cummings Publ., Menlo Park, CA.

201. Okada, T., and H. L. Carson. 1983h. Drosophilidae from banana traps over an altitudinal transect in Papua New Guinea. I. Descriptions of new species and notes on newly recorded species. Int. J. Entomol. 25:127–141.

202. Carson, H. L., and T. Okada. 1983i. Drosophilidae from banana traps over an altitudinal transect in Papua New Guinea. II. Frequency of species at eight collecting sites. Int. J. Entomol. 25:142–151.

203. Okada, T., and H. L. Carson. 1983j. The genus *Sphaerogastrella* DUDA (Diptera, Drosophilidae) of Papua New Guinea. Kontyû 51:367–375.

204. Carson, H. L. 1983k. Artificial selection of a secondary sexual character in *Drosophila silvestris* of the island of Hawaii. Proc. XV Int. Cong. Genet., Part II, p. 438 [abstract]. Oxford & IBH Publishing, New Delhi.

205. Carson, H. L., F. C. Val, and M. R. Wheeler. 1983l. Drosophilidae of the Galápagos Islands, with descriptions of two new species. Int. J. Entomol. 25:239–248.

206. Ashburner, M., H. L. Carson, and J. N. Thompson, Jr. (eds.). 1983m. The Genetics and Biology of Drosophila, Vol. 3d, 382 pp. Academic Press, London.

207. Carson, H. L., and W. B. Heed. 1983n. Methods of collecting *Drosophila*. In M. Ashburner, H. L. Carson and J. N. Thompson, Jr. (eds.), The Genetics and Biology of Drosophila, Vol. 3d, pp. 1–28. Academic Press, London.

208. Carson, H. L. 1983o. Genetical processes of evolution on high oceanic islands. In I. W. B. Thornton (ed.), Symposium on Distribution and Evolution of Pacific Insects. GeoJournal 7:543–547.

209. Okada, T., and H. L. Carson. 1983p. The genera *Phorticella* Duda and *Zaprionus* Coquillett (Diptera, Drosophilidae) of the Oriental region and New Guinea. Kontyû 51:539–553.

210. Chang, L. S., and H. L. Carson. 1984a. Metaphase chromosome comparisons among five species of Hawaiian *Drosophila*. Genetics 107:s18–19 [abstract].

211. Hunt, J. A., J. G. Bishop, III, and H. L. Carson. 1984b. Observation of a repetitive DNA element in five species of Hawaiian *Drosophila*. Genetics 107:s50 [abstract].

212. Carson, H. L., and J. V. Neel. 1984c. Harrison Dailey Stalker 1915–1982. Genetics 107:s138–140.

213. Carson, H. L., and L. T. Teramoto. 1984d. Artificial selection for a secondary sexual character in males of *Drosophila silvestris* from Hawaii. Proc. Natl. Acad. Sci. USA 81:3915–3917.

214. Herforth, R. S., H. L. Carson, and L. Chang. 1984e. A new arrival to the Hawaiian Islands: *Drosophila cardini*. Dros. Info. Serv. 60:124.

215. Carson, H. L. 1984f. Sexual dimorphism and sexual selection in Hawaiian *Drosophila*. Proc. Fifth Conference in Natural Sciences, Hawaii Volcanoes National Park, p. 51 [abstract]. Cooperative National Park Resources Study Unit, Department of Botany, University of Hawaii at Manoa, Honolulu.

216. Carson, H. L. 1984g. Genetics. In McGraw-Hill Yearbook of Science and Technology, pp. 193–194 McGraw-Hill Book Co., New York.

217. Carson, H. L., and A. R. Templeton. 1984h. Genetic revolutions in relation to speciation phenomena: the founding of new populations. Annu. Rev. Ecol. Systematics 15:97–131.

218. Carson, H. L., and R. Lande. 1984i. Inheritance of a secondary sexual character in *Drosophila silvestris*. Proc. Natl. Acad. Sci. USA 81:6904–6907.

219. Carson, H. L. 1984j. Speciation and the founder effect on a New Oceanic Island. In F. J. Radovsky, P. H. Raven, and S. H. Sohmer (eds.), Biogeography of the Tropical Pacific, pp. 45–54. B. P. Bishop Museum, Honolulu.

220. Hunt, J. A., J. G. Bishop, III, and H. L. Carson. 1984k. Chromosomal mapping of a middle-repetitive DNA sequence in a cluster of five species of Hawaiian *Drosophila*. Proc. Natl. Acad. Sci. USA 81:7146–7150.

221. Crews, D., L. T. Teramoto, and H. L. Carson. 1985a. Behavioral facilitation of reproduction in sexual and parthenogenetic *Drosophila*. Science 227:77–78.

222. Carson, H. L. 1985b. Genetic variation in a courtship-related male character in *Drosophila silvestris* from a single Hawaiian locality. Evolution 39:678–686.

223. Chang, L. S., and H. L. Carson. 1985c. Metaphase karyotype identity in four homosequential *Drosophila* species from Hawaii. Can. J. Genet. Cytol. 27:308–311.

224. Takenaka, J. H., H. L. Carson, and D. Crews. 1985d. Active courtship behavior and copulatory success by sterile XO males of *Drosophila mercatorum*. Genetics 110:s101 [abstract].

225. Titus, E. A., H. L. Carson, and R. G. Wisotzkey. 1985e. Another new arrival to the Hawaiian Islands: *Drosophila bryani* Malloch. Dros. Info. Serv. 61:171.

226. Carson, H. L. 1985f. *Evolution of Fish Species Flocks*, eds. A. A. Echelle and I. Kornfield, a review. Q. Rev. Biol. 60:347–348.

227. Carson, H. L. 1985g. Unification of speciation theory in plants and animals. Syst. Bot. 10:380–390.

228. Carson, H. L., C. B. Krimbas, and M. Loukas. 1985h. Slime fluxes, a larval niche of *Drosophila subobscura* Col. Le Congrès Int. sur la Zoogéographie et l'Écologie de la Grèce et des Régions Avoisinantes. Biol. Gallo-Hellen. 10:319–321.

229. Carson, H. L. 1985i. Genetic microdifferentiation due to sexual selection in *Drosophila* and man. In Y. R. Ahuja and J. V. Neel (eds.), Genetic Microdifferentiation in Human and Other Animal Populations, pp. 1–14. Indian Anthropological Association, University of Delhi.

230. Ashburner, M., H. L. Carson, and J. N. Thompson, Jr. (eds.). 1986a. The Genetics and Biology of Drosophila, Vol. 3e, 548 pp. London, Academic Press.

231. Carson, H. L. 1986b. Inversion heterozygosity is disproportionately high among copulating males of *Drosophila silvestris*. Genetics 113:s39 [abstract].

232. Wisotzkey, R. G., and H. L. Carson. 1986c. Sperm predominance in *Drosophila silvestris*. Genetics 113:s46 [abstract].

233. Carson, H. L. 1986d. Sexual selection and speciation. In S. Karlin and E. Nevo (eds.). Evolutionary Processes and Theory, pp. 391–409. Academic Press, London.

234. Carson, H., K. Kaneshiro, and W. P. Mull. 1986e. Natural hybridization between two species of Hawaiian *Drosophila*. Abstracts of the First Int. Cong. Dipterology, p. 35. Hungarian Academy of Sciences, Budapest.

235. Carson, H. L. 1986f. Drosophila populations in the Ola'a Tract, Hawaii Volcanoes National Park, 1971–1986. Proc. Sixth Conference in Natural Sciences, Hawaii Volcanoes National Park, pp. 3–9. Cooperative National Park Resources Unit, Department of Botany, University of Hawaii at Manoa, Honolulu.

236. Carson, H. L. 1986g. The origin of species. In R. J. Berry and A. Hallam (eds.), The Collins Encyclopedia of Animal Evolution, Collins, 8 Grafton St., London.

237. Carson, H. L. 1986h. Change in Isolation: speciation in the Hawaiian Islands. In R. J. Berry and A. Hallam (eds.), The Collins Encyclopedia of Animal Evolution, pp. 124–125. Collins, 8 Grafton St., London.

238. Carson, H. L. 1986i. Patterns of inheritance. Am. Zoologist 26:797–809.

239. Carson, H. L. 1987a. Population genetics, evolutionary rates, and Neo-Darwinism. In K. S. W. Campbell and M. F. Day (eds.), Rates of Evolution, pp. 209–217. Allen & Unwin, London.

240. Kaneshiro, K. Y., and H. L. Carson. 1987b. Sexual selection and the pomace fly. Nature Conservancy News 37:23–25.

241. Carson, H. L. 1987c. Tracing ancestry with chromosomal sequences. Trends Ecol. Evol. 2(7):203–207.

242. Carson, H. L. 1987d. High fitness of heterokaryotypic individuals segregating naturally within a long-standing laboratory population of *Drosophila silvestris*. Genetics 116:415–422.

243. H. L. Carson. 1987e. Colonization and speciation. In A. J. Gray, M. J. Crawley, and P. J. Edwards (eds.), Colonization, Succession and Stability, pp. 187–206. Blackwell, Oxford.

244. Craddock, E., and H. L. Carson. 1987f. Chromosomal heterogeneity, genetic structure, and recombination in populations of the island species *Drosophila silvestris* [abstract]. In Population Biologists of New England. Queens College, City University of New York, NY.

245. Carson, H. L. 1987g. The process whereby species originate. Bioscience 37(10):715–720.

246. Carson, H. L. 1987h. The genetic system, the deme, and the origin of species. Annu. Rev. Genet. 21:405–423.

247. Carson, H. L. 1987i. The contribution of sexual behaviour to Darwinian fitness. Behav. Genet. 17:597–611.

248. HuKai and H. L. Carson. 1987j. Interspecific copulation between far distant Drosophila species. Dros. Info. Serv. 66:75–76.

249. Tonzetich, J., T. W. Lyttle and H. L. Carson. 1988a. Induced and natural break sites in the chromosomes of Hawaiian Drosophila. Proc. Natl. Acad. Sci. USA 85:1717–1721.

250. Carson, H. L. and R. G. Wisotzkey. 1988b. Shift in a population of *Drosophila silvestris* following a bottleneck. Genome 30, Suppl. 1:386 (abstract).

251. Carson, H. L. 1988c. Genetic Biogeography of Island Species: Perspectives and Techniques. In: *The Biogeography of the Island Region of Western Lake Erie*. J. F. Downhower, Ed. Ohio University Press, Columbus. pp. 3–9.

252. Carson, H. L. 1989a. Gene Pool Conservation. In: *Conservation Biology in Hawaii*. C. P. and D. B. Stone, Eds. Cooperative National Park Resources Studies Unit. Honolulu: University of Hawaii Press. pp. 118–124.

253. Carson, H. L., K. Y. Kaneshiro and F. C. Val. 1989b. Natural hybridization between the sympatric Hawaiian species *Drosophila silvestris* and *Drosophila heteroneura*. Evolution 43:190–203.

2

Founder Effects and Genetic Revolutions in Microevolution and Speciation: An Historical Perspective

WILLIAM B. PROVINE

The evolutionary process produces both exquisite adaptation and amazing diversity in biological organisms. Some evolutionary biologists think that natural selection, which produces the adaptation, also produces the diversity by means of speciation. The only factor required for speciation beyond natural selection is geographical or ecological isolation. Richard Dawkins in his recent book, *The Blind Watchmaker* (1986), takes this view, as did R. A. Fisher in his famous book, *The Genetical Theory of Natural Selection* (1930). Other evolutionary biologists believe that the diversity of animals and plants is produced by a variety of mechanisms in addition to natural selection and isolation; they argue that speciation is often a process that is different and distinguishable from phyletic evolution in natural populations (Mayr, 1963; Eldredge and Gould, 1972).

Natural selection and speciation are both exceedingly difficult to study in action, but natural selection is easier. Charles Darwin believed that natural selection in action could not be measured in a period of less than 50 years. We know now that he was overly pessimistic. John Endler has published a careful listing and analysis of documented cases of natural selection in the wild (Endler, 1986). The surprise for someone who does not understand the great difficulties of documenting natural selection in action is that Endler's book is comprehensive, but only 336 pages long, and in many of his cases, only the *results* of natural selection were actually observed. The corresponding book on observed and documented cases of speciation in action would be very thin indeed, and some evolutionary biologists claim even those pages would be mostly blank. That is one good reason why, in the face of great mounds of evidence of the *results* of speciation processes, there is so much controversy about mechanisms of speciation.

Since Ernst Mayr invented the terms and popularized the concepts of "founder effects" (Mayr, 1942) and "genetic revolutions" (Mayr, 1954), both have played central roles in discussions of speciation. In the more than three decades since Mayr argued that some founder events led to genetic revolutions, which in turn led to speciation events, evolutionary biologists have debated the originality of Mayr's concepts and the evolutionary significance of founder effects and genetic revolutions. Mayr, Brown, Carson, Templeton, Lande, Charlesworth, Barton, and many others have extended the concepts and have evaluated them in the light of evidence from natural populations. At present, the diversity of opinion about founder effects and genetic revolutions is great and the debates are intense.

After a brief analysis of Mayr's views on founder effects and genetic revolutions, I will trace the historical development of concepts of founder effects in relation to genetic revolutions and speciation. In particular, I will address the question of the relationship of Sewall Wright's "peak shifts" on his adaptive landscape to Mayr's theory of founder effect and genetic revolution. Also, I will examine the continuing influence of theories of speciation in the period before 1954 upon those that have come after, with special reference to speciation in the Hawaiian Islands. The historical development is exceedingly complex, and this chapter is only a very preliminary analysis, especially for the period after 1954.

ERNST MAYR ON "FOUNDER EFFECTS" AND "GENETIC REVOLUTIONS"

From the beginning of his work in systematics under the influence of Stresemann and Rensch, even before he became a Darwinian (see Mayr and Provine, 1980, pp. 413–423), Mayr believed strongly that almost all speciation required geographical isolation. Physiological speciation, ecological speciation, and other versions of what Mayr later termed "sympatric" speciation have had, since the time of Darwin, strong advocates. But all of Mayr's efforts to understand the mechanisms of speciation are premised upon the belief that speciation is predominantly, though perhaps not exclusively, geographical. Thus the ideas of founder effect and genetic revolution refer, in Mayr's view, only to geographical speciation.

Founder Effect

Mayr's *Systematics and the Origin of Species* (1942), which contained his first statement of the founder principle, and his "Change of Genetic Environment and Evolution" (1954), where he first presented his concept of genetic revolutions, are two of his most widely read and influential works. A precise understanding of his arguments and evidence in these works is crucial for understanding their antecedents and later influence.

His first statement of the founder principle came in a long chapter entitled "The Biology of Speciation." Here Mayr classified the major factors influencing speciation into two interacting categories—the "internal" (physiological or genetical) and the "external" (environmental). Of the external factors, one of the most important was "restriction of random dispersal by geographic barriers," to which Mayr devoted a 14-page section. He clearly stated that geographical barriers, at least temporary ones, were a crucial requirement of the speciation process. And in the subheading "population size and variability," he argued further that population structure, especially population size, was an important variable in the genetic differentiation of populations separated by geographical barriers.

> Naturalists have known for a long time that island populations tend to have aberrant characteristics. Wright (1931, 1932, and elsewhere) found the theoretical basis for this by showing that in small populations the accidental elimination of genes may be a more successful process than selection. Furthermore, recessive mutations have a much better chance to become homozygous than in a large panmictic population. It is therefore very important to learn something about the actual size of distributional islands and of their populations. (Mayr, 1942, p. 234).

Mayr then offered examples of small isolated populations of cave animals, lizards, fish, birds, and mice, all of which exhibited aberrant characteristics. His analysis of these examples drew upon the work of Sewall Wright:

> An exact determination of the size of an isolated population is of importance, in view of Sewall Wright's work on gene loss in small populations. Owing to "accidents of sampling," small populations have a trend toward genetic homogeneity or at least toward a much-reduced variability. This is quite apparent in taxonomic work, although only a few systematists have taken the trouble to make careful measurements and to work out the coefficients of variation. (Mayr, 1942, p. 235)

Why exactly did population size have consequences for the speciation process? Mayr explained:

> The calculations of Sewall Wright (1931, 1932, and elsewhere) indicate that effective populations have to be rather small, in the order of several hundred individuals or less, before they can be expected to approach genetic homogeneity due to accidental gene loss. If the population size is larger (thousands to tens of thousands of individuals), there still may be rapid evolution owing to mutation pressure (in the absence of appreciable selection), but the population will remain much more variable. If the size of the effective breeding population is still greater, approaching panmixia in varying degrees, evolution will be slowed down considerably. The consequence of this consideration is that evolution should proceed more rapidly in small populations than in large ones, and this is exactly what we find. (Mayr, 1942, p. 236).

Clearly, Mayr relied upon Wright's theoretical formulations for analyzing reduced variability and increased rate of evolution in relatively small, geographically isolated populations. According to Mayr, however, random genetic drift was not the only way to produce reduced variability in isolated populations:

> The reduced variability of small populations is not always due to accidental gene loss, but sometimes to the fact that the entire population was started by a single pair or by a single fertilized female. These "founders" of the population carried with them only a very small proportion of the variability of the parent population. This "founder" principle sometimes explains even the uniformity of rather large populations, particularly if they are well isolated and near the borders of the range of the species. (Mayr, 1942, p. 237).

This was the first statement of the founder principle.

A corollary to Mayr's arguments on importance of population size in the speciation process came in the next section, "fluctuating population size." Here he argued that even very large populations that went through population bottlenecks could exhibit rapid evolution:

> Of greater importance than the seasonal changes, at least in a few species, are the cyclic changes such as have been described in some arctic animals (Elton 1930). These population fluctuations are, as Sewall Wright (1940a, b) and others have shown, of great importance in speeding up changes in the genetic make-up of a population. The effective population size of such a fluctuating species is much closer to the smaller than to the larger number. "Thus if the breeding population in an isolated region increases ten-fold in each of six generations during the summer (N_0 to $10^6 N_0$) but falls at the end of the winter to the same value N_0, the effective size of population ($N = 6.3 N_0$) is relatively small" (Wright 1940b, 242). It is obvious that the survival of only one N_0 out of one million N_0's must have a considerable effect on the genetic composition of the respective population, and this is true no matter whether accident or superior survival value accounts for the survival of the remaining few. (Mayr, 1942, p. 238)

In these passages, Mayr was attempting to explain the mechanisms of geographical speciation. He and other naturalists had observed that island populations exhibited characters different from those of mainland organisms and he concluded that the speciation process was accelerated by the isolation of relatively small populations. Mayr argued that evolution was faster in small populations than in large ones because inbreeding led to random genetic drift, accidental gene loss, and consequent increase of homozygosis and decrease of genotypic and phenotypic variability in the small populations. These mechanisms caused rapid, divergent evolution in the small populations. "The potentiality for rapid divergent evolution in small populations explains also why we have on islands so many dwarf or giant races, or races with peculiar color characters (albinism, melanism), or with peculiar structures (long bills in birds), or other peculiar characters (loss of special male plumage in birds)." (Mayr, 1942, p. 236).

Why or how random genetic drift led to "rapid divergent evolution," Mayr did not say, but he clearly was relying upon Sewall Wright's analysis of evolution in small populations. Later, I will turn to the question of whether Mayr was really following Wright's lead in this interpretation of the effects of random genetic drift.

Thus to Mayr in 1942, speciation required geographical isolation and was much accelerated by small population size, as in small island populations. Citing Wright, Mayr argued that even a very large population, if it underwent periodic crashes, would have a small effective population size that increased the probability of rapid evolution.

The "founder" principle, as introduced by Mayr in 1942, was an auxiliary mechanism (less important than random drift) for producing reduced variability in an isolated population started by a few individuals or even a single fertilized female. There was no hint in the 1942 book that the founder principle was a frequent factor in speciation. This founder principle, according to Mayr, could even explain the relative uniformity of "rather large" isolated populations. Although he did not say so directly, the implication was that the founder principle, like gene loss from random genetic drift, promoted rapid evolution and led to possible speciation. Again, as in the case of random drift, Mayr gave no indication of precisely how or why the founder principle led to speciation.

"Genetic Revolution"

Mayr's (1954) paper "Change of Genetic Environment and Evolution" begins where he left off, 12 years earlier, in *Systematics and the Origin of Species*. He had learned a great deal about genetics during this period, especially in the early 1950s, from extended discussions with Bruce Wallace at Cold Spring Harbor. Mayr was particularly impressed by Wallace's emphasis upon the high degree of integration and coadaptation of the genome. In the 1954 paper, Mayr again emphasized "the conspicuous difference of most peripherally isolated populations of species." But whereas in 1942 he left only a vague connection between the aberrant characteristics of these isolated populations and random genetic drift, he now argued that

> for such a striking dissimilarity of peripherally isolated populations two reasons are usually cited: difference of physical and biotic environment or genetic drift. It seems to me that neither of these factors nor a combination of the two can provide a full explanation, even though both may be involved. (Mayr, 1954, p. 158)

The answer to such aberrant characteristics, Mayr suggested, lay in understanding that the selective value of a single gene depended greatly upon the overall genetic environment. (Here he quoted from Wright, 1931.)

The genome of an individual was a highly interactive coadapted gene complex, Mayr argued. More importantly, the individuals in an entire species shared significant portions of that coadapted gene complex, which resisted changes: "Such a well-integrated, coadapted gene-complex constitutes an evolutionary unit in spite of its intrinsic variability. Any disharmonious gene or gene-combination which attempts to become incorporated in such a gene-complex will be discriminated against by selection" (Mayr, 1954, p. 165).

Therefore a major problem in speciation was to discover how was it possible to overcome the intertia of such a coadapted gene complex. One way was to send the gene-complex through a founder population:

> One of the obvious effects of the sudden reduction of population size in the founder population will be a strong increase in the frequency of homozygotes. As a consequence, homozygotes will be much more exposed to selection and those genes will be favoured which are specially viable in the homozygous condition. Thus, the "soloist" is now the favourite rather than the "good mixer."
>
> We come thus to the important conclusion that *the mere change of the genetic environment may change the selective value of a gene very considerably.* Isolating a few individuals (the "founders") from a variable population which is situated in the midst of the stream of genes which flows ceaselessly through every widespread species will produce a sudden change of the genetic environment of most loci. This change, in fact, is the most drastic genetic change (except for polyploidy and hybridization) which may occur in a natural population, since it may affect all loci at once. Indeed, it may have the character of a veritable "genetic revolution." Furthermore, this "genetic revolution," released by the isolation of the founder population, may well have the character of a chain reaction. Changes in any locus will in turn affect the selective values at many other loci, until finally the system has reached a new state of equilibrium. (Mayr, 1954, pp. 169–170)

Mayr emphasized that a genetic revolution was not to be expected in every founder population: "a 'genetic revolution' in the founder population is only a potentiality but does not need to happen every time a population is isolated, if the genetic constitution of the founders does not favour it" (p. 171). He also emphasized that "during a genetic revolution the population will pass from one well integrated and rather conservative condition through a highly unstable period to another new period of balanced integration. The new balance will be reached after a great loss of genetic variability" (p. 172). Many such populations, after undergoing a genetic revolution, would be severely depleted of genetic variability and subject to extinction should the environment change. Some populations, however, might find a new ecological niche after the genetic revolution and gradually accumulate genetic variability. This possibility Mayr illustrated with his famous diagram, illustrated in Figure 2.1.

Founder Effect and Genetic Revolution in Mayr's *Animal Species and Evolution* (1963)

In this book, Mayr discussed in detail his concepts of founder effect and genetic revolution. There was little significant conceptual change from the

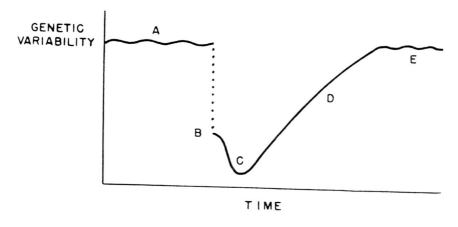

(from Mayr 1954, 174)

Figure 2.1. Mayr's illustration of the effect of a "genetic revolution" in a founder population on genetic variability. (Reproduced by permission from Mayr, 1954)

presentation in 1954, except that his earlier views on the coadaptedness of the genome had been strongly buttressed by Lerner's influential book, *Genetic Homeostasis* (1954); Mayr devoted a whole chapter to "the unity of the genotype." He also attempted to be more precise in specifying the preconditions necessary for a genetic revolution to occur in a founder population.

One particularly interesting question, which by 1963 had already attracted the attention of Hampton Carson and W. L. Brown (more to come on that), concerned the origin of the founders. Was it better that the founder come from the biological center of the species where, according to available evidence, genetic variability was greatest? (The biological center of a population, where genetic exchange is intense, can easily include the geographic periphery of a habitat—for example, an island.) Or, was the partial adaptation of a marginal population to its environment a precondition that gave the founders from such populations a greater probability of giving rise to a genetic revolution, despite apparent reduced genetic variability of the founders?

Mayr appeared to answer this question clearly.

> The border populations presumably barely maintain themselves and the new colonists beyond the species border (in mobile species such as birds and insects) come from farther inside the species range, where conditions permit a greater surplus of individuals and the increased population density in turn may stimulate emigration, but the gene complex is not adapted for the conditions of the border region. (Mayr, 1963, p. 524)

> The basic gene complex of the species (with all the species-specific canalizations and feedbacks) functions optimally in the area for which it had evolved by selection, usually somewhere near the center. Here it is in bal-

ance with the environment and here it can afford much superimposed genetic variation and experimentation in niche invasion. Toward the periphery this basic genotype of the species is less and less appropriate and the leeway of genetic variation that it permits is increasingly narrowed down until much uniformity is reached. Such an impoverished and yet dependent population is not the ideal starting point for speciation. (Mayr, 1963, p. 527)

In these passages, Mayr clearly said that the founders of populations likely to undergo genetic revolutions came from near the center of the parent population. But only a few pages later, he came to a different conclusion:

The reduction in polymorphism and heterozygous balances that we find in peripheral (and particularly isolated) populations reduces the genetic homeostasis and evolutionary inertia of these populations. They are far better capable of responding to new selection pressures and, consequently, to new evolutionary opportunities than populations from the "dead heart" of the species. They are more immediately in the position to utilize new gene combinations that are generated during the genetic revolution than are populations in which the genes are tightly knitted together by numerous balancing mechanisms. (Mayr, 1963, p. 534–535)

The origin of the founders was indeed a vexing problem. On the one hand, the founders had to carry with them as much variability as possible as a basis for flexible adaptation to the new environment and selective divergence. On the other hand, the founders also had to be capable of undergoing a genetic revolution, thus of overcoming the genetic inertia of their coadapted genome. Filling both requirements at the same time was a real problem. Mayr solved this problem in 1963 by inconsistency.

Concluding Comment on Mayr's Theory of Speciation

Mayr is generally known among evolutionary biologists as the premier "Neo-Darwinian." Most biologists now think of a Neo-Darwinian as one who believes that gradual natural selection dominates the evolutionary process. An important thing to understand about Mayr's conception of speciation is that the process involves far more than just natural selection. Population structure (isolated founders), random genetic drift (in small founder populations), and nonadaptive breakdowns of balanced genetic systems and creation of new ones through genetic revolutions are all central to Mayr's conception of speciation. His conception of speciation is not a simple extension of Darwin's views, nor is speciation for Mayr a mere extension of phyletic evolution within populations. He has constantly criticized Darwin's view of sympatric speciation as naive. If Fisher is to be called a "Neo-Darwinian," then I would prefer a different term to describe Mayr.

In a recent summary of his views on speciation (Mayr, 1982b), Mayr argued that before he championed founder effects and genetic revolutions in very small geographically isolated populations, the "classical" concept of allopatric speciation was the "dumbbell" model. This model posited a geographically widespread species, with a geographical barrier somewhere in the middle, in the pattern of a dumbbell. For Fisher and Wright, and for systematists generally, I think Mayr's thesis here is correct. Indeed, as late as 1949, Mayr's own presentation of allopatric speciation fit the dumbbell model, complete with a perfect drawing of a dumbbell (Mayr, 1949, p. 287). It is a measure of the influence of Mayr's peripatric model that Lande, Barton, B. Charlesworth, Slatkin, and many others regard Mayr's view as the dominant interpretation of speciation in the 1970s and 1980s.

A BRIEF HISTORICAL PERSPECTIVE ON FOUNDER EFFECTS AND GENETIC REVOLUTIONS

From the time of Darwin onward, evolutionary biologists have been familiar with the following four general steps of the speciation process, referring specifically to island populations or those separated by obvious geographical boundaries.

1. Founder effects—starting a new isolated population with relatively few individuals, perhaps only a single fertilized female. This category also includes the problem of the origin of the founder population.
2. Breakup of the old hereditary system.
3. Appearance of new hereditary constitutions.
4. Speciation events.

The four steps frequently merge into one another. In Mayr's concept of a genetic revolution, steps 2 and 3 happen simultaneously, with step 4 close behind.

Evolutionists and systematists before Mayr understood these steps and their sequence. No one has been more aware of this history than Mayr himself, whose detailed work on the history of geographical speciation may be found in his *Animal Species and Evolution* (1963, chapter 16), *Evolution and the Diversity of Life* (1976, pp. 117–220), and *Growth of Biological Thought* (1982). Mayr has never claimed to have invented the idea of founder effect, or the view that old hereditary constitutions have to be broken up and new ones formed for the speciation process. He has claimed to have strongly emphasized the importance of founder effects and genetic revolutions, his particular version of the breakup of old and creation of new genotypes so crucial to geographical speciation.

Tracing the history of concepts related to founder effect and genetic revolution in the late nineteenth and early twentieth centuries would require a large book. This book would examine the work of a great many biologists, among the more important of whom are Darwin, Moritz Wagner, John T. Gulick, G. J. Romanes, Alfred Russel Wallace, E. B. Poulton, David Starr Jordan, Vernon L. Kellogg, Karl Jordan, William Bateson, and Hugo de Vries, among many others. Such a survey is obviously impossible here. I will simply provide one example, Gulick, whose views on speciation were very influential in the late nineteenth and early twentieth centuries.

JOHN T. GULICK AND THE PROBLEM OF SPECIATION

For most of his active life, Gulick (1832–1923) was a missionary in China and Japan. He grew up in Hawaii where he developed a keen interest in natural history and was particularly struck by the almost incredible diversity of closely related species of snails, birds, and insects. As a naturalist, his primary interest was in the mechanisms of speciation (on J. T. Gulick, see A. Gulick, 1932; Lesch, 1975; Kottler, 1976; and Provine, 1986, pp. 217–220). In 1853, before he began his missionary service, Gulick gathered a huge collection of strikingly beautiful Achatinellid snails that he used for more than 50 years as his primary research material for studying the problems of speciation. His publications were sporadic, coinciding with breaks from his demanding missionary work and after his retirement. Even so, he published more on mechanisms of speciation than any other biologist of his time.

Darwin's thesis in the *Origin* was that speciation was primarily the result of geographical or ecological isolation in combination with altered forces of natural selection in the different habitats. Thus the divergence of two populations into varieties and later perhaps into species required no special mechanisms other than separation of the populations into situations in which the effects of the environment were different. Gulick's observations on Achatinellid snails led him to differ significantly from Darwin's view, which he considered a plausible hypothesis but one that did not fit direct observations in Hawaii. His views were first published in 1872, then amplified in long papers in 1888 and 1890, and in a major monograph in 1905.

> In my study of Sandwich-Island terrestrial mollusks my attention was early arrested by the fact that wide diversity of allied species occurs within the limits of a single island, and in districts which present essentially the same environment. As my observations extended, I became more and more impressed with the improbability that these divergences had been caused by differences in the environment. It was not easy to prove that sexual selection had no influence; but, owing to the very low grade of intelligence possessed by the creatures, it seemed impossible that the form and colouring of the shells should be the result of any such process. I was therefore led to search for some other cause of divergent transformation, the diversity of whose action is not dependent on differences in nature external to the organism.
>
> I found strong proof that there must be some such principle, not only in the many examples of divergence under uniform activities in the environment, but in the fact that the degrees of divergence between nearly allied

forms are roughly measured by the number of miles by which they are separated, and in the fact that this correspondence between the ratios of distance and the ratios of divergence is not perceptibly disturbed by passing
over the crest of the island into a region where the rainfall is much heavier,
and still further in the fact that the average size of the areas occupied by the
species of any group varies, as we pass from group to group, according as
the habits of the group are more or less favourable to migration. I perceived
that these facts could all be harmonized by assuming that there is some cause
of divergence more constant and potent than differences in nature external
to the organism; and that the influence of this cause was roughly measured
by the time and degree of separation. (Gulick, 1888, p. 189)

For Gulick, geographical separation and natural selection were insufficient to explain speciation in the Achatinellid snails; other mechanisms
were required.

Many of Gulick's views sound familiar to biologists working on the
problem of speciation today. He believed that geographical isolation was
necessary but not sufficient for speciation. He thought that after geographical isolation, many mechanisms contributed to continuing differentiation. Among them were random sampling, which resulted in changes
in mate recognition patterns, social behavior or habits, fecundity, tendency to migrate, time of fertility, ecological preference, utilization of
food sources, and many other factors. Small founder populations, which
often began with a nonrepresentative sampling of hereditary features,
would continue to differentiate from the parent population even if the environments were basically similar.

Gulick's widely read views on the process of speciation serve to illustrate my argument that the rough equivalent of founder effects, breakup
of old hereditary systems, and appearance of new ones were all crucial
for the understanding of the speciation process even for nineteenth-cen-
tury naturalists interested in geographical speciation. Gulick's evidence
for his views was inadequate; indeed, even by the standards of his time,
his understanding of the ecological settings of the snails in his collections
was very weak.

R. A. FISHER, NATURAL SELECTION, AND
THE FISSION OF SPECIES

Fisher wrote nothing about speciation before his *Genetical Theory of Natural Selection* in 1930. Speciation, or as he called it in 1930, "fission of
species," was for Fisher simply the effect of geographical or ecological
isolation of a part of a species and the continued differentiation of the two
parts under the influence of natural and/or sexual selection. In other
words, his theory of speciation was a direct extension of his theory of
phyletic evolution within populations.

He believed that species were closely tied together genetically, and he
referred to "the cohesive power of the species" (Fisher, 1930, p. 124).

The problem of speciation was to discover the stresses that could break up the shared genetic cohesion.

> The close genetic ties which bind species together into single bodies bring into relief the problem of their fission—a problem which involves complexities akin to those that arise in the discussion of the fission of the heavenly bodies, for the attempt to trace the course of events through intermediate states of instability, seems to require in both cases a more detailed knowledge than does the study of stable states. (Fisher, 1930, p. 125)

In principle, however, speciation was no mystery for Fisher. He clearly believed that natural selection of individual genes in relatively large populations was the primary determinant of the evolutionary process. The genetic cohesion of a species was no match for the power of natural selection operating differentially upon a species divided geographically or ecologically into two parts. Fisher pointed out that sexual preference was particularly subject to selection and could be a potent factor in promoting reproductive isolation (Fisher, 1930, pp. 125–131; in this connection, see especially the analysis of Lande, 1982, and the continuing work of Kaneshiro on speciation in Hawaiian *Drosophila*, represented by his article in this volume).

Fisher's view of speciation was very much like that of Darwin or Wallace (with the exception of sexual selection; see Wallace's *Darwinism*, 1889). Indeed, what Fisher had done was to provide a modern genetical interpretation of Darwin's view of speciation. This view continues to have strong proponents.

ECOLOGICAL PERSPECTIVE: CHARLES ELTON ON POPULATION FLUCTUATION AND SPECIATION

Elton, more than any other evolutionary biologist in the twentieth century, related observations on population fluctuations to the problem of speciation. At the time he wrote his first book on ecology (*Animal Ecology*, 1927), Elton believed that closely related species differed mostly by nonadaptive characters, a view held by Sewall Wright, Julian Huxley, and many other evolutionary biologists at the time. The problem of speciation for Elton was to find a mechanism that could produce nonadaptive differentiation and sterility between the divided populations. His research on fluctuations in animal numbers suggested an answer:

> Many animals periodically undergo rapid increase with practically no checks at all. In fact, the struggle for existence sometimes tends to disappear almost entirely. During the expansion in numbers from a minimum, almost every animal survives, or at any rate a very high proportion of them do so, and an immeasurably larger number survives than when the population remains constant. If therefore a heritable variation were to occur in the small nucleus of animals left at a minimum of numbers, it would spread very quickly and automatically, so that a very large proportion of numbers of individuals

would possess it when the species had regained its normal numbers. In this way it would be possible for non-adaptive (indifferent) characters to spread in the population, and we should have a partial explanation of the puzzling facts about closely allied species, and of the existence of so many apparently non-adaptive characters in animals. (Elton, 1927, p. 187)

Since 1927, Elton has frequently reiterated his argument that fluctuations in population size must be taken into account in explaining speciation, and that the character and intensity of natural selection is always dependent upon these fluctuations. His thesis has become part of the common sense of modern evolutionary biology.

SEWALL WRIGHT, SHIFTING BALANCE, AND SPECIATION

In a major paper, Charlesworth, Lande, and Slatkin state that Mayr's conception of the speciation process was that

> speciation events occur most often in small isolated populations. Random changes in genotypic composition caused by passage through a bottleneck in population size are supposed to induce a "genetic revolution," in which there is a shift in the state of a population from one set of "coadapted" genotypes to another. This idea is, in essence, a special case of Wright's theory of random genetic drift as a mechanism for triggering shifts from one stable equilibrium (adaptive peak) to another, when there are epistatic interactions in fitness effects between loci; such shifts could not occur under natural selection alone (Wright 1932, 1977 Ch. 3, 1978 Ch. 11). (Charlesworth, Lande, and Slatkin, 1982)

This is an interesting claim. Wright stated in our interviews for *Sewall Wright and Evolutionary Biology* (Provine 1986) that, during his career, he spent little of his energies thinking or writing about mechanisms of speciation. Yet the statement above suggests that by 1932, Wright had developed a general theory of the evolutionary process that covered speciation as well as changes of gene frequencies within populations; Mayr's theory of speciation, based upon three decades (by 1954) of studying species and speciation, turned out to be a special case of Wright's view, already developed in 1932.

An exchange of letters between Victor McKusick and Sewall Wright in 1977 brings out Wright's understanding of the claim made by Charlesworth, Lande and Slatkin. He agreed with them.

May 11, 1977

Dear Dr. Wright:
  Who introduced the concept of "founder effect"?
  Many thanks for your information and very best regards.

Sincerely,

Victor McKusick

May 24, 1977

Dear Dr. McKusick:

To the best of my knowledge Ernst Mayr introduced the term "founder" effect. I do not think that he added much if anything to the concept. He wrote in 1963 (in *Animal Species and Evolution*, p. 209), "The differences in the amount of variation of small and large colonies is presumably, in part, due to selection. This does not contradict the probability that many small colonies may have been originated by a single fertilized individual (Mayr 1942, 32) and may thus also demonstrate the founder principle."

Turning to Mayr 1942 *(Systematics and the Origin of Species)* he refers on page 32 to the probable frequent origin of colonies of the snail *(Cepaea hortensis)* by single fertilized females but the only citation of the "founder" principle in the index is to page 237:

> The reduced variability of small populations is not always due to accidental gene loss, but sometimes to the fact that the entire population was started by a single pair or by a single fertilized female . . . [reproduces the quotation from Mayr, 1942, p. 237, given above, under Mayr, Founder Effects].

I may note that the first sentence above seems to make a distinction where there is none.

Leading to this first statement of his "founder" principle were three references (pp. 234, 235, and 236) to my papers of 1931 (Wright 1931) and 1932 (Wright 1932). I did not refer to such extreme bottlenecks of population number as origin from a single pair, but went into the wide random drift of gene frequencies (leading sometimes to fixation or loss) resulting from small numbers and cited several reasons why the effective population number might be very much smaller than apparent.

Mayr probably based his "founder" effect more specifically on a paper that I presented at a 1939 AAAS symposium at which he presided [Wright, 1940b; actually, Dobzhansky, who organized the session, presided] which he lists in his bibliographies. I wrote:

> An important case arises where local populations are liable to frequent extinction with restoration from the progeny of a few stray immigrants. In such regions, the line of continuity of large populations may have passed repeatedly through extremely small numbers, even though the species has at all times included countless millions of individuals in its range as a whole.

This was followed at once by the following:

> Such mutations as reciprocal translocations that are strongly selected against until half fixed seem to require some such mechanism to become established.

I also referred to these phenomena in an earlier paper [Wright, 1938, p. 431]:

> In some cases (as when translocations become fixed) there is evidence of fixation against very strong selection likely to occur (in a sexually reproducing species) only if there are numerous outlying territories in which the populations frequently pass—through single stray individuals.

I discussed the conditions for fixation of a reciprocal translocation (semisterile if heterozygous) quantitatively in [Wright, 1941c].

The effects attributed to the "founder" principle by Mayr (gene loss, reduced variability) are the most obvious but the least important of the three that I had stressed. I attributed most significance to wide random variability of gene frequencies (*not fixation or loss*) expected to occur simultaneously in tens of thousands of loci at which the leading alleles are nearly neutral, leading to unique combinations of gene frequencies in each of innumerable different local populations and continually changing in each of these. I put great weight on different harmonious systems, corresponding to different peaks in the "surface" of fitness values of the multidimensional set of gene frequencies. With simultaneous stochastic variability of all nearly neutral pairs of alleles, some are likely to cross saddles (against weak selection) leading to superior selective peaks. Firm establishment of this superior peak by local individual selection, followed by selective diffusion predominantly from superior peaks, spreads the corresponding interaction system throughout the species, and gives an evolutionary process ("shifting balance") which can continue indefinitely even under unchanging conditions. Evolution by irregular shifts of control to ever more adaptive interaction system contrasts with orderly progress of rare favorable major mutations toward fixation emphasized by Haldane (Haldane 1924, 1932) and also with the equally regular decrease of the additive genetic variance with respect to fitness, under unchanging conditions emphasized by Fisher (1930).

Both Fisher and I stressed the major dependence of evolution on the multiple minor gene differences involved in quantitative variability (in which I followed Castle, East, and ultimately Darwin), in contrast with Haldane's emphasis on "sports." Fisher and I differed radically in his assumption that species are so nearly panmictic (except as differentiated by different conditions of selection) that there could be no significant local random differentiation. I assumed an abundance of such local differentiation, especially where conditions approached these on which Mayr later based the founder principle.

Both of the processes emphasized by Haldane and Fisher tend to exhaust the variability on which they depend under unchanging conditions and thus depend in the long run on restoration of such variability by changes of conditions giving an evolutionary process rather of the nature of a treadmill: the undoing of adaptation to one set of conditions, adaptation to another. This is not the case with the "shifting balance" process.

Thus I considered the favoring of this process to be the most important consequence of situations in which colonies are continually becoming extinct but being refounded by stray migrants from the more successful ones.

The next most important consequence was, in my view, the opportunity for fixation of a reciprocal translocation which, by producing only semisterile progeny from crosses with the neighboring colonies, makes a first step toward reproductive isolation of a new species.

Occasional losses of genes and reduced variability seemed to me to be relatively trivial consequences. My first statement of the "shifting balance" hypothesis was in the abstract of a paper in an AAAS symposium [Wright, 1929]. This paper reached publication as the 1931 paper referred to above. In this abstract I also stated:

> In too small a population, there is nearly complete random fixation, little variation, little effect of selection, and thus a static condition, modified

occasionally by chance fixation of a new mutation, leading to degeneration and extinction.

This refers to a species which remains very small in numbers for many generations. Essentially the same statement appeared in my 1931 and later papers.

Mayr missed all but the least important aspect.

Sincerely yours,

Sewall Wright

June 3, 1977

Dear Dr. Wright,

My heartiest thanks for your detailed letter of May 24. Very best regards.

Sincerely,

Victor McKusick

This classic exchange of letters (Wright's answers were almost always much longer than letters of inquiry) reveals Wright's belief that Mayr's concept of founder effect was derivative from Wright's earlier work.

An immediate word of caution is necessary. What Wright compares in this letter is Mayr's founder effect in relation to speciation with his own shifting balance theory of evolution in relation to speciation. The reasonable comparison is between Mayr's founder effect *and* genetic revolution, taken together, in relation to Wright's shifting balance theory of evolution, which Wright thought applied to both evolution within populations and to speciation. In other words, we should compare Mayr's theory of the mechanisms of speciation with Wright's theory of the mechanisms of speciation.

Wright's shifting balance theory of evolution in nature was based upon his belief that the genome was highly interactive. His research on color inheritance in guinea pigs impressed upon him the importance of epistasis. From very early, Wright viewed biological organisms as composed of complex interaction systems of genes rather than being a mosaic of unit characters, each determined by a single gene. Thus, for Wright, selection was far more effective when applied to entire interaction systems of genes, rather than single genes. The basic question for the breeder was, how could entire interaction systems of genes be exposed to selection? The answer was to inbreed enough to reveal the systems, as had occurred in the inbred families of guinea pigs in Wright's charge for the decade 1915–25 when he was at the USDA. For breeders this was relatively simple. For evolution in nature, population structure was the key.

By 1932, Wright's shifting balance theory was comprised of four phases. First came the required inbreeding, produced by a population structure in which the population was divided and subdivided into relatively small demes, mostly but not completely isolated, small enough for a significant inbreeding effect but large enough that the debilitating effects of inbreeding (primarily the lowering of fertility) were minimal. Second was the resulting kaleidoscopic shifting (not fixation) of gene frequencies due to random genetic drift, and the consequent production of novel interaction systems of genes. Third was natural selection acting within each deme of the population (intrademic selection), frequently taking place in a patchy environment. And finally there was what Wright called "intergroup" or "interdemic selection," which was really selective diffusion from the more successful demes. This shifting balance process was effective even in a constant environment, and was capable of changing a population much faster than straight mass selection. (For a more detailed historical account of Wright's shifting balance theory of evolution see Provine, 1986, chapter 9.)

Wright presented the theory in 1931 as a way of envisioning the interrelations of the mechanisms of phyletic evolution. But obviously he was aware that evolution also involved the splitting of species, and he applied his shifting balance theory directly, but almost as an afterthought in one sentence: "Complete isolation [of the subpopulations] . . . originates new species differing for the most part in nonadaptive respects but is capable of initiating an adaptive radiation as well as of parallel orthogenetic lines, in accordance with the conditions" (Wright, 1931, p. 159). In 1932, he expanded this sentence to a paragraph:

> It need scarcely be pointed out that with such a mechanism complete isolation of a portion of a species should result relatively rapidly in specific differentiation, and one that is not necessarily adaptive. The effective intergroup competition leading to adaptive advance may be between species rather than races. Such isolation is doubtless usually geographic in character at the outset but may be clinched by the development of hybrid sterility. The usual difference of the chromosome complements of related species puts the importance of chromosome aberration as an evolutionary process beyond question, but, as I see it, this importance is not in the character differences which they bring (slight in balanced types), but rather in leading to the sterility of hybrids and thus making permanent the isolation of two groups. (Wright, 1932, p. 363)

In short, the process of speciation for Wright was simply a continuation of the shifting balance process, but with complete or nearly complete isolation of some of the subpopulations. Because the shifting balance process yielded rapid evolutionary change, speciation would result from *any* complete barrier to breeding of any two portions of a species. Indeed, the speciation process would occur far more rapidly if the isolated portions were both large subdivided populations, rather than one of them being a single deme; speciation was even less likely if one of the portions was an extremely small isolate, which, in Wright's scheme, generally went extinct. Wright specifically argued that in a species the size of a single deme (4Nu, 4Ns medium) the rate of adaptive evolutionary change was extremely slow (Wright, 1932, pp. 362–363).

All that Wright's shifting balance theory required to be a theory of speciation was complete isolation of one group of subpopulations from the rest of the species. Because the shifting balance process automatically broke apart coadapted gene complexes and produced new ones in normal phyletic evolution (i.e., produced peak shifts), speciation required no mechanisms other than those of phyletic evolution (beyond the initial isolation).

Wright never changed this basic view of the speciation process, but added some elaborations during the next decade. In 1940, he published two papers directed to the issue of speciation: "The Statistical Consequences of Mendelian Heredity in Relation to Speciation" (Wright, 1940a) and "Breeding Structure of Populations in Relation to Speciation" (Wright, 1940b). Neither paper contained much about speciation; each was devoted to the shifting balance theory, which was then briefly applied to the problem of speciation.

One elaboration particularly important to Mayr's later founder effect related to the question of how species came to differ by reciprocal translocations. Wright suggested that for this relatively infrequent occurrence a possible mechanism was a combination of what later would be termed a series of founder–flush cycles reminiscent of Elton's theory of speciation. If local populations of a species were subject to frequent extinction and if restoration of the populations resulted from the offspring of a few stray immigrants, then even highly improbable combinations (including ones sterile with individuals in the rest of the species, as with reciprocal translocations) might evolve. Should this occur, speciation was almost certain. In the theoretical example given by Wright in the paper, he showed a large population that had been through six major bottlenecks in its immediate ancestry. In this paper, Wright also described in nonmathematical terms his new theory of isolation by distance (later much elaborated in Wright 1943a, b, 1946). By itself in a continuous habitat, isolation by distance was a mechanism perhaps important in evolution. Combined with already existing subdivisions of the population and patchiness of the environment, isolation by distance became an important additional part of the shifting balance theory and also, of course, of the shifting balance theory of speciation. Isolation by distance tended to speed up the evolutionary process within a phyletic line, and to speed up speciation if the necessary geographical isolation were established.

Two other insights, neither original with him, completed Wright's theory of speciation by 1942. The first was the distinction between *general* and *specializing* adaptations (Wright, 1941a, p. 347). If a species by the shifting balance process happened upon a new adaptation of general importance, then many ecological niches would be opened up and the new branch of the species could expand and diversify very rapidly, always by the shifting balance mechanism. The other possibility came when migrants found a relatively unoccupied territory (Wright, 1941b, p. 167). Wright summarized these possibilities in one paragraph in 1942:

> Under certain conditions the multiplication and diversification of species may be a very rapid process. These include a relaxation of the general selection pressure on the species permitting great increase in widely distinct ecological niches associated with almost complete isolation of the groups seizing these opportunities and with subdivision of these groups into partially isolated local populations. A species that is the first of its general kind to reach unoccupied territory finds most at least of these conditions realized. This is also the case with a species that by any means acquires an adaptation of first rate general significance which gives its subgroups an advantage over species already established in various ecological niches, that more than compensates for the initial lack of special adaptations for these niches. (Wright, 1942, p. 244).

As an example of the latter situation, Wright offered the case of the honeycreepers (Drepanidae) of Hawaii. Clearly arguing for the possibility of very rapid speciation, Wright again invoked his shifting balance theory as the mechanism.

MAYR, WRIGHT, AND "PEAK SHIFTS"

Now we can address the issue of whether or not Mayr's view of speciation reduces to a Wrightean peak shift. In the first volume of the journal *Evolution*, which he founded and edited, Mayr wrote a major paper, "Ecological Factors in Speciation." Here he stated:

> The process of speciation results in ecological diversification and consequently in an ever-increasing efficiency in the utilization of the environment. One aspect of gradual speciation is therefore that it is an ecological process. Every species lives on an adaptive peak and the problem of speciation is how to reach new, not previously occupied, adaptive peaks. A species might do this either (1) by becoming locally more euryoecous (ecologically tolerant) or (2) by invading new areas with different ecological conditions. (Mayr, 1947, p. 265)

Consider also the following. In 1939, Mayr heard Wright deliver a paper on his shifting balance theory and speciation. In 1940, Mayr credited Wright with having presented (in Mayr's presence) the most robust conception of a species yet given (Mayr, 1940, pp. 255). In *Systematics and the Origin of Species* (1942), Mayr credited Wright with being the architect who had clarified the genetic consequences of founder populations. In his 1954 paper introducing the concept of genetic revolution, Mayr cited Wright as one who had long argued for the coadaptedness and cohesion of the genome. And finally, in the same 1940 paper that Mayr heard Wright deliver and later cited often, Wright had clearly introduced the idea that a succession of founder events could induce a peak shift over a deep valley on the fitness surface. Taken together with the 1947 quotation above, in which Mayr closely followed Wright, even to the extent of invoking the peak-shift heuristic, one might be led to conclude that Mayr's theory of speciation was indeed just a special case of Wright's shifting balance theory of evolution. But this conclusion is unwarranted, as I will attempt to show in the following section.

MAYR, WRIGHT, AND FISHER ON GEOGRAPHICAL SPECIATION

I will try to state clearly the relationships between the views of Mayr, Wright, and Fisher on the process of geographical speciation.

Geographical Speciation

All three believed that geographical isolation was the primary stimulus for speciation. Wright gave some minimal weight to ecological speciation, Fisher somewhat more, whereas Mayr gave none or almost none.

Genetic Cohesion of Parent Populations

Wright first, and then Mayr, believed strongly in the genetic cohesion of natural populations. Both believed that the genome was highly interactive, and that coadapted gene complexes characterized populations. Both believed that for speciation to happen, something would have to release the genetic inertia of some part of the population. Mayr's belief in the cohesion of the genotype was even stronger than Wright's. Fisher was impressed by the genetic cohesion of natural populations, but believed that natural selection was easily powerful enough to break apart any genetic cohesion of species. He thought that natural selection would keep the genetic pool of a population well adjusted to itself during the speciation process.

Population Structure Conducive to Speciation

For Fisher, the population structure most conducive to speciation was that preserving the greatest genetic variance in both portions of the population isolated by a geographical barrier; speciation would proceed most rapidly if both portions of the population were large and panmictic, thus preserving the maximum genetic variance upon which natural selection could act.

Wright thought that speciation would proceed most rapidly, and in particular faster than in the Fisherian model, if the parent population were split by a geographical barrier into large populations subdivided into relatively small demes in which random genetic drift caused a constant shifting, but not fixation, of gene frequencies. Both portions of the parent population were then most likely to undergo "peak shifts," resulting in eventual genetic isolating barriers. In those unusual cases where the speciation event involved a genetic change that would have been at a great selective disadvantage in the original parent population (such as reciprocal translocations), Wright advocated a model requiring successive founder–flush cycles in a subdivided population in which the demes often became extinct and were refounded by single individuals. This process could produce a "peak shift" despite the deep valley between the two peaks.

Mayr advocated a model that superficially looks like the Fisherian with the addition of small peripheral populations around the edge of the species distribution. But Mayr also believed that central population was structured, not just large and panmictic. In the "dead heart" of the species was the greatest genetic variability and cohesion of the genotype; both began to weaken going toward the periphery of the parent population. The small geographically peripheral populations were often founded by a single individual. The fate of most founder populations was extinction, but, rarely, a founder population underwent a genetic revolution, which was at the same time a speciation event.

Microevolution in Relation to Speciation

For Fisher and Wright, speciation was generally a continuation of microevolution, with the crucial added factor of geographical or perhaps ecological isolation. Their views of microevolution differed significantly according to their views of population structure, but speciation was for both a continuation of the microevolutionary process.

For Mayr, prevalent mechanisms of microevolution were poor producers of speciation, even with geographical isolation of portions of a species. The problem was the genetic cohesion of the parent population. To disturb this cohesion, he thought that some mechanisms in addition to those usually operating in microevolution were required. In Mayr's view, the population subdivision necessary for efficient operation of Wright's shifting balance theory was not generally found in natural populations, and thus he looked for another mechanism to break up the old gene complex. His belief was that founder populations were likely (only in an evolutionary sense—founder populations usually went extinct) to break up the genetic cohesion of the parent population and undergo a transition to a new genetic constitution. Fisher and Wright both argued that microevolutionary processes (dominated by natural and sexual selection for Fisher, by shifting balance process for Wright) were sufficient to unravel the genetic cohesion of the parent population. In this sense, the speciation theories of Fisher and Wright are far closer in spirit than either is to Mayr's theory of speciation, which could not be considered a simple continuation of microevolution and geographical isolation. For this reason, I think that viewing Mayr's theory of speciation as merely a special case of Wright's shifting balance theory is misleading.

LATER CHANGES IN THE VIEWS OF FISHER, WRIGHT, AND MAYR ON SPECIATION

The leading ideas about mechanisms of speciation have always been the-oretical because such mechanisms must be inferred indirectly. The indi-rect evidence has increased substantially, especially since the 1950s. How has this increase of information about natural populations affected the theories discussed in this paper thus far? There can be no doubt that Fisher, Wright, and Mayr all modified their views on speciation in the light of newer evidence and theory. This has been occasionally frustrating for the younger workers, who of course wish to make their own original contributions distinct from those of the earlier greats.

Fisher's participation in the research and controversy over the moth *Panaxia dominula* (see Provine, 1986, pp. 420–437) and his knowledge of the field research of his friend E. B. Ford led him to modify his earlier notion that most natural populations were both large and random breed-ing. He became aware that many natural populations are actually quite small. Still, as the work on *Panaxia* indicated to him, natural selection was the key mechanism to explain fluctuations in gene frequency from year to year, even in populations as small as 1000 individuals. For popu-lations much smaller than this, Fisher suspected that extinction would be the evolutionary fate. Thus speciation did not require the reservoirs of variability found only in large populations; small populations still carried enough variability for natural selection to push them to the stage of spe-ciation. Fisher and Ford were very skeptical about Mayr's founder theory of speciation (apparently requiring random genetic drift), and they offered natural selection in relatively small populations as an alternative expla-nation of speciation involving isolated peripheral populations. No founder effects or genetic revolutions were necessary (see Ford, 1964, pp. 38–41, 63–65, 168–171).

Through the 1950s and 1960s, Wright argued that extremely small pop-ulations generally became extinct. His shifting balance theory of evolu-tion, in contrast to the belief of some of his critics (especially Fisher and Ford), was not a theory of evolutionary change under the direct influence of random genetic drift. Wright was keen to shed the image of a "random drifter" during this time. Yet as evidence grew that laboratory popula-tions could be founded by single fertilized females, remain small for sev-eral generations, and then expand without major detrimental effects, Wright began to believe that he had saddled himself too severely with the rejection of evolutionary importance of small isolated populations. Thus when I was interviewing Wright in 1982, he agreed with Lande's (1980, p. 475) assessment that progressive evolution was possibly significant in fully isolated small populations.

Moreover, in his chapter on speciation in Volume 4 of *Evolution and the Genetics of Populations* (1978), Wright gave a revealing response to

M. J. D. White's assessment that over 90% of speciation events are accompanied by karyotypic changes (White, 1978). White suggested, along with Hedrick (1981), that meiotic drive might explain how chromosomal alterations (disadvantageous in heterozygous combination) might come to dominate in some populations and thus lead to speciation. Wright suggested an alternative:

> Fixation by accidents of sampling is, however, another possibility. It can occur only in extremely small colonies (Wright 1940, 1941). The most favorable situation is that in a region in which there are numerous colonies, subject to frequent extinction and refounding by stray fertilized females from the more successful ones. Since this is also a situation especially favorable to peak-shifts, there should be a strong correlation between speciation by the above process and favorable character changes (Wright, 1982, p. 438)

Wright had proposed this mechanism in 1940 to cover what he then thought were rare speciation events. But in 1982, he was prepared to offer the mechanism to cover a wide range of cases of chromosomal speciation in which the heterokaryons were selectively disadvantageous. He was more willing to be a drifter!

Mayr changed his views in two ways after 1963. First, he settled the question of the origin of founders most likely to undergo a genetic revolution in favor of the "dead heart" of the species, not the ecological periphery. And in response to Templeton's "genetic transilience" theory, a revision of Mayr's founder effect followed by genetic revolution, Mayr quickly agreed that the "genetic revolution" might result from a relatively few genes of major effect rather than being a revolution affecting all genes. I will return to this point shortly.

MAYR, WRIGHT, FISHER, ELTON, AND THE RECENT DEBATES ON FOUNDER EFFECT AND GENETIC REVOLUTION

Evolutionary biology has many persistent controversies. Indeed, when I wrote a survey of adaptation and mechanisms of evolution after Darwin, the most reasonable subtitle I could think of was "a study in persistent controversies" (Provine, 1985). Founder effects in relation to genetic revolutions and speciation are no exception to the persistent controversy label.

For any generation of investigators, however, there is always an understandable tendency for investigators to mark their achievements off from those of the previous generation or two, and for those of the previous generation to stress the continuity of their conceptions through the newer investigations. The task of the historian is to see the issues in broad historical perspective to the extent possible, but not on so large a scale as to lose the detail that makes scientific work understandable.

From this perspective, I would argue that the basic terms for the recent debate on geographical speciation were set by Elton, Fisher, Wright, and Mayr; as yet, Brown, Carson, Templeton, Lande, Charlesworth, Barton, and the many others involved are still debating the basic theories raised by the previous generations.

William L. Brown, Jr. and Centrifugal Speciation

Brown was a student at the Museum of Comparative Zoology at the same time as E. O. Wilson and Thomas Eisner. He was intimately familiar with Mayr's views on speciation, and was also a field taxonomist working primarily on ants. By the mid-1950s, Brown had developed his view of "centrifugal speciation" (Brown, 1957). He was skeptical that what Mayr later called peripatric speciation was a primary mechanism of mainland speciation or of speciation of major bifurcations because insular fauna and flora were notoriously fragile. Animals and plants on the Hawaiian Islands were examples. Brown thought that primary speciation was far more likely if built upon variations produced in the biological center of the species (not necessarily the geographical center). The question was, how could such variants become isolated and transformed into new good species?

He hypothesized first that natural populations often fluctuated dramatically in size (he does not mention Elton in this regard, but rather other ecologists who were influenced by Elton): "Within the maximum range, the populations of the species normally undergo successive expansions into the less favorable peripheral areas, alternating with contractions into more favorable refuges" (Brown, 1957, p. 248). The resulting population structure at crucial stages of both expansion and contraction was just that of Wright's shifting balance theory of evolution, and Brown argued that Wright's "interdemic selection" was the probable cause of the actual differentiation that resulted in speciation. Brown specifically denied that the genetically depauperate founder populations advocated by Mayr in 1954 had the potential to become major new species. Brown's centrifugal speciation may be seen as a combination of the views of Elton and Wright.

Hampton L. Carson on Founder Effect, Genetic Revolution, Founder–Flush Cycling, and Speciation

Brown had already written his 1957 paper before reading Carson's first paper on the theory of speciation, given at the 1955 Cold Spring Harbor Symposium (for a background on Carson, see his 1980 paper). Stimulated by reading Mayr's 1954 paper on founder effects and genetic revolution (as he describes in the interview with Wyatt Anderson in this volume), Carson set to work in the field to see if marginal populations of *Drosophila robusta* had genetic characteristics in accordance with Mayr's theory. He discovered that the marginal populations were relatively homozygous for gene arrangement (as measured by inversions), and the central populations had extensive chromosome polymorphism. He concluded that the more homozygous peripheral populations allowed greater crossing-over and therefore were the best candidates for originating new species. Indeed, Carson suggested that successful cosmopolitan species like *Drosophila melanogaster* originated from peripheral populations with few inversions.

Brown, of course, disagreed strongly and wrote to Carson arguing that evolutionary novelties were most likely to arise in the center of a species rather on the periphery. Carson replied:

> Your contention that the center of a species has special properties with respect to the origin of evolutionary novelties is certainly a challenging one. My first thought would be that it sounds a little too inclusive and as if you are denying peripheral speciation altogether rather than proposing another possibility. I am also not clear as to where gene flow (from a center outward) ends and centrifugal speciation begins. (Carson to Brown, February 4, 1957)

At the Cold Spring Harbor Symposium of 1959, Carson firmly rejected Brown's centrifugal speciation:

> Brown (1957) has assembled some evidence, most of which is indirect, that marginal populations are relatively depauperate genetically. This evidence is adduced, however, to support the idea that the central part of the species is "the principle source of evolutionary change leading to 'potent' new species. . . ." This view is not in accord with the hypothesis being presented in this paper. (Carson, 1959, p. 91)

> The writer strongly disagrees with the conclusion of Brown that the "centrifugal" flows which he discusses are causally related to the formation of new species. (Carson, 1959, p. 99)

Like Brown, however, Carson at the same time attributed similar mechanisms for the speciation process itself:

> Marginal populations tend to be isolated and inbred. They usually exist in areas which offer to the species a limited series of environmental niches to be conquered. It seems inescapable that, when a subspecies becomes progressively adapted to these niches as the margin is approached, homoselection should play the dominant role. Marginal populations indeed come close to fulfilling the requirements for the small, semi-isolated populations to which Wright (1932) attributed the greatest potential for progressive change. (Carson, 1959, p. 92)

Thus for both Brown and Carson, the process that actually transformed the isolated populations into another species was Wright's shifting balance.

Carson had built his view of speciation upon the best available evidence from direct field research, a method to which he has firmly adhered during the years following. As the data from field research on founder effects and genetic revolutions have changed substantially since 1959, so did Carson's theories of speciation.

By 1967, he was convinced by the evidence that speciation in Hawaiian *Drosophila* was not in any way dependent upon karyotypic change (Carson, et al., 1967). A year later, he introduced the "population flush" as a possible major factor in breaking up the coadapted genome of the central population. His argument is very similar to that of Elton in 1927 except that following the flush and later crash, a population structure very conducive to Wright's shifting balance process is produced. "Thus, a new species is born under circumstances where adaptation is not the guiding force" (Carson, 1968, p. 135). One requirement of the flush–founder theory was that the initial population must possess much genetic variability. Thus relatively homozygous peripheral populations had little potential for producing new species with this mechanism. The flush–founder thesis also began to look more like Brown's centrifugal speciation, with the founders coming from the outcrossed central population during a population crash.

In 1970, Carson published two papers altering the sequence of events in speciation that he had presented in 1968. The genetic evidence indicated that a population flush from a central population had little chance by itself to disorganize the coadapted genome. The sequence (Carson, 1970a, b) was now that the central population produced a founder that ended up in an isolated but favorable environment. The founding event was followed by a few generations of random drift in turn followed by a population flush. The inevitable crash produced isolated demes where Wright's shifting balance mechanism completed the speciation process. After the speciation event, most of which so far was nonadaptive according to Carson, adaptive evolution dominated by natural selection characterized the evolutionary process.

In the late 1960s, the first wave of work on allozyme variability (e.g., by Lewontin and Hubby) began to appear. One of the early papers (Prakash et al., 1969) indicated clearly that marginal and central populations of *Drosophila pseudoobscura* did not differ significantly in their allozyme variability. And Carson himself had reported that marginal populations respond to selection as well as or better than central ones (Carson, 1959). In a 1971 statement of his position on speciation given at the Stadler Symposium, Carson still held to his earlier belief that founders came from the peripheral populations, but in the case of Hawaiian *Drosophila*, where species differed by few karyotypic changes, the reasons for believing that founders came from the marginal populations were beginning to weaken (Carson, 1971). In this paper, he still distinguished clearly between a nonadaptive speciation stage, which could be extremely rapid, and an adaptive phyletic stage that followed.

By 1975, Carson had made two further changes in his model of speciation in Hawaiian *Drosophila*. Here he stated: "Speciation is considered to be initiated when an unusual forced reorganization of the epistatic supergenes of the closed variability system occurs" (Carson, 1975, p. 88). In other words, he now believed the founders came from the center of the population, not the periphery. Moreover, from both experiment and theory he now believed that after a population crash, reduction only to deme size would not by itself engender much disorganization of the coadapted gene complexes of the parent population. A more drastic reduction was required. Thus in 1975, Carson's model of speciation in Hawaiian *Drosophila* looked like this: A central population produces a founder individual that migrates to a favorable isolated environment. Random drift is followed by a population flush, then a crash again to the level of single founders (what Carson called the "flush–crash–founder" cycle, which could be repeated any number of times). The selection that followed these disorganizational processes then forged new closed genetic systems typical of species centers.

Thus by 1975, Carson no longer looked upon speciation as a primarily nonadaptive process: "The formation of these new coadaptive relationships, according to my view, represents the essence of the speciation process" (Carson, 1975, p. 90). The number of generations required for the speciation process was correspondingly increased.

Carson's interpretation of the speciation process, so far as I can see from his published papers, has not changed in a major way since this 1975 paper. To see how much Carson's views on speciation had changed in the 20 years before this, all that is necessary is to compare Carson's 1955 (pp. 284–285) and 1975 (p. 92) summaries. These changes suggest several conclusions.

Carson was a highly accomplished naturalist long before he started work on speciation. His career has been characterized throughout by painstaking attention to field research, laboratory experimentation, and concrete data. His theorizing has always been very closely tied to experimental evidence. He began thinking about founder effects and genetic revolutions at a time when direct evidence from natural populations was almost entirely absent. My first conclusion, therefore, is that the evidence from natural populations and from laboratory experiments concerning founder effects and genetic revolutions grew and changed dramatically between 1955 and 1975, in large part from Carson's own researches and those stimulated by him. His theorizing about speciation was based closely upon this changing evidence.

The other primary influence upon his theorizing about speciation was the ideas of Wright, Mayr, and Elton. In a major paper summarizing his views on speciation in 1982, Carson specifically credited Wright (Carson, 1982, p. 426) and Mayr (p. 430) as the major influences upon his thinking about speciation. I would argue that Carson's theories of speciation between 1955 and 1975 were all combinations of the theories of Wright, Mayr, and Elton, and all closely tied with the latest experimental data.

Alan Templeton's Theory of "Genetic Transilience"

Drawing explicitly from the theoretical foundations of Mayr and Wright on founder effects and genetic revolutions, from the mounting experimental evidence from the field and laboratory on *Drosophila*, and from computer simulations, Templeton presented his theory of genetic transilience in 1980. This theory differed from Mayr's founder effect followed by genetic revolution chiefly in Templeton's careful analysis of probable genetic characteristics of the parent population (must be highly polymorphic for loci controlling development, physiology, and behavior), and by his argument that the revolution itself depended upon rapid changes in a relatively few genes of major effect (a genetic transilience), rather than upon Mayr's "veritable genetic revolution" that might "affect all loci at once."

In March 1980, Mayr received a preprint of Templeton's 1980 paper and immediately agreed with both his characterization of the parent population and with his argument that only a relatively few genes needed to be involved in the initial speciation process. He disagreed, however, that the new term, "genetic transilience," was needed because Mayr had no trouble incorporating Templeton's amendments under the term "genetic revolution," as the following letter from Mayr to Templeton suggests:

> Owing to Allan Wilson's kindness I have seen preprints of your forthcoming papers on the theory of speciation via the founder principle. I would like to congratulate you, not only on being the first geneticist to tackle the genetics of genetic revolutions, but also for having come up with a series of reasonable explanations, particularly the stress on major genes controlling multilocus systems, as well as the importance of the genetic structure of the ancestral population supplying the founders.
>
> When I proposed genetic revolutions in 1954 (I had been lecturing about it since 1951), I had abundant material on the actual occurrence of startling evolutionary changes in peripherally isolated populations. They did not respond at all to either of two earlier models, that of gradual evolutionary changes by the gradual replacement of alleles, or else Goldschmidt's model of systemic mutations, affecting a single gene or chromosome segment.
>
> I realized that a multilocus phenomenon was involved. I think historians will give me much credit for having recognized some of the major properties of such genetic revolutions. I did recognize the crucial fact that such a reduced population size would permit the rapid transgression through the bottleneck of heterozygosity. This is equally true for the chromosomal mutations. I recognized and emphasized the importance of the increased frequency of homozygotes in such small inbreeding populations. I also recognize the importance of drastic shifts in epistatic balances.
>
> Where I was apparently wrong, as much of the electrophoretic work has since demonstrated, was in assuming that such a genetic revolution would affect all variable loci. Actually, I never said that it would, but I did say that it might. I simply didn't know how many loci such a genetic revolution would affect.
>
> What I didn't understand at all was why such genetic revolutions occurred in some founder populations, but not in others. Birds, with their high chromosome number, are of course presumably particularly suited for genetic revolutions, even though their easy dispersal may counteract the inbreeding.
>
> The original assumption, implied by me and made articulate by some who followed me, that such genetic revolutions would affect all variable loci, became less and less likely as the heterogeneity of the DNA became known in the last 10 years, and in the wake of much electrophoretic work. It is now

evident that most enzyme genes do not participate in genetic revolutions. The work on the Bogota *D. pseudoobscura*, of course, also showed the importance of a limited number of genes.

In 1954 all I could do was to describe the phenomenon and try to speculate on its genetic basis. I am delighted that finally the geneticists are following up. However I feel somewhat hit below the belt by your endeavor to replace my term "genetic revolution" by your own term "genetic transilience." To be sure, my interpretation was in part incorrect, even though it contained most of the major explanatory components. Hence I do not see any particular need for a replacement of terms. More importantly, such replacement is not sanctioned by scientific tradition. I do not know of a single major concept in biology on which the underlying explanation has not been changed again and again. Just compare Johannsen's gene with our modern interpretations, or the species of Ray and Linnaeus with the modern species, or Dobzhansky's isolating mechanisms (which included geographical barriers) with the modern definition, or the mutation of De Vries with our modern definition, etc. It is taken for granted that the interpretation of phenomena described by new terms will change over time. I can speak with authority, since I am in the process of finishing a major volume on the history of concepts of biology. It is only when terms are given to actual objects, like nucleus or chromosome that no such changes occur. I would strongly urge you not to push your endeavor to insert your own term. . . .

Once more, my warmest congratulations on having made such a major contribution toward the solution of a rather puzzling phenomenon. I hope you are not upset that I am somewhat critical of your terminological endeavors. (Mayr to Templeton, March 24, 1980)

Perhaps because most evolutionary biologists tended to view both their theories of speciation by founder effects and genetic revolutions to be mere variants of Mayr's theory, Carson and Templeton collaborated on a long paper in 1984 in which they drew the lines of difference between their theories and Mayr's as sharply as possible (see especially table 1 in Carson and Templeton, 1984, p. 118). They concluded that "genetic revolution is not only different from the other two models [founder–flush and genetic transilience], it is incompatible with them" (Carson and Templeton, 1984, p. 118). Yet if the changes in Mayr's views after 1963 are taken into account, the differences between his version of genetic revolution and Templeton's genetic transilience are minimal. As I view the issue from the perspective provided in this paper, it appears from table 1 that Carson and Templeton have put flesh on the bones of Mayr's original theory, which from the beginning was at best vague on genetic details. (For the differences between Carson's founder–flush model and Templeton's genetic transilience model, which they describe as mostly matters of emphasis, see Carson and Templeton, 1984.)

Lande's Analysis of Allopatric Speciation

Lande, who views Mayr's 1954 statement of the founder effect and genetic revolution as a special case of Wright's shifting balance theory, provided in 1980 a spirited defense of the shifting balance theory as an explanation of allopatric speciation. In the paper, Lande developed a quantitative phenotypic version of Wright's adaptive landscape. Arguing that Mayr's conception of the genetic consequences of founder effects and genetic revolutions was flawed, Lande's analysis indicated that, with two additions, Wright's shifting balance theory fit both population genetics theory and data from natural populations on allopatric speciation. The additions were that Lande's version of shifting balance included the possibility that evolution in fully isolated small populations could sometimes be progressive and lead to speciation, and that the phenotypic version of the adaptive landscape allowed one to trace greater phenotypic changes than Wright's gene frequencies version. Wright was happy to accept both additions.

Barton and Charlesworth on Genetic Revolutions, Founder Effects, and Speciation

Although some version of Mayr's founder effect and genetic revolution has been the favored explanation for at least island speciation since 1954, some evolutionists have rejected this approach. Foremost among these dissenters have been Fisher, Ford, Cain, Sheppard, Clarke, Antonovics, and in general what is called the Oxford or British school of ecological genetics. Their primary objection is that the available evidence does not require postulating the existence of founder effects and genetic revolutions in order to explain speciation phenomena even on islands. They argue for a simple explanation, requiring fewer assumptions: that only the usual mechanisms of phyletic evolution are required for speciation.

Barton and Charlesworth, in a paper juxtaposed with that of Carson and Templeton in the 1984 *Annual Review of Ecology and Systematics*, have argued forcefully that neither data from the field or laboratory, nor theoretical arguments compel the conclusion that rapid speciation requires founder populations. With Fisher, they present evidence and theoretical models indicating that natural selection sufficient to cause speciation is not impeded by the genetic cohesion of a geographically divided species. Indeed, they argue that natural selection in a population of moderate size can cause more rapid divergence toward isolation that can random genetic drift in a very small population. The paper concludes:

> The strongest biogeographic evidence for founder effect speciation comes from a general correlation between peripheral isolation and phenotypic divergence, and in particular from the Hawaiian *Drosophila*, in which colonization of new islands is almost invariably associated with speciation. The generally small chance of achieving reproductive isolation or marked phenotypic change in a single founder event means that founder effects themselves probably do not provide the explanation. It is impossible to separate the effects of isolation, environmental differences, and continuous change by

genetic drift from the impact of population bottlenecks in these cases. Since all of these factors promote divergence by a variety of processes, it is not clear that the additional influence of founder events need be invoked. (Barton and Charlesworth, 1984, p. 158)

I am confident that Fisher would have liked this paper very much.

CONCLUSION

The history of ideas about founder effect and genetic revolution in speciation is extremely complicated. This paper is a mere scratch on the surface.

There has been much continuity of theory about founder effect and genetic revolution, going back into the time well before Mayr coined these names. I have argued that the speciation theories of Mayr, Elton, Fisher, and Wright were all distinguishable, and all have played major roles in the modern debates. Their theories have been combined and recombined by others.

Despite this historical continuity, Mayr's peripatric theory of speciation with its founder effects and genetic revolutions undoubtedly has had an enormous influence upon evolutionary biologists and systematists. One measure of its importance is the effort expended by its critics.

Although much more is now known about speciation in nature than in was known 1954, in general the evidence is still insufficient to discriminate between revised versions of the basic schemes of Fisher, Wright, or Mayr. Even in the most studied cases in animals, including Hawaiian *Drosophila* and *Geospizinae* of the Galapagos Islands (Grant, 1986), both Mayr and Wright can and have concluded with justification that their views on speciation are consistent with recent research, and a good case has been made for the consistency of Fisher's views with the same evidence.

My belief is that controversies about founder effect and genetic revolution (and about modes of speciation generally) cannot be settled with assurance until we have a greater understanding of speciation processes happening now. True, speciation is a slow process measured against the academic career of a single biologist. But biologists have a great many examples of introduced populations, both natural and mediated by humans. Many of these have been monitored. I think that if such geographically separated populations are studied over perhaps no more than ten generations of academics (five by Sewall Wright's standards), we might well begin to have the evidence to settle long-standing controversies about particular instances of speciation. Combined with the proliferation of techniques to study differences in existing closely related species, laboratory simulations of crucial aspects of speciation, and refinements of theory, convincing accounts of an increasing number of examples of speciation should emerge. The accumulation of such examples will provide a basis for more robust generalizations about the process of speciation in general.

ACKNOWLEDGMENT

This chapter is respectfully dedicated to Hampton L. Carson.

LITERATURE CITED

Barton, N. H., and B. Charlesworth. 1984. Genetic revolutions, founder effects, and speciation. Annu. Rev. Ecol. Systematics 15:133–164.

Brown, W. L., Jr. 1957. Centrifugal speciation. Q. Rev. Biol. 32:247–277.

Carson, H. L., 1955. The genetic characteristics of marginal populations of *Drosophila*. Cold Spring Harbor Symp. Quant. Biol. 20:276–287.

Carson, H. L. 1959. Genetic conditions which promote or retard the formation of species. Cold Spring Harbor Symp. Quant. Biol. 24:87–105.

Carson, H. L. 1968. The population flush and its genetic consequences. In R. C. Lewontin (ed.), Population Biology and Evolution, pp. 123–137. Syracuse University Press, Syracuse, NY.

Carson, H. L. 1970a. Chromosomal tracers of founder events. Biotropica 2(1): 3–6.

Carson, H. L. 1970b. Chromosome tracers of the origin of species. *Science* 168:1414–1418.

Carson, H. L. 1971. Speciation and the founder principle. Stadler Symp. 3:51–70.

Carson, H. L. 1975. The genetics of speciation at the diploid level. Am. Natur. 109:83–92.

Carson, H. L. 1980. Cytogenetics and the Neo-Darwinian synthesis. In E. Mayr and W. B. Provine (eds.), The Evolutionary Synthesis, pp. 86–95. Harvard University Press, Cambridge, MA.

Carson, H. L. 1982. Speciation as a major reorganization of polygenic balances. In C. Barigozzi (ed.), Mechanisms of Speciation, pp. 411–433. Alan R. Liss, New York.

Carson, H. L., and A. R. Templeton, 1984. Genetic revolutions in relation to speciation phenomena: the founding of new populations. Annu. Rev. Ecol. Systematics 15:97–131.

Carson, H. L., F. E. Clayton, and H. D. Stalker. 1967. Karyotypic stability and speciation in Hawaiian *Drosophila*. Proc. Natl. Acad. Sci. USA 57:1280–1285.

Charlesworth, C., R. Lande, and M. Slatkin, 1982. A Neo-Darwinian commentary on macroevolution. Evolution 36(3):474–498.

Eldredge, N., and S. J. Gould. 1972. Punctuated equilibria: an alternative to phyletic gradualism. In T. J. M. Scopf (ed.), Models in Paleobiology, pp. 82–115. W. H. Freeman, San Francisco.

Dawkins, R. 1986. The Blind Watchmaker. Norton, New York.

Elton, C. S. 1927. Animal Ecology. Macmillan, New York.

Elton, C. S. 1930. Animal Ecology and Evolution. Oxford University Press, Oxford.

Endler, J. A. 1986. Natural Selection in the Wild. Princeton University Press, Princeton, NJ.

Fisher, R. A. 1930. The Genetical Theory of Natural Selection. Oxford University Press, Oxford.

Ford, E. B. 1964. Ecological Genetics. Methuen, London.

Grant, P. R. 1986. Ecology and Evolution of Darwin's Finches. Princeton University Press, Princeton, NJ.

Gulick, A. 1934. John Thomas Gulick: Evolutionist and Missionary. University of Chicago Press, Chicago.

Gulick, J. T. 1872. On the diversity of evolution under one set of external conditions. J. Linn. Soc. Lond. (Zoology) 11:496–505.

Gulick, J. T. 1888. Divergent evolution through cumulative segregation. J. Linn. Soc. Lond. (Zoology) 20:189–274.

Gulick, J. T. 1890. Intensive segregation, or divergence through independent transformation. J. Linn. Soc. Lond. (Zoology) 23:312–380.

Gulick, J. T. 1905. Evolution: Racial and Habitudinal. Carnegie Inst. of Washington Publ. No. 25. Washington, DC.

Haldane, J. B. S. 1924. A mathematical theory of natural and artificial selection. I. Trans. Cambridge Philos. Soc. 23:19–41.

Haldane, J. B. S. 1932. The Causes of Evolution. Longmans, London.

Hedrick, P. W. 1981. The establishment of chromosomal variants. Evolution 35:322–332.

Kottler, M. J. 1976. Isolation and speciation, 1837–1900. Ph.D. dissertation, Yale University.

Lande, R. 1980. Genetic variation and phenotypic evolution during allopatric speciation. Am. Natur. 116:463–479.

Lande, R. 1982. Rapid origin of sexual isolation and character divergence in a cline. Evolution 36(2):213–223.

Lerner, I. M. 1954. Genetic Homeostasis. John Wiley, New York.

Lesch, J. T. 1975. The role of isolation in evolution: George J. Romanes and John T. Gulick. Isis 66:483–503.

Mayr, E. 1940. Speciation phenomena in birds. Am. Natur. 74:249–278.

Mayr, E. 1942. Systematics and the Origin of Species. Columbia University Press, New York.

Mayr, E. 1947. Ecological factors in speciation. Evolution 1:263–288.

Mayr, E. 1949. Speciation and systematics. In G. L. Jepson, E. Mayr, and G. G. Simpson. (eds.), Genetics, Paleontology, and Evolution, pp. 281–298. Princeton University Press, Princeton.

Mayr, E. 1954. Change of genetic environment and evolution. In J. Huxley, A. C. Hardy, and E. B. Ford (eds.), Evolution as a Process, pp. 157–80. Allen & Unwin, London.

Mayr, E. 1963. Animal Species and Evolution. Harvard University Press, Cambridge, MA.

Mayr, E. 1976. Evolution and the Diversity of Life. Harvard University Press, Cambridge, MA.

Mayr, E. 1982a. The Growth of Biological Thought. Harvard University Press, Cambridge, MA.

Mayr, E. 1982b. Processes of speciation in animals. In C. Barigozzi (ed.), Mechanisms of Speciation, pp. 1–19. Alan R. Liss, New York.

Mayr, E., and W. B. Provine (eds.) 1980. The Evolutionary Synthesis. Harvard University Press, Cambridge, MA.

Prakash, S. R., R. C. Lewontin, and J. L. Hubby, 1969 A molecular approach to the study of genic heterozygosity in natural populations. IV. Patterns of genic variation in central, marginal and isolated populations of *Drosophila pseudoobscura*. Genetics 61:841–858.

Provine, W. B. 1985. Adaptation and mechanisms of evolution after Darwin: a study in persistent controversies. In D. Kohn (ed.), *The Darwinian Heritage*, pp. 825–866. Princeton Univ. Press, Princeton, NJ.

Provine, W. B. 1986. Sewall Wright and Evolutionary Biology. University of Chicago Press, Chicago.

Templeton, A. R. 1980. The theory of speciation via the founder principle. Genetics 94:1011–1038.

Wallace, A. R., 1889. Darwinism. Macmillan, London.

White, M. J. D. 1978. Modes of Speciation. W. H. Freeman, San Francisco.

Wright, S. 1929. Evolution in a Mendelian population. Anat. Rec. 44:287.

Wright, S. 1931. Evolution in Mendelian populations. Genetics 16:97–159.

Wright, S. 1932. The roles of inbreeding, crossbreeding and selection in evolution. In Proc. Sixth Int. Cong. Genet., Vol. 2, pp. 356–366.

Wright, S. 1938. Size of population and breeding structure in relation to evolution. Science 87:430–431.

Wright, S. 1940a. The statistical consequences of Mendelian heredity in relation to speciation. In Julian S. Huxley (ed.), The New Systematics, pp. 161–183. Oxford University Press, Oxford.

Wright, S. 1940b. Breeding structure of populations in relation to speciation. Am. Natur. 74:232–248.

Wright, S. 1941a. Review of The "Age and Area" Concept Extended, by J. C. Willis. Ecology 22:345–347.

Wright, S. 1941b. Review of The Material Basis of Evolution, by R. Goldschmidt. Sci. Monthly 53:165–170.

Wright, S. 1941c. On the probability of fixation of reciprocal translocations. Am. Natur. 75:513–522.

Wright, S. 1942. Statistical genetics and evolution. Bull. Am. Math. Assoc. 48:223–246.

Wright, S. 1943a. Isolation by distance. Genetics 28:114–138.

Wright, S. 1943b. Analysis of local variability of flower color in *Linanthus parryae*. Genetics 28:139–156.

Wright, S. 1946. Isolation by distance under diverse systems of mating. Genetics 31:29–59.

Wright, S. 1977. Evolution and the Genetics of Populations, Vol. 3. University of Chicago Press, Chicago.

1978. Evolution and the Genetics of Populations, Vol. 4. University of Chicago Press, Chicago.

Wright, S. 1982. Character change, speciation, and the higher taxa. Evolution 36:427–443.

II

PLANT SPECIATION AND THE FOUNDER PRINCIPLE

3

Adaptive Radiation of the Hawaiian Silversword Alliance (Compositae–Madiinae): A Comparison with Hawaiian Picture-Winged *Drosophila*

GERALD D. CARR, ROBERT H. ROBICHAUX,
MARTHA S. WITTER, AND DONALD W. KYHOS

The Hawaiian silversword alliance comprises a total of 28 species in the three genera *Argyroxiphium*, *Dubautia*, and *Wilkesia* (Carr, 1985a). These genera are endemic to the major islands of the Hawaiian archipelago, and representative species occur from Kauai in the northwest to Hawaii in the southeast. The silversword alliance is related to members of the tarweed subtribe (Madiinae) of the sunflower tribe (Heliantheae). The tarweed subtribe is highly developed in California and is represented there by more than a dozen genera whose species are mostly restricted to the west coast of North America.

The Hawaiian tarweeds are exceedingly diverse in growth form, habitat preferences, and leaf morphology and anatomy. They also exhibit substantial variation in floral morphology, chromosome number and structure, and physiological tolerances. The purpose of this paper is to describe and substantiate this diversity further, and to provide evidence that the entire silversword alliance has evolved from a single founder population. In this respect, the adaptive radiations of the Hawaiian Madiinae and the Hawaiian picture-winged *Drosophila* are strikingly similar. Comparisons of the Hawaiian Madiinae and Hawaiian picture-winged *Drosophila* will underscore additional similarities between these unique Hawaiian groups.

Figure 3.1.   Habits and habitats of Hawaiian Madiinae. (**A**) *Argyroxiphium sand-wicense* subsp. *macrocephalum,* cinder cone within Haleakala Crater, Maui. (**B**) *Wilkesia gymnoxiphium,* dry scrub on margin of Waimea Canyon, Kauai. (**C**) *Du-bautia scabra* subsp. *scabra,* pumice deposit of Kilauea Iki, Hawaii. (**D**) *D. reticulata,* upper Hana rain forest, Haleakala, Maui.

HABIT

As is the case with Hawaiian *Drosophila*, an impressive array of growth forms is found among the Hawaiian Madiinae (Carr, 1985a, b). Among the five species of the genus *Argyroxiphium*, habital types range from unbranched silvery rosette shrubs that flower once and die (e.g., the Haleakala silversword, *A. sandwicense;* see Fig. 3.1A) to branched green rosette shrubs that may flower repeatedly (e.g., the greensword, *A. grayanum*). The two species of the genus *Wilkesia* are also rosette shrubs similar to *Argyroxiphium* in habit, but have stems that are woody and more highly developed. One of the species, *W. gymnoxiphium* (Fig. 3.1B), is typically unbranched and monocarpic (i.e., flowers once and dies), whereas the other, *W. hobdyi*, is branched from the base and is polycarpic (i.e., flowers repeatedly). Habital variation among the 21 species of *Dubautia* is much more extreme, the forms ranging from nearly herbaceous mat-forming subshrubs in *D. scabra* (Fig. 3.1C), to compact cushion plants in *D. waialealae*. Still other forms include large sprawling shrubs (e.g., *D. raillardioides*), erect woody shrubs (e.g., *D. laxa*), very large shrubs or small trees (e.g., *D. knudsenii*), and large, woody trees (e.g., *D. reticulata*, Fig. 3.1D). Another striking growth form in *Dubautia* is that of the vine, *D. latifolia*.

HABITAT

Another similarity between Hawaiian Madiinae and Hawaiian *Drosophila* is the great diversity of habitats in which both groups occur. Species of Hawaiian Madiinae grow in habitats as varied as exposed lava, dry scrub, dry woodland, mesic forest, wet forest, and bog (cf. Fig. 3.1). These habitats constitute an unusually wide array of moisture environments, with annual rainfall varying from less than 400 mm in the dry scrub habitat to more than 12,300 mm in the wet forest and bog habitats. Indeed, the latter habitats are among the wettest terrestrial environments on Earth (Grosvenor, 1966). These habitats also span a wide range of elevations, from less than 75 m to more than 3750 m. At the higher elevations, species are often exposed to marked temperature fluctuations during the day, and may experience minimal daily temperatures below 0°C for extended periods of time. At the lower elevations, in contrast, temperatures are quite moderate, with species rarely experiencing minimal daily temperatures below 15°C.

Species in this group often exhibit localized habitat partitioning at sites of sympatry. A particularly striking example of this partitioning is exhibited by *Dubautia scabra* and *D. ciliolata* at a site of sympatry on the upper slopes of Mauna Loa, Hawaii (Carr and Kyhos, 1981; Robichaux,

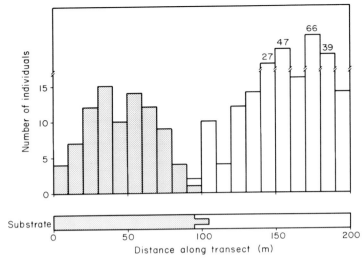

Figure 3.2. Localized habitat partitioning in *Dubautia scabra* and *D. ciliolata* at a site of sympatry on the upper slopes of Mauna Loa, Hawaii (Robichaux, Stemmermann, and Liebman, unpublished data). Species distributions were determined along a 200-m transect that crossed two lava flows. The first 100 m crossed a 1935 flow of pahoehoe lava, whereas the second 100 m crossed a prehistoric flow of pahoehoe and aa lava. In the zone of interdigitation at 95–105 m, fingers of the younger flow covered sections of the older flow. Where the two flows were in contact in this zone, the boundary between them was always abrupt. The transect was 10 m wide. For the number of individuals along the transect, the shaded bars denote *D. scabra*, whereas the unshaded bars denote *D. ciliolata*. For the substrate distribution along the transect, the shaded section denotes the younger lava flow, whereas the unshaded section denotes the older lava flow.

1984). This site consists of a mosaic of two different lava flows. An extensive flow produced by a 1935 eruption of Mauna Loa covers most of the site. This 1935 flow is discontinuous in several places, such that lava from an older, prehistoric flow is exposed. These pockets, or *kipukas*, of older lava vary in size from 4 to 4000 m². Wherever the older flow is exposed, the boundary between the two flows is abrupt, such that the transition from one flow to the other occurs over a distance of a few millimeters. The two flows differ not only in age but also in physical structure. The younger flow consists entirely of smooth, billowy, pahoehoe lava, whereas the older flow consists of a mixture of pahoehoe lava and rough, chunky, aa lava. Though both *Dubautia* species are common at this site, they are differentially restricted to the two substrates. *Dubautia scabra* grows exclusively on the younger flow, whereas *D. ciliolata* grows only on the older flow (Fig. 3.2). This differential restriction means that the pattern of distribution of the two species at this locality is a distinctive mosaic.

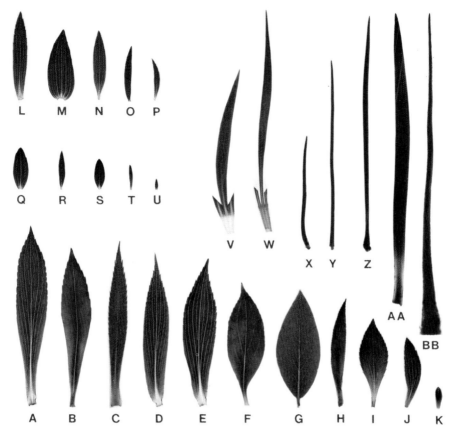

Figure 3.3. Leaf profiles of Hawaiian Madiinae. (**A**) *Dubautia raillardioides*; (**B**) *D. pauciflorula*; (**C**) *D. laevigata*; (**D**) *D. microcephala*; (**E**) *D. plantaginea* subsp. *plantaginea*; (**F**) *D. knudsenii* subsp. *knudsenii*; (**G**) *D. latifolia*; (**H**) *D. imbricata* subsp. *imbricata*; (**I**) *D. laxa* subsp. *laxa*; (**J**) *D. paleata*; (**K**) *D. waialealae*; (**L**) *D. dolosa*; (**M**) *D. platyphylla*; (**N**) *D. sherffiana*; (**O**) *D. reticulata*; (**P**) *D. herbstobatae*; (**Q**) *D. arborea*; (**R**) *D. linearis* subsp. *linearis*; (**S**) *D. menziesii*; (**T**) *D. scabra* subsp. *scabra*; (**U**) *D. ciliolata* subsp. *ciliolata*; (**V**) *Wilkesia hobdyi*; (**W**) *W. gymnoxiphium*; (**X**) *Argyroxiphium caliginis*; (**Y**) *A. kauense*; (**Z**) *A. virescens*; (**AA**) *A. grayanum*; (**BB**) *A. sandwicense* subsp. *sandwicense*. All, × 1/4.

LEAF MORPHOLOGY AND ANATOMY

Leaf form is very diverse among the Hawaiian Madiinae (Fig. 3.3) and in any given instance appears to be highly correlated with the nature of the habitat. The leaves of *Argyroxiphium* are long and narrow, succulent, and often exceedingly hairy. These features are of obvious adaptive significance in the high-elevation, open cinder or bog habitats where these plants are found. In both of these situations, lack of water (or restricted uptake of water under bog conditions) and very high insolation may be two strong selective factors accounting for these leaf types. As noted by Carlquist (1957, 1959a, b), the succulent leaves of *Argyroxiphium* are further distinguished from those of the other two genera of Hawaiian Madiinae by the presence of prominent pectic channels and by a unique three-tiered vascular system. The leaves of *Dubautia* are more variable, ranging from thin-textured, expansive forms in the forest species to small, firm types in bog, dry scrub, or cinder habitats. Leaves of the forest species are adapted for low light conditions where water is generally not a limiting factor, whereas leaves of the species of open habitats are adapted for conditions of low water availability and high insolation. Anatomically, the small leaves of species adapted to dry sites are thick and have compact tissue and thick cuticles to reduce water loss (Fig. 3.4A), whereas the large leaves of species adapted to wet forest conditions are thin and have loosely packed tissue and thin cuticles (Fig. 3.4B). The vascular patterns of leaves of *Dubautia* are also highly variable, but little is known of the adaptive significance of this variation.

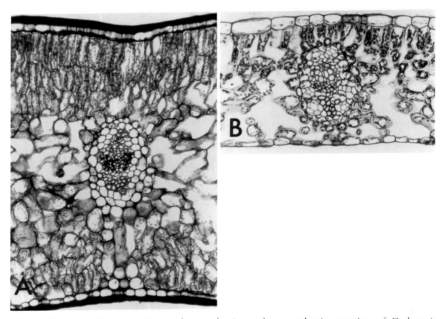

Figure 3.4. Leaf transections of xerophytic and mesophytic species of *Dubautia*. (**A**) *D. menziesii* of open, alpine desert scrub habitat. Note thick leaf with compact organization of tissue and very highly developed cuticle on upper and lower epidermis. (**B**) *D. knudsenii* subsp. *knudsenii* of mesic forest habitat. Note thin leaf with loose organization of tissue and thin cuticle on upper and lower epidermis. Magnification, ×100.

FLORAL MORPHOLOGY AND BREEDING SYSTEMS

There is considerable variation in the flowering heads among species of Hawaiian Madiinae (Carr, 1985a, b). A whorl of peripheral ray flowers distinguishes the heads of *Argyroxiphium* from those of *Dubautia* and *Wilkesia*. In *Argyroxiphium*, the heads are up to 3.5 cm in diameter and may contain over 600 florets. In contrast, the rayless head of *Dubautia pauciflorula* has only two or three florets and is less than 2 mm broad. Other species of *Dubautia* and *Wilkesia* have flower heads of sizes between these extremes. Flower colors include deep wine-red, yellow-orange, cream, and white. Other features of the flowers, notably the morphology and micromorphology of the pappus, also vary considerably among species of this alliance (Carr, 1985a). In *Argyroxiphium*, the pappus is absent or consists of irregular scales. In *Wilkesia* it consists of lanceolate scales, whereas in *Dubautia* it varies from scales to naked bristles to plumose bristles.

In accordance with the observations of Baker (1955), one would expect the highly isolated flora of Hawaii to be depauperate in self-incompatible species. Indeed, until recently, there appear to have been no reports of genetically based self-incompatibility among native Hawaiian plant taxa. However, Carr, Powell, and Kyhos (1986) have documented the occurrence of self-incompatibility in all three genera of Hawaiian Madiinae. Insofar as they are obligate outcrossers, these species are reproductively similar to Hawaiian *Drosophila*. However, other species of Hawaiian Madiinae are known to be self-compatible. Because it seems unlikely that self-incompatibility would have arisen de novo after dispersal, we suggest that the original founder may have been pseudo–self-compatible (sensu Mulcahy, 1984). Alternatively, the founder may have been a long-lived rhizomatous or otherwise asexually propagating self-incompatible species, the sexual reproductive system of which gradually became restored through the accumulation of diversified incompatibility alleles produced by mutation subsequent to the dispersal event.

CYTOGENETICS AND HYBRIDIZATION

Another biological similarity between the Hawaiian Madiinae and the Hawaiian *Drosophila* is the structural repatterning of chromosomes that has occurred in both groups. Paracentric inversions commonly distinguish mainland species of *Drosophila*, and, although many homosequential taxa are known, this mechanism of chromosomal evolution has also been active among Hawaiian species of the genus (Carson, 1983). Likewise, chromosomal evolution involving reciprocal translocations has been documented in several mainland tarweeds (Clausen, 1967; Carr and Carr, 1983; Tanowitz, 1985), and has also been detected among Hawaiian representatives (Carr and Kyhos, 1981, 1986). Furthermore, two groups of Hawaiian Madiinae may be distinguished on the basis of chromosome number. *Argyroxiphium, Wilkesia,* and 12 species of *Dubautia* have 14 pairs of chromosomes, whereas 9 species of *Dubautia* have 13 pairs.

In spite of the overwhelming diversity exhibited by the silversword alliance, its genetic cohesiveness is amply demonstrated by the occurrence of 35 different spontaneous hybrid combinations in nature (Carr, 1985a). Furthermore, 55 additional hybrid combinations have been produced artificially in the laboratory (Carr and Kyhos, 1986). Figure 3.5 summarizes these hybrids according to generic and habital categories. It can be readily seen that all categories have been involved in one or more hybrid combinations.

A great advantage in the study of hybridization in this group is the fact that in many instances the observation of chromosome pairing during meiosis allows positive confirmation of the hybrid nature of a given individual. This is true in cases where the parents have an aneuploid relationship or where they differ with respect to one or more reciprocal chromosome translocations. For example, the simplest meiotic chromosome pairing configuration in hybirds between aneuploid species with 14 and 13 pairs of chromosomes is 12 pairs and a linear chain of three chromosomes (Carr and Kyhos, 1981). Hybrids between *Dubautia scabra* ($n = 14$) and any 13-paired species of *Dubautia* consistently give this result. In fact, the synthesis of chromosomal, distributional, ecological, and morphological data suggests that *D. scabra* is the nearest living ancestor of the 13-paired cytotype.

Hybrids between 13-paired species have consistently exhibited normal meiosis and essentially normal fertility. These results and the uniformity of the meiotic pairing configuration of chromosomes in all hybrids involving *D. scabra* and any 13-paired species of *Dubautia* suggest that the genomes of species with 13 pairs of chromosomes are structurally uniform (Carr, 1985a; Carr and Kyhos, 1986).

However, hybrids among 14-paired species give different results, depending on the combination. The meiotic chromosome configuration in these hybrids ranges from 12 pairs and a simple chain or ring of four chromosomes to 11 pairs and a chain of six chromosomes (Carr, 1985a; Carr and Kyhos, 1986). Collectively, the cytogenetic evidence indicates that there are at least seven distinctive genomic arrangements present in the assemblage of 14-paired species (Carr and Kyhos, 1986). Unlike the situation in Hawaiian picture-winged *Drosophila,* these genomes are structurally differentiated from one another by reciprocal chromosome translocations. Although critical data are presently still lacking in some instances, it appears that these genomes are largely if not wholly differentiated from one another by single steps of chromosome evolution (interchanges). Another feature of chromosome evolution in this group is the redundancy of translocations repeatedly involving a limited number of chromosomes (Carr and Kyhos, 1986).

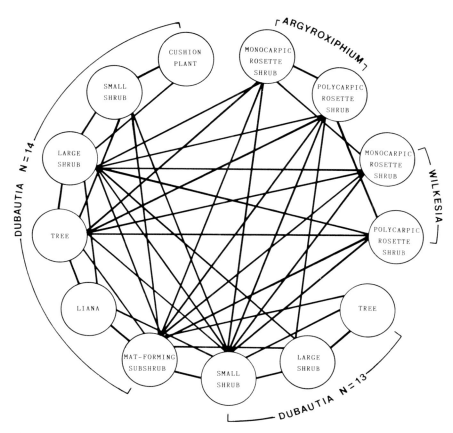

Figure 3.5. Crossing polygon summarizing artificial and natural hybrids that have been produced among generic and habital categories of Hawaiian Madiinae (cf. Carr, 1985a).

Of the two aneuploid cytotypes, the 14-paired cytotype includes the greater number of species, the three most widely ranging species, and the greater morphological, ecological, chromosomal, and isozymic diversity (see below). In addition, it is the only cytotype found on Kauai, the geologically oldest island occupied by the group. All of these observations suggest that the 14-paired cytotype occupies an ancestral position in the evolution of the Hawaiian Madiinae.

The hybrids that have been analyzed have mean pollen stainabilities ranging from 9% to 99%. In virtually all cases, the reduction in pollen stainability of hybrids can be ascribed to chromosomal differentiation of the parents resulting from aneuploidy and/or reciprocal chromosome translocations. Moreover, even the most sterile of hybrid combinations are capable of producing vigorous backcross progenies. For example, progeny tests in progress reveal that as many as 5% of the fruits from the spontaneous intergeneric hybrid combination *Argyroxiphium sandwicense* ($n = 14$) × *Dubautia menziesii* ($n = 13$) have viable embryos that germinate and grow vigorously.

PHYSIOLOGY

The evolutionary diversification of the Hawaiian Madiinae has been accompanied by a significant degree of change at the physiological level. Among the *Dubautia* species, for example, significant interspecific variation exists in tissue elastic properties (Robichaux, 1984; Robichaux and Canfield, 1985). The pattern of variation in these properties appears to be correlated with the pattern of variation in the habitats and chromosome numbers of these species. Species that grow in dry habitats and have 13 pairs of chromosomes exhibit significantly lower tissue elastic moduli near full hydration (E_i) than species that grow in mesic to wet habitats and have 14 pairs of chromosomes. Values of E_i range from approximately 2 to 4 megapascals (MPa) among the former species and from 9 to 18 MPa among the latter species (Table 3.1). Hybrids between the species in these

Table 3.1. Tissue elastic modulus near full hydration (E_i) in nine Hawaiian *Dubautia* species that differ in habitat and chromosome number

Species	Habitat	E_i (MPa)[a]
13-paired species		
D. ciliolata	exposed lava[c]	2.22 (0.20)
D. platyphylla	dry scrub	2.94 (0.30)
D. menziesii	dry scrub	3.51 (0.30)
D. linearis	dry scrub	3.93 (0.32)
14-paired species		
D. raillardioides	wet forest	9.57 (0.61)
D. scabra	exposed lava[c]	10.23 (0.78)
D. paleata	bog	13.33 (0.65)
D. plantaginea	mesic forest	14.47 (0.36)
D. knudsenii	mesic forest	18.23 (0.42)
Hybrid		
D. ciliolata × D. scabra[b]	exposed lava	6.67 (0.41)

[a]The *tissue elastic modulus* is defined as the change in tissue turgor pressure for a fractional change in tissue water content. The units are megapascals. Standard errors are given in parentheses. All pairwise comparisons between species with different chromosome numbers are significantly different at $p < .001$.

[b]Data for the natural hybrid between *D. ciliolata* and *D. scabra* are also provided.

[c]Though both *D. ciliolata* and *D. scabra* grow on exposed lava at their site of sympatry, the environment of the former species is significantly drier (Robichaux, 1984).

Source: Adapted from Robichaux (1984) and Robichaux and Canfield (1985).

two groups, such as between *D. ciliolata* and *D. scabra*, exhibit intermediate values of E_i (Table 3.1).

As a result of these differences in tissue elastic properties, the capacity for maintaining high turgor pressures as tissue water content decreases is much greater in the 13-paired species from dry habitats than in the 14-paired species from mesic to wet habitats. This is illustrated in Figure 3.6 for the 13-paired *Dubautia menziesii* and the 14-paired *D. knudsenii*. Although the maximal values of turgor pressure in these two species are similar, the initial rates at which turgor pressure declines with decreasing tissue water content differ markedly. Turgor pressure declines much more slowly in *D. menziesii* than in *D. knudsenii*, which is reflected in the significantly lower value of E_i in the former species (Table 3.1). The effect of this difference on the magnitude of tissue turgor pressure at moderate tissue water contents is very pronounced. At a relative water content of 0.93, for example, the value of turgor pressure is 0.76 MPa in *D. menziesii*, but only 0.08 MPa in *D. knudsenii*.

With their greater capacity for turgor maintenance, the 13-paired *Dubautia* species may be able to tolerate conditions of low moisture availability to a greater extent than the 14-paired species. This stems from the

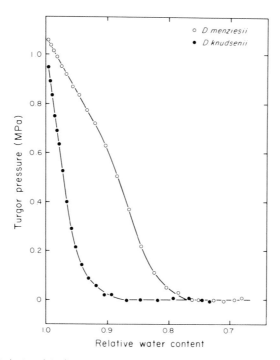

Figure 3.6. Relationship between tissue turgor pressure and tissue relative water content for the 13-paired *Dubautia menziesii* and the 14-paired *D. knudsenii* (Robichaux and Canfield, 1985). Tissue relative water content is defined as: (fresh weight − dry weight) ÷ (saturated weight − dry weight).

fact that a variety of physiological processes, such as stomatal opening and photosynthetic carbon assimilation, appear to exhibit a significant turgor dependence (Robichaux and Canfield, 1985).

ENZYMES

Enzyme electrophoresis has been used to evaluate genetic differentiation and the genetic organization of the Hawaiian Madiinae and to provide additional biosystematically useful characters. Species from all three genera have been surveyed for 11 enzyme systems presumed to be coded by a total of 26 loci.

Although the Hawaiian Madiinae exhibit normal diploid meiotic chromosomal behavior, the pattern of enzyme expression indicates that they are genetically tetraploid. The polyploid nature of the Hawaiian genomes is seen in the enzyme systems phosphoglucoisomerase (PGI, EC 5.3.1.9), phosphoglucomutase (PGM, EC 2.7.5.1), and triose phosphate isomerase (TPI, EC 5.3.1.1) and is expressed by fixed, nonsegregating heterozygote banding patterns typical of organisms of known polyploid origin (e.g., Roose and Gottlieb, 1976; Gottlieb, 1982). Potentially, the duplicate nature of the founding genome may have provided a significant source of variation for adaptive evolution in island plant groups. Of the five plant groups in which insular adaptive radiation has been studied—*Bidens* (Helenurm and Ganders, 1985; Ganders, Chapter 4, this volume), *Dendroseris* (Crawford et al., 1985), *Euphorbia* (Pearcy, personal communication), *Tetramolopium* (Lowrey and Crawford, 1985), and the Hawaiian Madiinae—all except *Tetramolopium* exhibit polyploid gene expression.

The proposed link between the mainland and Hawaiian Madiinae, based largely on morphology and anatomy (Carlquist, 1959b; Carr, 1985a), is further supported by enzyme data. A unique gene marker has been found in a cytoplasmic PGM duplication that occurs throughout the mainland genera (Gottlieb, personal communication; Warwick and Gottlieb, 1985). The Hawaiian Madiinae exhibit a minimum of four cytoplasmic and two plastid PGM isozyme bands. We provisionally interpret this to be the expression of the same PGM duplication found in the mainland genera, with the additional isozymes resulting from the effects of polyploid gene expression.

The enzyme data are congruent with other evidence that the Hawaiian tarweeds are a cohesive group of common origin and are the result of radiation subsequent to only a single founder event. There are genetic similarities among even the most distinctive species. All Hawaiian taxa share a single allele at the malic enzyme (ME) locus (EC 1.1.1.40), whereas glutamic oxaloacetic transaminase (GOT, EC 2.6.1.1, 3 loci), PGI (2 loci), TPI (3 loci), and PGM (6 loci) all have a predominant allelic form that occurs in common among various species from all three genera. The close genetic relationships of the Hawaiian Madiinae are qualitatively similar to other plant groups in island ecosystems (Crawford et al., 1985; Helenurm and Ganders, 1985; Lowrey and Crawford, 1985; Ganders, Chapter 4, this volume), all of which show much lower levels of genetic differentiation than their mainland counterparts. Presumably, this is a reflection of the recent and common origin of the insular groups.

The pattern of morphological differentiation and the geographic distribution of species strongly support the hypothesis that the 14-paired cytotype occupies the ancestral position in the evolution of the Hawaiian Madiinae. Using genetic data based on enzymes, one would predict that the more ancient taxa would exhibit a greater degree of interspecific genetic differentiation, reflecting a longer history of separation and isolated evolution within the Hawaiian archipelago. To test this hypothesis, the percentages of 14-paired and 13-paired species of *Dubautia* exhibiting unique alleles or polymorphic variants at each of five loci were calculated and compared (Table 3.2). On average, the percentage of species with fixed unique alleles was six times greater for the 14-paired species than for the 13-paired species (30% vs. 5%). The same trend is clear for the polymorphic variants: The percentage of species showing polymorphic variation was twice as great for the 14-paired species as for the 13-paired species (45% vs. 22%).

The Hawaiian tarweeds are of interest in that a great number of morphologically unique species have arisen rapidly from a common gene pool. The prerequisite to understanding the origin of this diversity is a phylogeny of accurate species relationships superimposed on the current and historical geographical distributions of these taxa. Furthermore, it is important to be able to distinguish those instances in which speciation has been coincident with interisland migration from those in which speciation has resulted from differentiation of populations on a single island. Electrophoretic data have been helpful in assessing one such situation involving two 13-paired species, *Dubautia sherffiana* and *D. herbstoba-*

Table 3.2. Frequency (%) of *Dubautia* species of $n = 14$ and $n = 13$ cytotypes with unique or polymorphic variant alleles at five enzyme loci

Enzyme locus	% Species with unique alleles		% Species with polymorphic variant alleles		Sample size (No. species)[a]	
	$n = 14$	$n = 13$	$n = 14$	$n = 13$	$n = 14$	$n = 13$
PGI-1	56	25	44	12.4	9	8
PGI-2	56	0	44	62.5	9	8
GOT-1	12.5	0	37.5	0	8	6
GOT-2	0	0	75	12.5	8	8
GOT-3	25	0	25	25	4	4
Means	29.9	5	45.1	22.5	7.6	6.8

[a]For each species, a minimum of 20 individuals were assayed.

tae, on Oahu. The majority of the species in this 13-paired group and also the presumed nearest 14-paired relative, *D. scabra*, are centered on the islands of the Maui complex and the island of Hawaii. The existence of the two 13-paired species on Oahu is best accounted for by colonization of Oahu from Maui by either one or two independent founder events. However, because these two Oahu species share a unique, derived PGI-1 allele, we propose a common ancestry involving a single founder event on Oahu followed by subsequent in situ speciation.

FOUNDER EVENTS

Two recent papers by Carson (1983, 1984) include a map showing the geographical distribution of Hawaiian picture-winged *Drosophila*. The map also indicates the proposed founder events that account for the present-day distributions of species. These papers provided inspiration to construct a similar scheme for the Hawaiian Madiinae and to compare it with the situation in the Hawaiian picture-winged *Drosophila*. The hypothetical founder events for Hawaiian Madiinae are summarized in Table 3.3 and are compared graphically with the picture-winged *Drosophila* in Figure 3.7. The patterns are remarkably similar in the two groups. Kauai is perceived as the home of the ancestral species complex for both groups. In both instances the number of founder events on Hawaii is greatest, whereas the number of species produced per founder on Hawaii is lowest. Likewise, in both groups the "Maui complex" has the greatest ratio of species to interisland founders: 3.0 for the plants and 3.3 for the insects. Whereas "back migration" from the "Maui complex" to Oahu is per-

Table 3.3.   Hypothetical founder events for the Madiinae in the Hawaiian archipelago

Island	Taxon and genome[a]	Inferred origin
Kauai	*Dubautia knudsenii* (DG1)	
	D. laxa (DG1)	
	D. microcephala (DG1)	
	D. plantaginea (DG1)	
	D. paleata (DG2)	
	D. laevigata (DG3)	
	D. latifolia (DG3 or DG4)	North America[b]
	D. imbricata	
	D. pauciflorula	
	D. raillardioides	
	D. waialealea	
	W. gymnoxiphium (WG)	
	W. hobdyi	

Table 3.3.   (cont.)

Island	Taxon and genome[a]	North America[b]
Oahu	D. laxa (DG1)	Kauai[c]
	D. plantaginea (DG1)	Kauai[c]
	D. herbstobatae (DG5)	
	D. sherffiana (DG5)	Maui complex[d]
Maui complex	Argyroxiphium caliginis	
	A. grayanum (AG1)	
	A. sandwicense (AG2)	Kauai[e]
	A. virescens	
	D. laxa (DG1)	Oahu[c]
	D. plantaginea (DG1)	Oahu[c]
	D. scabra (DG4)	
	D. dolosa (DG5)	
	D. linearis (DG5)	
	D. menziesii (DG5)	Kauai[c]
	D. platyphylla (DG5)	
	D. reticulata (DG5)	
Hawaii	A. kauense	
	A. sandwicense (AG2)	Maui complex[e]
	D. plantaginea (DG1)	Maui complex[c]
	D. scabra (DG4)	Maui complex[c]
	D. arborea (DG5)	Maui complex[c]
	D. ciliolata (DG5)	Maui complex[c]
	D. linearis (DG5)	Maui complex[c]

[a] AG1, AG2 = *Argyroxiphium* genomes 1 and 2; DG1–DG5 = *Dubautia* genomes 1–5; WG = *Wilkesia* genome; see Carr and Kyhos (1986).
[b] Anatomy, morphology, and cytology (see Carlquist, 1959a; Carr, 1985a).
[c] Cytogenetics and morphology (see Carr, 1985a; Carr and Kyhos, 1981, 1986).
[d] Cytogenetics and isozymes (see text and Carr and Kyhos, 1986).
[e] The distinctive morphology of *Argyroxiphium* is approached only by *Wilkesia* from Kauai. It appears likely that *Argyroxiphium* was derived from some member of the ancestral plexus that was dispersed from Kauai.

ceived to have occurred in each group, the plant group shows no evidence of dispersal to Kauai from any other island. Similarly, whereas there is evidence of direct dispersal from Kauai to Hawaii and from Oahu to Hawaii among the *Drosophila*, no such evidence exists among the species of Hawaiian Madiinae. Both groups, however, exhibit evidence of direct dispersal from Kauai to the "Maui complex." Indeed, the paucity of species of Hawaiian Madiinae on the intervening island of Oahu is somewhat perplexing.

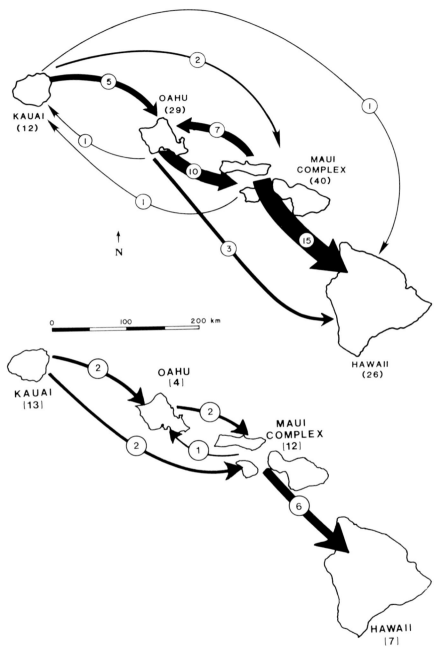

Figure 3.7. Geographical summary of the hypothetical founder events invoked to account for the origin of the picture-winged *Drosophila* fauna (above) and the madiinoid flora (below) on each of the major Hawaiian Islands. The width of the arrows is proportional to the number of proposed founders. The number of species found on each island is given in parentheses. (*Drosophila* summary adapted from Carson, 1983)

SUMMARY OF ADAPTIVE RADIATION

The Hawaiian Madiinae appear to have originated from a single madiinoid colonizing ancestor with characteristics similar to extant mainland genera such as *Adenothamnus* and *Raillardella*. The dramatic radiation into forms as varied as cushion plants, rosette shrubs, mat-forming subshrubs, erect shrubs, trees, and lianas may have been facilitated by the close juxtaposition of highly diversified habitats within the archipelago. While allopatry and ecological isolation have probably been important during the evolution of these taxa, partial internal isolation through reciprocal chromosome translocations may also have played a role in numerous instances. In addition, major adaptive shifts, as in the case of *Argyroxiphium* on the younger islands, may have been facilitated by founder effects associated with long-distance dispersal between islands.

Diversification into new habitats may also have been aided by the evolution of novel physiological traits. At least one major evolutionary event, the aneuploid reduction in chromosome number, appears to have been accompanied by a major shift in tissue elastic properties. This physiological shift may have aided the derived 13-paired species in exploiting significantly drier habitats than those occupied by the ancestral 14-paired species.

Once underway, rapid diversification may have been further promoted by self-incompatibility, frequent outcrossing, and hybridization between differentially adapted genotypes. Recurrent volcanic activity, which repeatedly disrupted stable communities, may have opened vast areas for potential colonization by the novel products of hybridization. The resulting species of Hawaiian Madiinae now occupy habitats that range in elevation from near sea level to 3750 m and in annual rainfall from less than 400 mm to more than 12,300 mm. Indeed, these species grow in virtually every terrestrial habitat in the archipelago except coastal strand. As a consequence, they appear to be the most spectacular example of adaptive radiation in the plant kingdom.

ACKNOWLEDGMENTS

This work was partially supported by National Science Foundation grants to Gerald D. Carr (DEB-7822819, BSR-8306917) and Robert H. Robichaux (DEB-8206411). We thank In Sun Kim for making the anatomical preparation utilized in Figure 3.4B.

LITERATURE CITED

Baker, H. G. 1955. Self-compatibility and establishment after "long-distance" dispersal. Evolution 9:347–349.

Carlquist, S. 1957. Leaf anatomy and ontogeny in *Argyroxiphium* and *Wilkesia* (Compositae). Am. J. Bot. 44:696–705.

Carlquist, S. 1959a. Vegetative anatomy of *Dubautia, Argyroxiphium*, and *Wilkesia* (Compositae). Pacific Sci. 13:195–210.

Carlquist, S. 1959b. Studies on Madinae: anatomy, cytology, and evolutionary relationships. Aliso 4:171–236.

Carr, G. D. 1985a. Monograph of the Hawaiian Madiinae (Asteraceae): *Argyroxiphium, Dubautia*, and *Wilkesia*. Allertonia 4:1–123.

Carr, G. D. 1985b. Habital variation in the Hawaiian Madiinae (Heliantheae) and its relevance to generic concepts in the Compositae. Taxon 34:22–25.

Carr, G. D., and D. W. Kyhos. 1981. Adaptive radiation in the Hawaiian silversword alliance (Compositae–Madiinae). I. Cytogenetics of spontaneous hybrids. Evolution 35:543–556.

Carr, G. D., and D. W. Kyhos. 1986. Adaptive radiation in the Hawaiian silversword alliance (Compositae–Madiinae). II. Cytogenetics of artificial and natural hybrids. Evolution 40:959–976.

Carr, G. D., E. A. Powell, and D. W. Kyhos. 1986. Self-incompatibility in the Hawaiian Madiinae (Compositae): an exception to Baker's rule. Evolution 40:430–434.

Carr, R. L., and G. D. Carr. 1983. Chromosome races and structural heterozygosity in *Calycadenia ciliosa* Greene (Asteraceae). Am. J. Bot. 70:744–755.

Carson, H. L. 1983. Chromosomal sequences and interisland colonizations in Hawaiian *Drosophila*. Genetics 103:465–482.

Carson, H. L. 1984. Speciation and the founder effect on a new oceanic island. In F. J. Radovsky, P. H. Raven, and S. H. Sohmer (eds.), Biogeography of the Tropical Pacific, pp. 45–54. B. P. Bishop Museum Special Publ. No. 72.

Clausen, J. 1967. Stages in the Evolution of Plant Species. Hafner Publishing, New York.

Crawford, D. J., T. F. Stuessey, and M. Silva O. 1985. Allozymic divergence and evolution of *Dendroseris* (Compositae) on the Juan Fernandez Islands. Am. J. Bot. 72:947–948 [abstract].

Gottlieb, L. D. 1982. Conservation and duplication of isozymes in plants. Science 216:373–380.

Grosvenor, M. B. (ed.). 1966. National Geographic Atlas of the World. National Geographic Society, Washington, DC.

Helenurm, K., and F. R. Ganders. 1985. Adaptive radiation and genetic differentiation in Hawaiian *Bidens*. Evolution 39:753–765.

Lowrey, T. K., and D. J. Crawford. 1985. Allozyme divergence and evolution in *Tetramolopium* (Compositae: Astereae) on the Hawaiian Islands. Syst. Bot. 10:64–72.

Mulcahy, D. L. 1984. The relationships between self-incompatibility, pseudo-compatibility, and self-compatibility. In W. F. Grant (ed.), Plant Biosystematics, pp. 229–235. Academic Press, New York.

Robichaux, R. H. 1984. Variation in the tissue water relations of two sympatric Hawaiian *Dubautia* species and their natural hybrid. Oecologia 65:75–81.

Robichaux, R. H., and J. E. Canfield. 1985. Tissue elastic properties of eight Hawaiian *Dubautia* species that differ in habitat and diploid chromosome number. Oecologia 66:77–80.

Roose, M. L., and L. D. Gottlieb. 1976. Genetic and biochemical consequences of polyploidy in *Tragopogon*. Evolution 30:818–830.

Tanowitz, B. D. 1985. Systematic studies in *Hemizonia* (Asteraceae: Madiinae): hybridization of *H. fasciculata* with *H. clementina* and *H. minthornii*. Syst. Bot. 10:110–118.

Warwick, S., and L. D. Gottlieb. 1985. Genetic divergence and geographic speciation in *Layia* (Compositae). Evolution 39:1236–1241.

4

Adaptive Radiation in Hawaiian *Bidens*

FRED R. GANDERS

Continental species of *Bidens* produce achenes with barbed awns that are well adapted for dispersal by adhesion to mammals and birds. The ancestor of Hawaiian species of *Bidens* probably arrived on the islands by accidental long distance dispersal by birds. Several species of shorebirds annually migrate between Hawaii and continental areas or other Polynesian islands. Furthermore, several North American species of birds are sighted each decade as accidental arrivals in the Hawaiian Islands (Shallenberger, 1981).

Evidence from chromosomes (Gillett and Lim, 1970), breeding experiments (Gillett and Lim, 1970; Gillett, 1975; Ganders and Nagata, 1983b, 1984), isozymes (Helenurm and Ganders, 1985), polyacetylenes (Marchant et al., 1984), and flavonoids (McCormick, Bohm, and Ganders, unpublished) clearly indicates that all Hawaiian species evolved from a single common ancestor. From this single immigrant species have evolved a morphologically and ecologically diverse group of 19 species and 8 subspecies endemic to the Hawaiian Islands (Ganders and Nagata, 1983a, 1984, in press). Hawaiian species of *Bidens* exhibit a greater range of morphological diversity than does the rest of the genus. Morphological differentiation among the Hawaiian species (Fig. 4.1) includes characters of growth habit, leaf shape, inflorescence structure, flowers and heads, and achenes (Gillett and Lim, 1970; Ganders and Nagata, 1984, in press). Hawaiian species have evolved character states not found in continental species, such as completely fused inner involucral bracts that split irregularly at flowering in *B. amplectens;* achenes enclosed by receptacular bracts and elongated styles with pollen presentation on style branches in *B. cosmoides;* coiled achenes in several species, with the extreme in *B. torta;* gynodioecy in nine species; and indeterminate growth in nine species. Hawaiian *Bidens* occur in a diversity of habitats, including coastal bluffs and lithified sand dunes, cliffs, cinder cones, and montane ridges and bogs. They occur from sea level to over 2200 m elevation, in areas with annual rainfall ranging from 0.3 m to over 7.0 m.

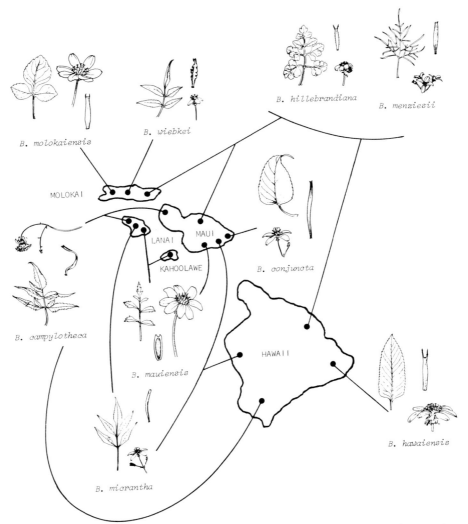

Figure 4.1. (*Above and at right*) Distribution of endemic species of *Bidens* on the Hawaiian Islands. Illustration shows typical flower heads, achenes, and leaves of each species.

INTERSPECIFIC DIFFERENTIATION

In order to understand the process of adaptive radiation in a taxon, it is necessary to know both the types of characters that have diverged and the extent of the divergence. Formal genetic analysis of character differences has not been completed, but Hawaiian *Bidens* have been as thoroughly studied as any example of adaptive radiation in plants.

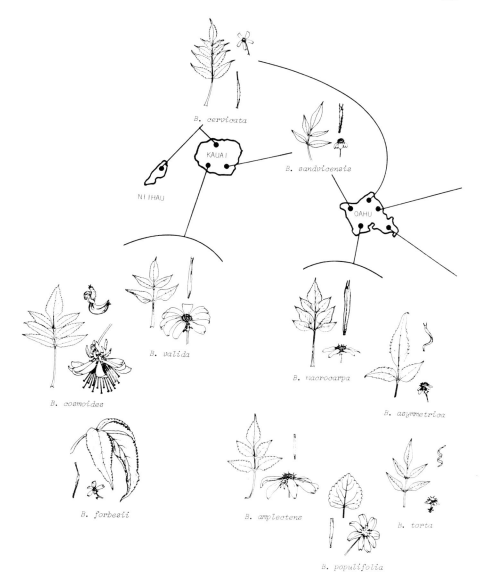

Breeding Behavior and Chromosomes

Despite their morphological divergence, all Hawaiian species of *Bidens* are interfertile (Ganders and Nagata, 1983b, 1984). As predicted in Ganders and Nagata (1983b), hybrids between *B. cosmoides* and other Hawaiian species are completely pollen and seed fertile, even though *B. cosmoides* was previously classified in a different section of the genus. Adaptive radiation in Hawaiian *Bidens* has occurred without the evolution of crossability barriers or intrinsic postzygotic isolating mechanisms.

All Hawaiian *Bidens* that have been counted have a somatic chromosome number $2n = 72$, show diploid pairing at meiosis, but are presumably ancient hexaploids based on $x = 12$ (Gillett and Lim, 1970). Isozyme banding phenotypes resulting from duplicated loci confirm the polyploid status of Hawaiian *Bidens* (Helenurm and Ganders, 1985). If there has been any chromosomal differentiation, it has been so minor that it does not affect interspecific F_1 hybrid fertility. However, detailed cytological studies have not been done.

Despite interfertility, natural hybridization is relatively uncommon. Of the 171 interspecific hybrid combinations possible, less than 6% are known. Obvious F_1 hybrids between *B. valida* and *B. forbesii* subsp. *kahiliensis* have been found recently on Mt. Kahili, Kauai (*Ganders 84-01*, UBC), and nine other interspecific hybrids have been documented (Ganders and Nagata, 1984). Natural interspecific hybridization is rare because most species are allopatric.

Other Hawaiian examples of adaptive radiation in flowering plants that have been investigated biosystematically include *Lipochaeta* (Gardner, 1976, 1979; Rabakonandrianina and Carr, 1981); *Tetramolopium* (Lowrey and Crawford, 1985); *Dubautia, Argyroxiphium,* and *Wilkesia* (Carr and Kyhos, 1981; Carr, 1985) in the Asteraceae; and *Scaevola* (Gillett, 1966) in the Goodeniaceae. In the Macaronesian Islands (the Azores, Canary Islands, Madeira, and Cape Verde Islands), *Sinapodendron* and *Lobularia* (Borgen, 1984) in the Brassicaceae, *Scrophularia* (Dalgaard, 1979) in the Scrophulariaceae, and *Asteriscus* (Borgen, 1984), and *Argyranthemum* (Humphries, 1976, 1979) in the Asteraceae have been studied biosystematically. All are characterized by allopatric speciation, absence of crossability barriers, absence of hybrid sterility except where chromosome number or structure has changed, yet only infrequent natural hybridization because distributions are allopatric, except in *Dubautia* where hybridization is rather frequent. The evidence is overwhelming that intrinsic isolating mechanisms are not necessary for divergent evolution in plants on islands.

Isozymes

The electrophoretically detectable polymorphic isozymes typically surveyed in population genetic studies are mostly enzymes that function in primary metabolic pathways. Consequently, they are a sample of a fraction of the genome that is different from that controlling morphological characters. Isozyme studies also have the advantage that allele and genotype frequencies can be measured at several loci. Genetic distances or identities between populations or taxa calculated from this data provide quantitative measures of genetic divergence for this portion of the genome.

Helenurm and Ganders (1985) surveyed 12 species and three additional subspecies of Hawaiian *Bidens*, as well as four American species, for eight enzyme systems controlled by 21 loci. Average genetic diversity within populations is high, essentially the same as the average for all outcrossing flowering plants surveyed (Table 4.1). Adaptive radiation has not resulted in genetically depauperate or monomorphic populations.

Furthermore, it appears that founder effects have not reduced genetic variability at isozyme loci in populations of Hawaiian *Bidens*. If they had, average genetic variability should be lower on younger islands, because the Hawaiian Islands are a linear chain both geographically and chronologically. There is no correlation between age of island and average genetic variability in populations (Fig. 4.2). Likewise, there has been no reduction in genetic variability during subspeciation following colonization of younger islands or volcanoes, or colonization of different elevations. Ecogeographical evidence suggests that *B. sandvicensis* subsp. *sandvicensis* is ancestral to subsp. *confusa*, and that *B. micrantha* subsp. *micrantha* is ancestral to both subsp. *ctenophylla* and subsp. *kalealaha*. In all cases the derived subspecies are at least as polymorphic as are the ancestral subspecies (Table 4.2).

Genetic identities among populations of Hawaiian *Bidens* are high. Identities between different species are as high as between different subspecies of a species, and as high as between populations of the same taxon (Table 4.3). The Hawaiian species are as similar to each other at isozyme loci as are intraspecific populations of most outcrossing flowering plants. However, they are differentiated from American species, and the American species are differentiated from each other, so that high interspecific genetic identities are a distinctive feature of Hawaiian *Bidens* rather than a characteristic of the genus. Despite diversity within populations, adaptive radiation in Hawaiian *Bidens* has not involved differentiation at isozyme loci.

Genetic identities are equally high among species of *Tetramolopium* in Hawaii (Lowrey and Crawford, 1985), the only other case of adaptive radiation in plants on oceanic islands in which isozymes have been investigated. Adaptive radiation in Polynesian land snails of the genus *Partula* (Johnson, Clarke, and Murray, 1977) and in Hawaiian *Drosophila* (Carson and Johnson, 1975; Johnson et al., 1975; Sene and Carson, 1977; Crad-

Table 4.1. Average gene diversity for 21 loci within Hawaiian *Bidens* populations compared to average for outcrossing plants

	Percent loci polymorphic	Mean alleles per polymorphic locus	Gene diversity (H_s)
Hawaiian *Bidens*	39	3.1	0.10
Outcrossing plants*	37	2.9	0.09

Source: Gottlieb (1981).

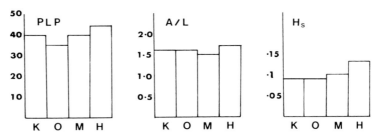

Figure 4.2. Average genetic variability in *Bidens* populations on each island. Islands are arranged by decreasing age from left to right: K, Kauai; O, Oahu; M, Maui and Molokai; and H, Hawaii. PLP, percent loci polymorphic; A/L, alleles per locus; H_s, gene diversity—each calculated from data in Helenurm and Ganders (1985).

Table 4.2. Genetic diversity within ancestral and derived subspecies of *B. sandvicensis* and *B. micrantha*.

Subspecies	Ancestral (a) or derived (d)	Percent loci polymorphic	Mean alleles per locus	Gene diversity (H_s)
B. sandvicensis				
subsp. *sandvicensis*	a	36	1.6	0.07
subsp. *confusa*	d	48	1.6	0.12
B. micrantha				
subsp. *micrantha*	a	43	1.6	0.12
subsp. *kalealaha*	d	53	1.8	0.13
subsp. *ctenophylla*	d	48	1.6	0.16

Table 4.3. Genetic identities among Hawaiian *Bidens* compared to American species and averages for all plants

	Genetic identity		
Comparison	Mean	Standard deviation	Range
Hawaiian taxa			
Within taxa	.98	.01	.96–1.00
Between subspecies	.96	.02	.93–.99
Between species	.96	.03	.89–1.00
Between Hawaiian and American species	.60	.05	.51–.73
Between American species	.80	.11	.65–.96
Average interspecific*	.67	.07	
Average intraspecific*	.95	.02	

Note: Genetic identities were determined among 22 populations of 11 species and three subspecies of Hawaiian *Bidens* and four populations of four species of American *Bidens*.
Source: Gottlieb (1977).

dock and Johnson, 1979) is also characterized by little differentiation at isozyme loci. Although there have been rather few isozyme studies in taxa that have undergone adaptive radiation on islands, they are all consistent with the hypothesis that adaptive radiation does not involve differentiation in this portion of the genome.

Flavonoids and Polyacetylenes

Diversity in two groups of so-called secondary metabolites (flavonoids and polyacetylenes) have been investigated in Hawaiian *Bidens*. A large number of flavonoids occur within each individual plant, and there is considerable diversity among plants and taxa. However, most of the compounds are chalcones and aurones, which differ in degrees of acetylation and methylation. Some species are differentiated from each other, as are subspecies within species. However, within species such as *B. macrocarpa*, *B. sandvicensis*, and *B. torta*, there are greater differences among individuals in the same population than there are between any of the species. There is variability and diversity of flavonoids within species of Hawaiian *Bidens*, but little significant differentiation correlated with speciation or adaptive radiation. Adaptive radiation has not been characterized by significant divergence in flavonoids among the species (McCormick, Bohm, and Ganders, unpublished).

Flavonoids have been investigated in two other Hawaiian genera that have undergone adaptive radiation: *Lipochaeta* and *Scaevola*. *Lipochaeta* is actually a biphyletic genus as presently circumscribed, consisting of a polyploid section and a diploid section, each independently evolved from *Wedelia*, and both have radiated on the Hawaiian Islands (Rabakonandrianina and Carr, 1981). The polyploid section of *Lipochaeta* is variable for flavonoids within taxa and shows some variation among taxa (Gardner, 1976), like Hawaiian *Bidens*, which are also polyploid. The diploid section of *Lipochaeta* is depauperate in flavonoids, with only three quercetin glycosides and little variation or differentiation among species (Gardner, 1976). The diploid species of *Scaevola* in the Hawaiian Islands also exhibit low diversity and little differentiation in flavonoids (Patterson, 1984). Flavonoid differentiation has not been a significant feature of adaptive radiation in any of these Hawaiian plants.

Polyacetylenes have also been thoroughly studied in Hawaiian *Bidens* (Marchant et al. 1984). In fact, Hawaiian *Bidens* are the only group in which these compounds have been systematically studied. Roots and leaves of Hawaiian *Bidens* accumulate a moderate diversity of polyacetylenes which may all be biosynthetically related. Two compounds are ubiquitous and unique to Hawaiian *Bidens*, which is consistent with other evidence that all are derived from a single ancestral immigrant species (Marchant et al., 1984).

Polyacetylenes are usually constant within taxa, and many taxa can be distinguished by their distinctive arrays of compounds, but species-specific compounds are rare. However, in *B. torta* there is as much difference among populations as there is among any of the species. Above the level of species or subspecies, polyacetylenes are not correlated with relationships based on morphology. Adaptive radiation has produced a group of species that combine an assortment of polyacetylenes that occur in a large number of combinations. This pattern might be expected in a case of multiple divergences from a common ancestor. There has been less evolutionary divergence in polyacetylenes than in morphology, but more than in isozymes. The situation is comparable to that in flavonoids. Adaptive radiation has involved only minor divergence among species in production and accumulation of secondary compounds.

MORPHOLOGICAL EVOLUTION AND ADAPTIVE RADIATION

Adaptive radiation in Hawaiian *Bidens* has involved considerable morphological differentiation but little or no differentiation in secondary compounds, isozymes involved in primary metabolism, or chromosomes. Interspecific isolating mechanisms have not evolved. Furthermore, all available evidence suggests that adaptive radiation in *Bidens* is typical of adaptive radiation in plants on oceanic islands.

Morphological differentiation in many plant characters may be controlled by relatively few gene loci. Hilu (1983) reviewed the literature on Mendelian inheritance of morphological mutations in plants and concluded that taxonomically important changes could involve only one or two loci. Gottlieb (1984) concluded from his survey of the literature that differences within and between species of plants in structure, shape, and presence or absence of characters are controlled by only one or two loci, although these loci are often duplicated in polyploids.

The morphological differences that distinguish Hawaiian species of *Bidens* are characters of growth habit, leaves, inflorescences, flowers, and fruits (Table 4.4). The growth habit and leaf characters of species are usually correlated with their habitat. Inflorescence and flower characters

Table 4.4. Types of morphological differences between Hawaiian species of *Bidens*, and examples where these differences are controlled by one or two loci

Character	Adaptation to habitat (h), pollinators (p), or dispersal (d)	Controlled by 1 or 2 loci in:[a]
Growth habit		
Height	h	*Pisum, Oryza*
Determinate or not	h	*Nicotiana, Phaseolus*

Table 4.4. (cont)

Character	Adaptation to habitat (h), pollinators (p), or dispersal (d)	Controlled by 1 or 2 loci in:[a]
Branching pattern	h	*Layia, Potentilla*
Decumbent or erect	h	*Crepis, Hemizonia*
Leaves		
Leaflet number	h	*Potentilla*
Leaf size	h	
Leaflet size	h	
Leaf dissection	h?	*Tropaeolum, Carthamnus*
Leaf pubescence	h	*Crepis, Gossypium*
Inflorescence		
Open or dense	p	*Potentilla*
Heads per inflorescence	p	*Helianthus*
Peduncle pubescence	?	*Crepis*
Peduncle length	p?	*Nicotiana*
Head size	p	
Phyllary size	?	
Phyllary fusion	d?	
Erect or pendant	p	*Aquilegia, Crepis*
Flowers		
Color	p	many (Hilu, 1983)
Ray number	p	
Disk floret number	p	
Size	p	*Viola, Senecio*
Style length	p	*Eschscholzia*
Male sterility	?	*Zea, Triticum*
Fruits		
Size	d	
Curvature	d	*Medicago, Pisum*
Awns	d	*Layia, Carthamnus*
Color	?	*Cucurbita*
Wings	d	*Coreopsis, Plectritis*
Pubescence	d	*Dithyrea, Valerianella*
Accessory bracts	d?	
Shape	d?	*Cucurbita*

[a]From Gottlieb (1984).

may be related to pollination strategies or specific pollinators, and the fruit differences are nearly all adaptations for dispersal in the absence of vectors such as mammals and birds. Most of the morphological differentiation between species of *Bidens* involves characters that are known to be controlled by only one or two loci in other plant species, and hence could be under similar simple genetic control in *Bidens* (Table 4.4, and see Gottlieb, 1984). In other examples of adaptive radiation in island plants such as *Argyranthemum* in the Canary Islands (Humphries, 1976) and *Lipochaeta* in the Hawaiian Islands (Gardner, 1979), the morphological differences between species are also characters that are known to be controlled by one or two loci in other plants.

Preliminary analysis suggests that several interspecific differences in *Bidens* may be controlled by only one or two loci; these include presence or absence of pubescence on leaves, peduncles, and achenes, indeterminate versus determinate growth, presence or absence of awns or wings on achenes, and coiling of achenes. Other characters such as leaflet number, number of heads per inflorescence, leaf and head size, and number of ray and disk flowers per head show a greater range of character states. I hypothesize that these multistate characters are controlled by a small number of loci. They are probably controlled by one or two loci that have been duplicated by polyploidy. Hawaiian *Bidens* are diploidized hexaploids (Gillett and Lim, 1970).

Multilocus Inheritance of Leaflet Number

One example of a multistate character is leaflet number. Different species have leaves that vary from always simple to bipinnately or tripinnately compound with numerous leaflets. The number of leaflets is clearly under genetic control, and parental phenotypes can be recovered from F_2 progenies of crosses between extremely different parents (Mensch and Gillett, 1972). This suggests that only a few loci are involved.

The simplest genetic model assumes additive inheritance of alleles at duplicated, essentially homologous, loci. With this system, plus (+) alleles produce additional leaflets or leaves. A simple leaved species may have no + alleles, a species with three leaflets may be homozygous + at two loci, a species with five leaflets may be homozygous + at three loci, etc. (Table 4.5). With enough + alleles, the leaves are bipinnately compound. If species groups A through F are ranked by decreasing number of leaflets and hypothesized + alleles, the F_1 interspecific hybrids also show a uniformly decreasing number of leaflets and + alleles to the right and downward in the crossing matrix (Table 4.5). This pattern is analogous to the phenomenon called "constant ranking" in host–parasite genetics for nonspecific resistance by plant hosts to fungal disease (Fleming and Person, 1982). Constant ranking also occurs in genetic crosses within species (Person et al., 1982, 1983), and appears to be a general feature of multilocus systems with additive effects. More complex modes of inheritance involving dominance or epistasis do not produce constant ranking.

Table 4.5. Leaflet number and number of hypothetical + alleles in F_1 hybrids among Hawaiian species of *Bidens*

		Parental species groups					
		A	B	C	D	E	F
Leaflet number		bi	7-bi	5–7	3–5	1(3)	1
Number of + alleles		12	10	8	6	4	0
Leaflet number and	A	bi 12	bi 11	bi 10	bi 9	7-bi 8	3–5 6
number of + alleles	B	bi 11	bi 10	7-bi 9	5–7 8	5–7 7	1–3 5
(underscored) in F_1	C	bi 10	7-bi 9	5–7 8	5–7 7	3–5 6	1(3) 4
hybrids	D	bi 9	5–7 8	5–7 7	3–5 6	1–3 5	1 3
	E	7-bi 8	5–7 7	3–5 6	1–3 5	1 4	1 2
	F	3–5 6	1–3 5	1(3) 4	1 3	1 2	1 0

Note: Groups B–E show constant ranking. See text for discussion. Bipinnate leaves are indicated by "bi."

The F_1 progeny of crosses between groups B through E show constant ranking, and fit the model of additive inheritance. From this I hypothesize that each of the parental species groups differs from adjacent groups by fixation for leaflet alleles at one locus. Groups A and F have a stronger influence on F_1 phenotype (Table 4.5). Group A may differ from group B by fixation at two loci, or there may be different, incompletely dominant alleles fixed in group A. Group F may also differ from group E in either of these ways. These models can be tested further by F_2 segregations from F_1 hybrids between adjacent groups.

All the evidence indicates that adaptive radiation in *Bidens* has involved significant differentiation primarily at loci controlling morphology. Preliminary evidence suggests that these interspecific differences could be controlled by only a few loci. Furthermore, there are no qualitative differences between subspecies and species. They differ only in degree of differentiation.

FOUNDER POPULATIONS AND SPECIATION

Geographical distributions of taxa indicate that speciation has been allopatric and is frequently associated with founder events on different islands, or at different elevations on the same island. Most species or their subspecies are endemic to a single island. Furthermore, 85% of the possible species pairs are allopatric, and an additional 4% are geographically contiguous but their ranges are separated by differences in elevation (Ganders and Nagata, 1984).

Interisland colonization has been a rare event. Only eight taxa occur on more than one island (Table 4.6). In all cases except *B. campylotheca* subsp. *campylotheca*, these taxa occur only on adjacent islands. Assum-

Table 4.6. Distribution of *Bidens* taxa that occur on more than one island

Taxon	Islands
B. *campylotheca* subsp. *campylotheca*	Oahu, Lanai, Hawaii
B. *cervicata*	Kauai, Niihau, Oahu
B. *hillebrandiana* subsp. *polycephala*	Molokai, Maui
B. *mauiensis*	Maui, Lanai, Kahoolawe
B. *menziesii* subsp. *menziesii*	Molokai, Maui
B. *micrantha* subsp. *kalealaha*	Maui, Lanai
B. *molokaiensis*	Oahu, Molokai
B. *sandvicensis* subsp. *sandivcensis*	Kauai, Oahu

ing that these taxa originally evolved on one of the islands in their present ranges, there have been only 11 cases in which a species has colonized another island and persisted to the present without differentiating into a taxonomically recognizable subspecies or different species.

The distribution of subspecies within species illustrates the importance of colonization by founder populations to the early stages of the speciation process. In four species—*B. campylotheca*, *B. hillebrandiana*, *B. menziesii*, and *B. micrantha*—colonization of a different island has resulted in subspeciation, with allopatric subspecies endemic to different islands. Thus, subspeciation has occurred in four out of 15 of the known cases of successful interisland colonization by species. Additional interisland colonizations have resulted in divergence to the point that the founders are now recognizable as different species. It is difficult to determine the exact number of cases, but certainly *B. mauiensis*, endemic to Maui, Lanai, and Kahoolawe, evolved from *B. molokaiensis*, endemic to Molokai and Oahu. The two species share the apomorphic characters of solitary heads on long terminal peduncles, decumbent growth habit, and short generation time. But the brown, glabrous, winged achenes of *B. mauiensis* are obviously advanced over the black, setose, wingless achenes of *B. molokaiensis*. Successful interisland colonization has frequently resulted in the evolution of new subspecies or species.

The other four cases of subspeciation have occurred on the same island. In the case of *B. micrantha* subsp. *micrantha* on West Maui and subsp. *kalealaha* on East Maui, the two subspecies are allopatric on different mountain ranges. In *B. campylotheca*, *B. forbesii*, and *B. sandivicensis*, pairs of subspecies on the same island are isolated by differences in elevation (Ganders and Nagata, 1984). Evolutionary divergence of founder populations, either on different islands or at different elevations on the same island, is characteristic of the speciation process in Hawaiian *Bidens*.

Speciation and adaptive radiation in Hawaiian *Bidens* have typically involved genetic changes at a small number of loci in founder populations. Stochastic processes are likely to be especially significant in such circumstances. However, plausible adaptive significance can be hypothesized for some of the morphological differences among species (Table 4.4). The great diversity of habitats in the Hawaiian Islands means that founder populations are likely to encounter selective forces different from those in the source population. Natural selection in founder populations has also probably been a factor of major importance in the evolution of *Bidens* in Hawaii.

LITERATURE CITED

Borgen, L. 1984. Biosystematics of Macaronesian flowering plants. In W. F. Grant (ed.), Plant Biosystematics, pp. 477–496. Academic Press Canada, Don Mills.

Carr, G. D. 1985. Monograph of the Hawaiian Madiinae (Asteraceae): *Argyroxiphium, Dubautia,* and *Wilkesia.* Allertonia 4:1–123.

Carr, G. D., and D. W. Kyhos. 1981. Adaptive radiation in the Hawaiian silversword alliance (Compositae–Madiinae) I. Cytogenetics of spontaneous hybrids. Evolution 35:543–556.

Carson, H. L., and W. E. Johnson. 1975. Genetic variation in Hawaiian *Drosophila.* I. Chromosome and allozyme polymorphism in *D. setosimentum* and *D. ochrobasis* from the island of Hawaii. Evolution 29:11–23.

Craddock, E. M., and W. E. Johnson. 1979. Genetic variation in Hawaiian *Drosophila.* V. Chromosomal and allozymic diversity in *Drosophila silvestris* and its homosequential species. Evolution 33:137–155.

Dalgaard, V., 1979. Biosystematics of the Macaronesian species of *Scrophularia.* Opera Botanica 51:1–64.

Fleming, R. A., and C. O. Person. 1982. Consequences of polygenic determination of resistance and aggressiveness in nonspecific host: parasite relationships. Can. J. Plant Pathol. 4:89–96.

Ganders, F. R., and K. M. Nagata. 1983a. New taxa and new combinations in Hawaiian *Bidens* (Asteraceae). Lyonia 2:1–16.

Ganders, F. R., and K. M. Nagata. 1983b. Relationship and floral biology of *Bidens cosmoides* (Asteraceae). Lyonia 2:23–31.

Ganders, F. R., and K. M. Nagata. 1984. The role of hybridization in the evolution of Hawaiian *Bidens.* In W. F. Grant (ed.), Plant Biosystematics, pp. 179–194. Academic Press Canada, Don Mills.

Ganders, F. R., and K. M. Nagata. in press. *Bidens.* In W. L. Wagner, D. R. Herbst, and S. H. Sohmer (eds.), A Manual of the Flowering Plants of the Hawaiian Islands. Bishop Museum Press, Honolulu.

Gardner, R. C. 1976. Evolution and adaptive radiation in *Lipochaeta* (Compositae) of the Hawaiian Islands. Syst. Bot. 1:383–391.

Gardner, R. C. 1979. Revision of *Lipochaeta* (Compositae: Heliantheae) of the Hawaiian Islands. Rhodora 81:291–343.

Gillett, G. W. 1966. Hybridization and its taxonomic implications in the *Scaevola gaudichaudiana* complex of the Hawaiian Islands. Evolution 20:506–516.

Gillett, G. W., 1975. The diversity and history of Polynesian *Bidens* section *Campylotheca.* Univ. Hawaii, H. L. Lyon Arbor. Lect. 6:1–32.

Gillett, G. W., and E. K. S. Lim, 1970. An experimental study of the genus *Bidens* in the Hawaiian Islands. Univ. Calif. Publ. Bot. 56:1–63.

Gottlieb, L. D., 1977. Electrophoretic evidence and plant systematics. Ann. Missouri Bot. Gard. 64:161–180.

Gottlieb, L. D., 1981. Electrophoretic evidence and plant populations. In L. Feinhold et al. (eds.), Progress in Phytochemistry, Vol. 7, pp. 1–46. Pergamon Press, New York.

Gottlieb, L. D. 1984. Genetics and morphological evolution in plants. Am. Natur. 123:681–709.

Helenurm, K., and F. R. Ganders. 1985. Adaptive radiation and genetic differentiation in Hawaiian *Bidens*. Evolution 39:753–765.

Hilu, K. W. 1983. The role of single-gene mutations in the evolution of flowering plants. Evol. Biol. 16:97–128.

Humphries, C. J. 1976. A revision of the Macaronesian genus *Argyranthemum*. Webb ex Schultz Bip. (Compositae–Anthemideae). Bull. Br. Mus. Nat. Hist. Bot. 5:147–240.

Humphries, C. J. 1979. Endemism and evolution in Macaronesia. In D. Bramwell (ed.), Plants and Islands, pp. 177–199. Academic Press, New York.

Johnson, M. S., B. Clarke, and J. Murray. 1977. Genetic variation and reproductive isolation in *Partula*. Evolution 31:116–126.

Johnson, W. E., H. L. Carson, K. Y. Kaneshiro, W. W. M. Steiner, and M. M. Cooper. 1975. Genetic variation in Hawaiian *Drosophila*. II. Allozymic differentiation in the *D. planitibia* subgroup. Isozymes 4:563–584.

Lowrey, T. K. and D. J. Crawford. 1985. Allozyme divergence and evolution in *Tetramolopium* (Compositae: Astereae) on the Hawaiian Islands. Syst. Bot. 10:64–72.

Marchant, Y. Y., F. R. Ganders, C.-K. Wat, and G. H. N. Towers. 1984. Polyacetylenes in Hawaiian *Bidens* (Asteraceae). Biochem. Syst. Ecol. 12:167–178.

Mensch, J. A., and G. W. Gillett. 1972. The experimental verification of natural hybridization between two taxa of Hawaiian *Bidens* (Asteraceae). Brittonia 24:57–70.

Patterson, R. 1984. Flavonoid uniformity in diploid species of Hawaiian *Scaevola* (Goodeniaceae). Syst. Bot. 9:263–265.

Person, C. O., R. Fleming, and L. Cargeeg. 1982. Non-specific interaction based on polygenes. In H. M. Heybroek et al. (ed.), Resistance to Diseases and Pests in Forest Trees, pp. 318–325. Centre for Agricultural Publishing and Documentation, Wageningen.

Person, C. O., R. Fleming, and L. Cargeeg. 1982. Non-specific interaction based on polygenes. In H. M. Heybroek et al. (eds.), Resistance to Diseases and Pests in Forest Trees, pp. 318–325. Centre for Agricultural Publishing and Documentation, Wageningen.

Rabakonandrianina, E., and G. D. Carr. 1981. Intergeneric hybridization, induced polyploidy, and the origin of the Hawaiian endemic *Lipochaeta* from *Wedelia* (Compositae). Am. J. Bot. 68:206–215.

Sene, F. M., and H. L. Carson. 1977. Genetic variation in Hawaiian *Drosophila*. IV. Allozymic similarity between *D. sylvestris* and *D. heteroneura* from the island of Hawaii. Genetics 86:187–198.

Shallenberger, R. J. (ed.). 1981. Hawaii's Birds. Hawaii Audubon Society, Honolulu.

5

Plant Speciation and the Founder Principle

G. LEDYARD STEBBINS

As is true of many other fundamental problems in biology, that of the origin of biological species via reproductive isolation is difficult to solve for one major reason. It consists not of a single one, but of several different problems. Solutions obtained for one kind of animal or plant will not necessarily be valid for other groups. Nevertheless, principles that have been developed from observations and experiments carried out on one group, given sufficient knowledge and discrimination, can often be generally applied. One of these is the founder principle, which is the subject of the present symposium. Extrapolation from animals to plants and vice versa is possible and useful if carried out with due consideration of biological and ecological differences.

With respect to the origin of reproductive isolating mechanisms, evidence from plants is particularly valuable. The reasons for this are as follows:

1. Carefully analyzed examples of speciation are more numerous and diverse in plants than in animals. Cytogenetic information on animal species exists only for isolated genera, such as mice, mole rats, *(Spalax)*, and higher primates among mammals; a few gallinaceous birds; anuran amphibia; a few fishes; and among insects only *Drosophila* and a few other Diptera, a handful of Lepidoptera, and some Orthoptera. Among plants, the relative ease of culture and rearing has in many families enabled biosystematists to analyze complete species groups in several related genera on the basis of data from chromosomes, genetic segregation of morphological differences, and variation with respect to biochemical differences among proteins and secondary organic compounds.

2. Even within the same comparium of intercrossable plant species, great differences can exist with respect to ecological preferences, length of life cycles, overall size of the organism, self- versus cross-fertilization, and other characteristics that affect the rate and nature of speciation. Situations exist in which only one of these factors differs between two

groups. The effect of this particular factor can be estimated by comparing the two groups because other factors are similar in them.

3. Ecological factors that affect speciation are fewer and easier to analyze in plants than in animals. Population size and degree of spatial isolation can be estimated with relative ease. Complications such as food chains, courtship patterns, and social dominance are lacking. Plant evolutionists, therefore, can more easily focus attention on factors that affect directly the origin and spread of a smaller number of reproductive isolating mechanisms. Complications raised by widespread polyploidy are a drawback to formulating general principles. Nevertheless, many predominantly diploid genera have been analyzed cytogenetically so that comparisons can be made between species having the same chromosome number, or that form aneuploid series similar to those found in animals. In the present article, only genera of this kind will be considered.

DEGREE OF MORPHOLOGICAL DIFFERENCE IS NOT ASSOCIATED WITH STRENGTH OF REPRODUCTIVE ISOLATION

Beginning with the earliest analyses of genetic segregation in the offspring of semisterile interspecific hybrids (Müntzing, 1930), the fact has now become firmly established that chromosomal and genic differences that are responsible for sterility segregate independently of genes that affect morphological differences between the parental species (Stebbins, 1950, 1982; Grant, 1971). Moreover, one can find in related groups belonging to the same genus two contrasting situations: (1) species that differ strongly from each other with respect to morphological characteristics, and are partly subdivided into morphologically recognizable subspecies, but which can be easily crossed to form fertile hybrids; and (2) clusters of sibling species, that the taxonomist can separate only with difficulty, if at all, but are difficult or impossible to hybridize, and/or form highly sterile F_1 hybrids. In examples of the first kind, reproductive isolation is based chiefly upon differences in habitat and ecological adaptations that either prevent the species from growing near each other or greatly restrict the ability of the F_1 hybrids to become established in suitable habitats. In the Compositae, tribe Madiinae, both of these contrasting situations exist within the same genus (Clausen, 1951; see next section). Why should this be so?

The difference can be explained largely on the basis of the founder principle. Morphological and ecological divergence can be promoted by diversifying natural selection in large populations, whereas hybrid sterility in plants, caused chiefly by chromosomal repatterning, results from the action of chance as populations pass repeatedly through bottlenecks of small size and undergo simultaneously strong but temporary divergent selection pressures. If this postulate is correct, then the difference between the two situations should be based both upon the ecological nature

of adaptation and upon population structure. First, the tendency to produce polytypic species that are poorly isolated from each other should be
associated either with ecological dominance or with the capacity for rapid
colonization of newly invaded habitats. Dominant species usually form
large populations. If colonization is rapid, populations do not remain
small for periods long enough to permit chance fixation of adaptively neutral differences, such as the majority of chromosomal rearrangements.
Clusters of sibling species would be expected chiefly in pioneer species
that are adapted to temporary habitats having patchy distributions, such
as are often associated with specialized, restrictive soil types. These clusters should be particularly characteristic of genotypes and populations
that are ecologically specialized, being unable to establish themselves in
a variety of habitats. They are more likely to evolve in annuals, but exist
also in perennials, such as the bunch grass *Elymus glaucus* (Snyder,
1951). Among adaptive generalists that are adapted to generalized habitats, morphological and ecological divergence should evolve more rapidly
than reproductive isolation, whereas populations consisting of genetic and
ecological specialists should break up into clusters of sibling species.

Other factors that should contribute to the difference are those of the
genetic system. Obligate outcrossing, plus free genetic recombination
based upon a high chromosome number and unrestricted crossing-over,
should greatly retard the evolution of reproductive isolation, whereas both
self-fertilization and the restriction of genetic recombination caused by
low chromosome number and low frequencies of crossing-over should
promote reproductive isolation and the evolution of clusters of sibling species. To what extent are these predictions borne out?

EXAMPLES FROM INDIVIDUAL GENERA

Facts supporting independent divergence of morphology and reproductive isolation have already been reviewed in several publications (Stebbins 1950, 1982; Grant 1971, 1981). Among woody plants that dominate
their ecosystems, such as oaks (*Quercus*), poplars (*Populus*), blueberries
(*Vaccinium*), the genus *Ceanothus*, and the woody species of *Mimulus*,
great morphological and ecological differences are unaccompanied by reproductive isolation barriers. The same is true of some cross-pollinating
perennial herbs that dominate grasslands, such as diploid species of *Dactylis* (Stebbins and Zohary, 1959). On the other hand, sibling species, or
pairs of species that in spite of morphological differences are ecologically
similar but well isolated reproductively, are well known in annual genera
such as *Clarkia* (Lewis and Roberts, 1956; Lewis and Raven, 1958;
Lewis, 1962, 1966).

Particularly revealing are examples of generalized plus specialized species existing side by side in the same genus. Three such examples exist
in the Madiinae. Two of them are reviewed by Clausen (1951). *Madia*

elegans is an outcrossing generalist within which taxonomists can recognize four interfertile subspecies, one of which differs so much from the others that it was originally assigned to a different genus (*Hemizonia Wheeleri Gray*). *Madia nutans* and *M. Rammii* are restricted to volcanic substrates of a particular kind, are morphologically homogeneous, and relatively local geographically. They are isolated from each other by strongly developed sterility barriers, that cannot be overcome even by allopolyploidy (Clausen et al., 1945). Three other species of *Madia*—*M. yosemitana, M. Hallii,* and the recently described *M. Stebbinsii*—are likewise ecologically restricted and morphologically homogeneous. Although they have not been hybridized with other species, they can be expected to behave like *M. nutans* and *M. Rammii.*

In still another genus of the tribe Madiinae, Carr (1976, 1977, 1980; Carr and Carr, 1983) has obtained data suggesting a similar contrast between the widespread *Calycadenia multiglandulosa* and the more restricted species, largely confined to serpentine: *C. ciliosa, C. Fremontii, C. pauciflora,* and *C. villosa.* In this latter group, within which species are hard to define even on the basis of recognized barriers of reproductive isolation, the process of speciation may be actively taking place.

A similar example from another family is the species group that centers about *Clarkia unguiculata,* as analyzed by Vasek (1958, 1964, 1968). The principal species, *C. unguiculata* sensu stricto, is a widespread outcrossing generalist, adapted to a variety of habitats, distributed almost throughout lowland California. It consists of many populations that are adapted to different habitats, but are interfertile. In the southern end of the San Joaquin Valley and neighboring foothills, it is replaced by different species, all of which are normally self-fertilizing, smaller in stature, and with smaller flowers. These species—*C. exilis, C. temblorensis, C. springvillensis,* and *C. "Caliente"*—are isolated from each other and from *C. unguiculata* by well-developed internal barriers, including chromosomal rearrangements, even though the habitats that they occupy are similar to each other.

These examples form promising material for investigating the role of population–environment interactions in the origin of species. Three postulates can be tested:

1. The generalized species is either directly ancestral to the more specialized ones or resembles most closely the common ancestor of the group.
2. The more specialized, restricted species, which in each example occur at, near, or beyond the range of the more generalized species, have originated by a combination of peripheral migration, divergent selection, and the action of the founder principle.
3. Genotypes belonging to the more widespread, generalized species possess a wider range of ecological tolerance than those of the more restricted species.

The elegant experimental techniques developed by J. L. Harper (1977) for investigating seedling development under a diverse array of controlled environments would be particularly valuable for the latter purpose.

AUTOGAMY AND SPECIATION

According to Mayr (1954, 1963), one of the most important factors of the founder principle is the shift from outcrossing to inbreeding. He postulates two kinds of genes or alleles: (1) "soloists," which apparently possess an adaptive advantage in the homozygous condition, having little interaction with other genes; and (2) "good mixers," which are more highly adaptive in the heterozygous condition, also displaying epistatic interactions with other genes. He has maintained that the shift from outcrossing to inbreeding can cause "a veritable genetic revolution" in which "changes in any locus will in turn affect the selective values at many other loci, until finally the system has reached a new state of equilibrium."

Anyone familiar with modern genetics can detect a serious flaw in Mayr's reasoning. Biochemical evidence indicates that epistatic interaction between alleles at different loci is far more important for harmonious development than are allelic interactions at a single locus. Many enzymes and other proteins that are vital for cell metabolism consist of two or more different polypeptide chains that are coded by different gene loci. Growth and molecular synthesis are based upon biosynthetic pathways, carried out by harmonious interaction between several enzymes coded by different nonallelic genes. The shift from heterozygosity to homozygosity that results from reduction in population size accompanied by inbreeding will have little effect on epistatic interactions, unless the process uncovers recessive alleles that interact poorly with genes at other loci. The practical experience of plant breeders has shown that such inferior alleles must be weeded out before an inbred homozygote can succeed. Given below are examples of hybrids between outcrossed and inbred populations that are both vigorous and highly fertile in both first and second generations. This could not occur if the outcrossed parent were heterozygous for a large number of deleterious recessive alleles.

Plants that have hermaphroditic or bisexual flowers are unusually favorable material for testing this hypothesis, since an outcrosser can be converted into a selfer by means of a few gene substitutions. The derived selfer that can start a new population from a single individual is particularly well adapted to colonizing new habitats that are far from its original home (Baker, 1955). One can therefore find a series of examples that either increase or decrease the probability of a prediction that follows logically from Mayr's hypothesis: Genetic differences between closely related outcrossed and inbred populations, particularly those that promote reproductive isolation and speciation, are likely to be concerned as much with internal adjustment and developmental harmony as with adaptation to new environments.

The first example is a single well-known species, within which the shift from outcrossing to predominant selfing, accompanied by adaptation to a new habitat, has not led to change any more drastic than the origin of conventional subspecies or races.

Within the species *Potentilla glandulosa* eight subspecies are recognized. Five of these are self-incompatible obligate outcrossers; the other three are known or presumed to be usually self-fertilizing. Two of the latter, subspp. *ewanii* and *globosa*, are localized and little known. The other, ssp. *reflexa*, is widespread in dry foothills and open forests at middle altitudes. It is distinguished from all of the other seven by characteristics similar to those that in other groups of *Potentilla* serve to distinguish species: (1) smaller flowers; (2) bright yellow petals that are shorter than the sepals and become reflexed in age; and (3) taller, more slender stems and branches that are only sparingly glandular. If it could retain its identity when growing sympatrically with other subspecies, it would be recognized as a species. That, however, is not the case. In many localities in the Sierra Nevada, subsp. *reflexa* grows near outcrossing populations of subsp. *Hansenii*, and intergrades with them (Clausen et al., 1940). Populations in the North Coast Ranges, where the ranges of subspp. *reflexa* and *glandulosa* overlap, have not been carefully studied, but casual observations suggest that integradation between these two subspecies is extensive. In this example, the shift from outcrossing to predominant inbreeding has been accompanied by changes in flower structure that might be expected, since in a selfer large, conspicuous flowers have less adaptive value than in an outcrosser, and in the relatively dry habitat occupied by subsp. *reflexa* small flowers would favor more rapid maturation during the relatively short favorable springtime. Because *Potentilla* plants are perennials that may live for tens or scores of years, one might argue that genetic shifts normally associated with autogamy have been delayed by its long generation time. The remaining examples, however, are all strict annuals.

The second example consists of two closely related species, or semispecies in the sense of Mayr (1963), *Lasthenia minor* and *L. maritima*, in the family Asteraceae. Both are diploid annuals ($n = 4$), restricted to the Pacific Coast of North America. Confined to mainland northern California, *L. minor* consists of a coastal and an inland race. It is self-incompatible, an obligate outcrosser. Its relative, *L. maritima*, is confined to islands, islets, and "bird rocks" at distances varying from a few hundred meters to several kilometers from the coast (Vasey, 1985). It is self-compatible and normally self-pollinating (Ornduff, 1966). It occurs in more than 70 widely separate localities extending from southern California, to Vancouver Island, British Columbia, a distance of 2000 km. The species differ with respect to floral morphology (fruit size and pubescense, number and shape of pappus scales, length of ligules), but form highly fertile hybrids. Although morphological differences exist between the widely scattered populations of *L. maritima*, their differences from *L. minor* are consistent enough so that the hypothesis of Ornduff (1966), supported by Vasey (1985), that *L. maritima* arose only once and was spread by long-distance dispersal through the action of sea bird vectors, is highly probable. Crawford et al. (1985) have found electrophoretic differences between the proteins of the two species. In this respect differences between populations of *L. maritima* are greater than between those of *L. minor*. Consequently, *interpopulational* divergence without speciation has been greater in the autogamous *L. maritima* than in the outcrossing species *L. minor*. The differences between the fruits of the two species all render those of *L. maritima* more susceptible to bird transport than are the fruits of *L. minor*, suggesting that morphological evolution in this example has been guided by natural selection.

Two important lessons can be drawn from this example: First, the shift from obligate outcrossing to predominant selfing is not accompanied by genetic reorganization of the genome, any more than it is in the origin of subspecies in *Potentilla glandulosa*. Second, once self-fertilization and spatial isolation have been achieved, repeated colonization of new habitats can take place, being accompanied by alteration of genotypes, but not by further speciation.

The third example, described by Gottlieb (1973, 1982), is *Stephanomeria malheurensis* and its probable ancestor, *S. exigua* subsp. *coronaria*. Although Gottlieb believes that this is an example of sympatric speciation, the evidence that he presents does not require this interpretation. Although *S. malheurensis* is known from only a single locality, where it is sympatric with *S. exigua* subsp. *coronaria*, this locality is in the middle of a large steppe-desert that is traversed by only a few roads. It has such an impoverished flora that plant collectors rarely visit it. The geological map of the area shows that the unusual tuffaceous sedimentary formation that has formed the distinctive soil on which *S. malheurensis* grows is widely scattered throughout the area, occurring often far from a highway. Explorers for rare plants in western North America have found consistently that in such a situation, careful search reveals additional localities for species that are at first believed to be confined to a single spot. Hence the hypothesis is highly plausible that *S. malheurensis* originated via a colonization of such a presently undiscovered spot, beyond the present range of *S. e.* subsp. *coronaria*, and that their present sympatry is secondary.

Whatever may have been its origin, *S. malheurensis* is similar to the two previous examples in that a shift from cross- to self-fertilization has accompanied speciation. It differs in that speciation has been accompanied also by chromosomal rearrangements, which are the principal cause of the hybrid sterility that, probably more than predominant selfing, maintains their reproductive isolation and enables them to "pass the test of sympatry" (Mayr, 1963). In this respect, the example resembles those in *Clarkia*, mentioned above.

My final example is a species complex in the genus *Linanthus* (Polemoniaceae) that is still imperfectly known. In many ways it resembles examples in the neighboring genus *Gilia*, discussed by Grant (1950, 1952, 1954, 1971). *Linanthus androsaceus* and *L. bicolor* occur sympatrically in many places in the North Coast Ranges of California, including the Stebbins-Cold Canyon Reserve, where their distinctive ecological preferences are being studied, and whence the plants were obtained that are discussed below. The F_1 hybrid between them is easily made, is highly vigorous, and varies greatly in fertility (Table 5.1). An F_2 generation of 106 plants, derived from several F_1 parents, included 92 that were mostly vigorous but highly sterile (Table 5.1). In 81 of these, sterility was accompanied by abortion of anthers, such that no pollen was formed, and in many of them growth of anthers was arrested before the beginning of meiosis. This was in striking contrast to the sterility of the F_1 plants, nearly all of which had anthers of normal size, and formed pollen grains with normal exine. Nevertheless, in most plants pollen grains were irregular in size and either empty or deficient in cytoplasmic content.

The sharp decrease in fertility from F_1 to F_2 is in striking contrast to the behavior of progeny from the hybrid *S. e.* subsp. *coronaria* × *S. malheurensis* in which mean fertility (pollen stainability) increased from 24% in F_1 to 62% in the F_2 generation, and 18% of the F_2 segregates had over 80% pollen stainability (Gottlieb, 1973).

With respect to its parentage, the example from *Linanthus* is similar to the previous ones. The species are sympatric and similar morphologically. One *(L. androsaceus)* is an obligate outcrosser, and the other *(L. bicolor)* is a predominant selfer. The F_1 hybrid is easily made and is highly vigorous. It differs from the first two in possessing genetically determined hybrid sterility, and from *Stephanomeria* in the greater degree of sterility, which is even greater in the F_2 generation.

The series of examples discussed is comparable to a series that can be recognized within the genus *Drosophila*. The subspecies of *Potentilla glandulosa* are comparable to those of *D. willistonii* and *D. equinoxialis* (Ayala, 1975). Except for its lack of prezygotic isolating mechanisms, the

Table 5.1. Pollen fertility of F_1 and F_2 generation hybrids, *Linanthus androsaceus* × bicolor

		% Stained pollen				
		0–10	11–30	31–50	51–70	Total
Abnormal anthers	+	−	−	−	−	+ −
No. plants, F_1	2	12	7	0	3	2 22
No. plants, F_2	81	11	5	6	3	81 25

Lasthenia example resembles *D. sylvestris–heteroneura* (Val, 1977). Those in *Clarkia* and *Stephanomeria* resemble borderline species belonging to the *D. repleta* group. Finally, the presence of developmental abnormalities in F_2 hybrids of the *Linanthus androsaceus* complex makes it comparable to those of *Drosophila willistonii* and *pseudoobscura–persimilis–miranda* (Dobzhansky, 1970).

DISCUSSION

The significant elements of the founder principle as postulated by Mayr (1954) are (1) a reduction in population size; (2) the elimination of gene flow via migration from other populations; (3) the shift from outcrossing to inbreeding; and (4) a "genetic revolution" brought about by gene substitution, that is, genes that are adaptive by virtue of interactions with other genes are replaced by those that contribute to adaptation by virtue of their own properties. Are all of these factors equally important for speciation, and are they, either separately or acting together, sufficient to complete the speciation process? On the basis of the examples summarized above, plus other examples from plants that resemble them, I shall try to answer these questions.

The importance of reduction in population size, already recognized by plant geneticists (Grant, 1981) as well as by the animal geneticists that have participated in this and many other symposia, can hardly be doubted. Every example known to me of speciation that on the basis of various lines of evidence can be judged as recent involves passage of populations through one or more bottlenecks of small size. The plant evidence is even stronger. It includes a highly significant correlation: Populations that occupy climax or dominant positions in closed ecosystems have poorly developed postzygotic or internal barriers of reproductive isolation even between species that differ from each other greatly with respect to morphology and ecological adaptation. Clusters of morphologically similar sibling species exist in groups that occupy patchy habitats and often colonize newly available habitats. This correlation is emphasized by Grant (1981) and myself, and is briefly reviewed in earlier sections of this paper.

The question of major importance is, therefore, which of the kinds of changes that usually accompany passage through bottlenecks contribute the most to the speciation process?

In a previous section of this paper, allelic substitution directly connected with size reduction has been relegated to a minor role. Because integration of the genotype is achieved chiefly via epistatic rather than interallelic interactions, and because epistasis can be as prevalent in homozygotes as in heterozygotes, a minimum amount of gene substitution can maintain integrated adaptive systems even if population size becomes drastically reduced, accompanied by inbreeding.

Two other changes that accompany reduction in population size are much more important: first, drift and random fixation of neutral alleles or chromosomal differences, and second, biased sampling of the gene pool of the ancestral large population in such a way that adaptive or potentially adaptive gene combinations that are rare in the ancestral large population become predominant in one or more of its smaller descendants. Which of these two factors is the most important?

Based upon the examples reviewed above plus several others, the following is the most logical answer to this question. Either one of these factors may be more important than the other, depending upon the gene and chromosomal pool of the ancestral large population, as well as the environment in which the new isolate finds itself. The morphological differences and genetic relationships between *Lasthenia minor* and *L. maritima* are best explained by assuming that divergent selection played the major role in the origin of *L. maritima*. The first successful colonizer of the *L. maritima* habitat was probably derived from a coastal population of *L. minor* in which some of the genes responsible for the achene and pappus differences, as well as adaptation to guano-impregnated soils, may well have been present already, though in low frequencies. The absence of chromosomal differences between the two species may reflect the chromosomal homogeneity found in *L. minor.*

On the other hand, the reverse condition may have been true of the origin of *Stephanomeria malheuresis* from *S. exigua* subsp. *coronaria.* The adaptive difference between them is relatively slight, but is maintained by reproductive isolation due to chromosomal differences that may have become fixed from a gene pool in which some degree of chromosomal polymorphism was already present.

When other examples from flowering plants are reviewed, the same conclusion is reached. The examples of the Hawaiian silverswords and *Bidens*, reviewed in other chapters of this volume, resemble *Lasthenia*, whereas those of *Holocarpha*, *Madia nutans-Rammii*, and various species complexes in *Clarkia*, mentioned in an earlier section, are more like *Stephanomeria*. On the other hand, plant taxonomists are familiar with scores of species that contain or consist entirely of small populations that are widely separated from each other. In such species, many bottlenecks of reduction to small size followed by expansion in a new habitat have probably occurred unaccompanied by speciation.

My final conclusion is as follows. Passage through bottlenecks accompanied by founder events are by far the commonest condition to elicit and accompany speciation. Speciation, therefore, is usually a relatively sudden and abrupt process, as Eldredge and Gould (1972) have maintained. It can be completed during a few hundred or thousand generations, without participation of genetic changes that individually are any more drastic than those responsible for differentiation of well-marked races or subspecies. All that is required is the establishment of particular differences that in various ways reduce gene flow between sympatric populations.

Nevertheless, bottlenecks and founder events do not by themselves bring about speciation. Those populations that respond to them by forming new species have obtained from their parental populations gene complexes or chromosomal differences that, when fixed in the derived population, cause it to be reproductively isolated from all others. Newly arisen mutations or chromosomal changes may also contribute to the process.

The hypothesis that speciation is relatively uncommon, even among populations that have been subjected to conditions that favor it, is compatible with the pattern of species numbers and species differences as we see them in the modern plant world. Felsenstein (1981) has pointed out that, given the number of opportunities for speciation and the great length, in genetic terms, of the evolutionary time scale, the more serious question facing evolutionists is not, why are there so many species? but rather, why aren't there more? A little calculation bears out his conclusion. Imagine an evolutionary line with respect to which bottlenecks and founder events occur on the average every 50,000 years, and each event requires 5000 years for speciation to become complete. This is by no means explosive speciation, since 90% of the time the populations are static with respect to speciation. Yet if no extinction of species occurred, and if every species present at each of the 5000-year periods that would favor speciation should split into two species, at the end of 1,000,000 years a single ancestral species would have given rise to 2^{20} or 1,049,376 descendant species. Although a million years is an inconceivably long time for the geneticist, it is short for the paleontologist. Probably the majority of plant species now living in the temperate northern hemisphere are more than a million years old.

These figures present a challenge to the evolutionist. Speciation events are not the inevitable result of processes that are taking place continually in populations. Some intraspecific races may be the ancestors of future species, but most of them are not. For an evolutionary line to achieve speciation, both unusual environmental changes and unusual favorable situations within the population are needed. Moreover, several different combinations of environmental and population genetic factors can accomplish speciation almost equally well. The facts reviewed in this and other chapters in this volume have pointed to the nature of some of the most favorable conditions for speciation. Identifying all of them is a task that will occupy evolutionists for some time to come.

SUMMARY

Higher plants are particularly favorable for studying processes of speciation since many diverse groups can be and have been analyzed and compared under controlled conditions. The literature on higher plant cytogenetics contains many more separate examples of well-analyzed genera than exist in animals. Both analyses of individual progenies and comparisons between groups have shown that morphological differences between species are to a great extent genetically independent of factors that promote reproductive isolation. Intergroup comparisons show that reproductive isolation develops most rapidly in groups that are adapted to patchy habitats and are successful pioneers in newly available habitats. Examples that illustrate these generalizations are reviewed, taken from woody genera *(Quercus, Ceanothus)* and annuals native to California and Oregon *(Madia, Layia, Calycadenia, Clarkia, Potentilla, Lasthenia, Stephanomeria,* and *Linanthus).* They suggest that although great reduction in population size usually accompanies speciation, it does not automatically bring it about. Among the factors that accompany such reduction, the shift from outbreeding to inbreeding has relatively little effect by itself, because in any genotype, genic interactions that promote harmonious integration are mostly epistatic in nature, and so can be equally effective in largely homozygous genotypes that result from inbreeding as in heterozygotes. More important factors are the effects of strongly divergent natural selection, and drift plus random fixation of chromosomal and other differences that promote reproductive isolation. In some plant examples, divergent selection is most important; in others, chromosomal repatterning plays a major role.

LITERATURE CITED

Ayala, F. J. 1975. Genetic differentiation during the speciation process. Evol. Biol. 8:1–78.

Baker, H. G. 1955. Self-compatibility and establishment after "long-distance" dispersal. Evolution 9:347–349.

Carr, G. D. 1976. Chromosome evolution and aneuploid reduction in *Calycadenia pauciflora* (Asteraceae). Evolution 29:681–699.

Carr, G. D. 1977. A cytological conspectus of the genus *Calycadenia* (Asteraceae), an example of contrasting modes of evolution. Am. J. Bot. 64:694–703.

Carr, G. D. 1980. Experimental evidence for saltational chromosome evolution in *Calycadenia pauciflora* Gray. Heredity 45:109–115.

Carr, R. L., and Carr, G. D. 1983. Chromosome races and structural heterozygosity in *Calycadenia ciliosa* (Asteraceae). Am. J. Bot. 70:744–755.

Clausen, J. 1951. Stages in the Evolution of Plant Species. Cornell University Press, Ithaca.

Clausen, J., Keck, D. D., and Hiesey, W. M. 1940. Experimental studies on the nature of species. I. Effect of varied environments on North American plants. Carnegie Inst. Wash. Publ. No. 520.

Clausen, J., Keck, D. D., and Hiesey, W. M. 1945. Experimental studies on the nature of species. II. Plant evolution through amphiploidy and autoploidy, with examples from the Madiinae. Carnegie Inst. Wash. Publ. No. 564.

Crawford, D. J., Ornduff, R., and Vasey, M. C. 1985. Allozyme variation within and between *Lasthenia minor* and its derivative species *L. maritima.* Am. J. Bot. 72:1177–1184.

Dobzhansky, T. 1970. Genetics of the Evolutionary Process. Columbia University Press, New York.

Eldredge, N., and Gould, S. J. 1972. Punctuated equilibria as alternative to phyletic gradualism. In T. J. M. Schopf (ed.), Models in Paleontology, pp. 82–115. San Francisco, Freeman, Cooper.

Felsenstein, J. 1981. Skepticism toward Sant Rosalia, or why are there so few kinds of animals? Evolution 35:124–138.

Gottlieb, L. D. 1973. Genetic differentiation, sympatric speciation and the origin of a diploid species of Stephanomeria. Am. J. Bot. 65:970–982.

Gottlieb, L. D. 1982. Does speciation facilitate the evolution of adaptation? In C. Barigozzi (ed.), Mechanisms of speciation, pp. 179–190. Alan R. Liss, New York.

Grant, V. 1950, 1952. Genetic and taxonomic studies in Gilia. I–III. El Aliso 2:239–316, 361–373, 375–388.

Grant, V. 1954. Genetic and taxonomic studies in Gilia. VI. Interspecific relationships in the leafy-stemmed gilias. El Aliso 3:35–49.

Grant, V. 1971. Plant Speciation. Columbia University Press, New York.

Grant, V. 1981. Plant Speciation, 2nd ed. Columbia University Press, New York.

Harper, J. L. 1977. Population Biology of Plants. Academic Press, New York.

Lewis, H. 1962. Catastrophic selection as a factor in speciation. Evolution 16:257–271.

Lewis, H. 1966. Speciation in flowering plants. Science 152:167–172.

Lewis, H., and Raven, P. H. 1958. Rapid evolution in Clarkia. Evolution 12:319–336.

Lewis, H., and Roberts, M. R. 1956. The origin of Clarkia lingulata. Evolution 10:126–138.

Mayr, E. 1954. Change of genetic environment and evolution. In J. Huxley, A. C. Hardy, and E. B. Ford (eds.), Evolution as a Process, pp. 157–180. George Allen & Unwin, London.

Mayr, E. 1963. Animal Species and Evolution. pp. 797, Harvard University Press, Cambridge, MA.

Müntzing, A. 1930. Outline to a genetic monograph of Galeopsis. Hereditas 13:185–341.

Ornduff, R. 1966. A biosystematic survey of the goldfield genus Lasthenia. Univ. Calif. Publ. Bot. 40:1–92.

Snyder, L. A. 1951. Cytology of interstrain hybrids and the probable origin of variability in Elymus glaucus. Am. J. Bot. 38:195–202.

Stebbins, G. L. 1950. Variation and Evolution in Plants. Columbia University Press, New York.

Stebbins, G. L. 1982. Plant speciation. In C. Barigozzi (ed.), Mechanisms of Speciation, pp. 21–39. Alan R. Liss, New York.

Stebbins, G. L., and Zohary, D. 1959. Cytogenetic and evolutionary studies in the genus Dactylis. I. The morphology, distribution, and interrelationships of the diploid subspecies. Univ. Calif. Publ. Bot. 31:1–40.

Val, F. C. 1977. Genetic analysis of the morphological differences between two interfertile species of Hawaiian Drosophila. Evolution 31:611–629.

Vasek, F. C. 1958. The relationship of Clarkia exilis to Clarkia unguiculata. Am. J. Bot. 45:150–162.

Vasek, F. C. 1964. The evolution of Clarkia unguiculata derivatives adapted to relatively xeric environments. Evolution 18:26–42.

Vasek, F. 1968. The relationships of two ecologically marginal, sympatric Clarkia populations. Am. Natur. 102:25–40.

Vasey, M. C. 1985. The specific status of Lasthenia maritima (Asteraceae), an endemic of seabird breeding habitats. Madrono 32:131–142.

III

FOUNDER EFFECTS AND CHROMOSOMAL EVOLUTION

6

Chromosomal Evolution and Speciation in Hawaiian *Drosophila*

Since the initiation of the Hawaiian Drosophilidae Project by the late Professor Wilson S. Stone of the University of Texas and Professor D. E. Hardy of the University of Hawaii in 1963, a number of senior investigators from various institutions have been involved in the project. Because of their successful research, more is known today about the taxonomy, ecology, behavior, and cytogenetics of Hawaiian Drosophilidae than of any other insect group of comparable size.

In a series of cytogenetic and evolutionary studies of Hawaiian Drosophilidae, we have studied the patterns of chromosome evolution and have determined the phylogenetic relationships among several groups of Hawaiian *Drosophila* species. Several new experiments on mechanisms of chromosomal evolution have been initiated as another step toward our long-range objectives of understanding the evolutionary process. During this study, a model of the "pseudochromocenter" was formulated as a possible major mechanism for chromosome evolution (Yoon et al., 1972a; Yoon and Richardson, 1978a).

The purposes of this paper are (1) to present a summary of karyological data on Hawaiian *Drosophila* thus far studied; (2) to give information as to the mechanisms of chromosome evolution and speciation in Hawaiian *Drosophila;* and furthermore, (3) to provide additional data regarding the patterns of chromosomal puffing, as indicators of gene function, in selected "homosequential" species, as a necessary step to understanding the process of speciation by which they arose.

SYSTEMATICS AND EVOLUTIONARY PATTERN

The family Drosophilidae is developed most remarkably in the Hawaiian Islands and is one of the most striking examples of rapid and adaptive radiation known in the world. Approximately one-fourth of the described species of *Drosophila* and one-half of those of *Scaptomyza* are endemic to the Hawaiian archipelago. Not only have numerous species developed in a limited area here, but also they exhibit the greatest diversity of form and behavior known for any taxonomically similar group of Drosophilidae. A great many structural peculiarities are also found, which apparently do not occur in other drosophilid faunas. Taxonomists conservatively estimate that the total fauna consists of at least 800 species, leaving an additional 150 species of *Drosophila* and over 100 of *Scaptomyza* remaining to be described in Hawaii (Carson and Kaneshiro, 1976).

In spite of their morphological (Hardy, 1965; Throckmorton, 1966; Hardy and Kaneshiro, 1979), anatomical (Kaneshiro, 1974, 1976a, b), ecological (Heed, 1968), and ethological diversity (Spieth, 1966, 1978), the Hawaiian drosophilids constitute a closely related evolutionary group. These studies indicated that there are only two major lineages in the evolution of Hawaiian Drosophilidae—the drosophiloids and the scaptomyzoids.

An analysis of the cytological studies (both polytene and metaphase karotypes of Hawaiian drosophiloids) also points to the same interpretation as do the anatomical and behavioral studies. On the basis of observations of the metaphase karyotypes, Clayton (1966, 1968) and my colleagues and I (Yoon et al., 1972a–c, 1973a, b, 1975; Yoon and Carson, 1973; Yoon and Wheeler, 1973; Yoon and Richardson, 1974, 1976a, b, 1977, 1978a–e) have found that the endemic species of Hawaiian Drosophilidae fall into two metaphase karyotype groups which correspond with the genus *Drosophila* and the genus *Scaptomyza*, but with infrequent intermediate types, suggesting a common ancestor to the two lineages. (From what is known, no intermediate type exists outside of Hawaii.)

Although the superspecies classification of Hawaiian *Drosophila* is not settled, operationally several superspecies taxa are presently used. They are (1) the picture-winged species group; (2) the modified-mouthparts species group; (3) the modified tarsi species group (bristle tarsi, split tarsi, and spoon tarsi); and (4) a miscellaneous group.

The typical evolutionary pattern of Hawaiian *Drosophila* observed in earlier studies is that the species were formed on different islands. Carson et al. (1970) found that almost all of the species (98%) are restricted to single islands, with the most closely related forms often on adjacent islands. For example, each species of the *crassifemur* complex is endemic to one of the four major islands of the Hawaiian archipelago. From a phylogenetic study of the complex reported in Yoon et al. (1975), we have shown that the four species in the complex appear to form a linear biogeographical sequence of accumulated inversions paralleling the age of the islands. This, in turn, suggests that the evolutionary sequence is from the primitive species found on Kauai, the oldest island, to the most derived species on Hawaii, the youngest island. Concomitant with speciation were changes in the amount and distribution of heterochromatin among the metaphase chromosomes. These metaphase karyotype studies, combined with the polytene chromosome homologies, made it possible to designate a common chromosome nomenclature for all genera of Hawaiian Drosophilidae (Yoon and Carson, 1973).

GENOMIC ORGANIZATION, HETEROCHROMATIN, AND CHROMOSOME EVOLUTION

The karyology of Hawaiian *Drosophila* has been carried out by several workers (Yoon et al., 1972a–c, 1973a,b, 1975; Yoon and Carson, 1973; Yoon and Wheeler, 1973; Yoon and Richardson, 1974, 1976a,b, 1977, 1978a–e; Clayton, 1976; Clayton and Wheeler, 1975). Determinations of metaphase and polytene karyotypes were made by the lacto-aceto-orcein techniques (Yoon et al., 1973b). Heterochromatic bands were identified by the fact that they stain darker than euchromatic bands. Polytene chromosomes are composed mainly of euchromatic bands.

Fortunately, we have been dealing with an organism for which powerful tools are now available for studying both chromosomal and molecular evolution in natural populations. For example, in the nuclei of salivary gland cells of *Drosophila*, there are giant polytene chromosomes which are exceptionally favorable for examining the structural changes in chromosomes (Yoon et al., 1973a, b, Yoon and Wheeler, 1973). Hawaiian *Drosophila* have six polytene elements, five long and one short, in their genome. The principal type of chromosomal variation among these Hawaiian species is the paracentric inversion, which is relatively easy to detect. Furthermore, most of these inversions have become fixed in the populations of the one or more species in which they are found. Therefore, the phylogenetic relationships among many species have been established in great detail. The gene order for *D. hystricosa* was selected as the standard reference order (Yoon et al., 1972a).

Combining our data (Yoon et al., 1972a–c, 1973a,b, 1975; Yoon and Carson, 1973; Yoon and Wheeler, 1973; Yoon and Richardson, 1974, 1976a,b, 1977, 1978a–e, Yoon, 1982, 1983; Carson and Yoon, 1982) brings to 326 the total number of inversions thus far detected in the 165 species of Hawaiian *Drosophila* that have been studied (see Fig. 6.1). This means that, on the average, two inversions per species have been fixed during their evolution. The maximum difference between subgroups within the picture-winged group is about 14 inversions. The two subgroups in the modified mouthparts groups are differentiated by 11 inversions. (Preliminary studies suggest that the picture-winged group is differentiated from the modified mouthparts group by 30–40 inversions.) We devised a phylogenetic measure utilizing the proportion of the polytene chromosome karyotype that could be homologized (Yoon et al., 1972b–d). Loss of detectable homology is partly a result of fragmentation and redistribution of blocks of polytene bands, and partly a result of accumulated changes in individual band morphology that occur either with changes in relative

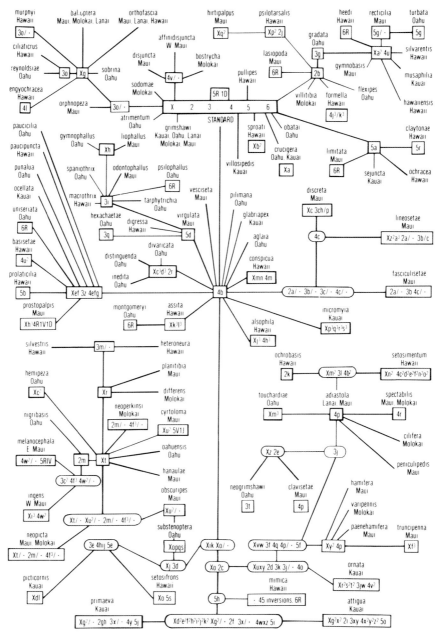

Figure 6.1. Chromosomal phylogeny of the picture-winged species group. For details, see text and also Carson and Yoon (1982).

time of puffing of nearby bands or with changes in chromatin within a band. With the use of this measure, chromosomal homologies were established among Hawaiian Drosophilidae (Fig. 6.2).

Chromosome inversions do not always differentiate species or species groups. Some inversions have remained as chromosomal polymorphisms in more than one species. For example, four species of the *hystricosa* subgroup (the modified mouthparts group) are polymorphic for some of the same inversions. These polymorphisms most likely arose in a common ancestor of these species (Yoon et al., 1972a,b). Carson et al. (1970) found similar instances in the picture-winged group, and several cases are known in non-Hawaiian *Drosophila*. We have also demonstrated cytogenetic evidence for a common ancestor for two genera of Hawaiian drosophilids Yoon et al. (1972d).

The function and effects of heterochromatin have been an enigma. Constitutive heterochromatin, for example, has been considered generally to have no discernible genetic function, although it has a "position effect" on the expression of euchromatic genes in *Drosophila* (Hannah, 1951; Baker, 1968). However, it has been suggested that centromeric heterochromatin plays a significant role in the evolution of the karyotype of both plants and animals, being associated with changes in chromosome morphology through breakage and refusion (Pathak et al., 1973; White, 1973, 1978; Hsu, 1979). Recently, it has been found that intercalary heterochromatin is closely associated with clusters of chromosomal breakpoints (Yoon and Richardson, 1976b). The functional and structural aspects of heterochromatin were reviewed by John and Miklos (1979).

We have observed phylogenetic changes of heterochromatic sites in polytene chromosomes of Hawaiian *Drosophila*, as well as in total heterochromatin associated with individual metaphase chromosomes (Yoon et al., 1975). Therefore, we have concluded that the distribution of heterochromatin is related also to speciation. Several phylogenies in the Hawaiian *Drosophila* exhibit an accumulation of heterochromatin on the "dot" (micro-) chromosome, thereby converting it to a rod. These large accumulations of heterochromatin on microchromosomes have been observed only in species that have been derived recently from their nearest relatives but show no differences in gene sequence (Yoon and Richardson, 1978c). Recently an interesting relationship between the microchromosomes and heterochromatin in Hawaiian *Drosophila* has been postulated in homosequential species (Yoon and Richardson, 1978c).

THE PSEUDOCHROMOCENTER: A MODEL
FOR GENOMIC CHANGE

In addition to the true chromocenter, an additional chromocenter-like configuration was found in polytene nuclei of several species of Hawaiian *Drosophila*. Only specific sites of certain chromosomes join to give the

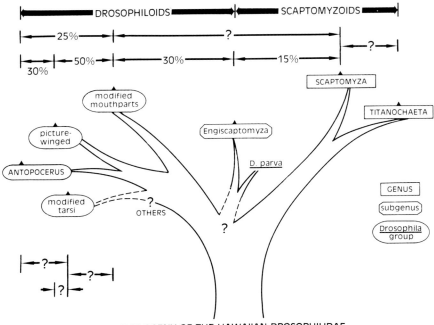

PHYLOGENY OF THE HAWAIIAN DROSOPHILIDAE
BASED ON PERCENT POLYTENE CHROMOSOME BANDING HOMOLOGY

A

Figure 6.2.   (**A**) Phylogeny of the Hawaiian Drosophilidae based on percent polytene chromosome banding homology. (**B**) Morphological diversity and size variation in several representative species of Hawaiian *Drosophilia*. 1, *D. planitibia*; 2, *D. grimshawi*; 3, *D. picticornis*; 4, *D. hystricosa*; 5, *D. biseriata*; 6, *D. mimica*; 7, *D. eurypeza*; and 8, *D. hawaiiensis.*

characteristic configuration, which was designated earlier as the "pseudochromocenter" (Yoon et al., 1972a). Apparently these sites are composed of intercalary heterochromatin and represent multiple points where nonhomologous association (i.e., ectopic pairing) occurs (Fig. 6.3).

Our earlier study also revealed that the pseudochromocenter is related to chromosome rearrangements that occur in salivary gland cells. Several inversions and reciprocal translocations have been observed in which breakpoints correspond to heterochromatic sites involved in the pseudochromocenter. In addition, ring chromosomes corresponding to the loop seen in the formation of the pseudochromocenter configuration, and chromosomes with the deleted segment corresponding to the ring have been observed in numerous cells. Recently, we found a new chromosomal rearrangement that occurred in an interspecific hybrid involving one of these lines. For example, the centromeric ends of the second and fifth

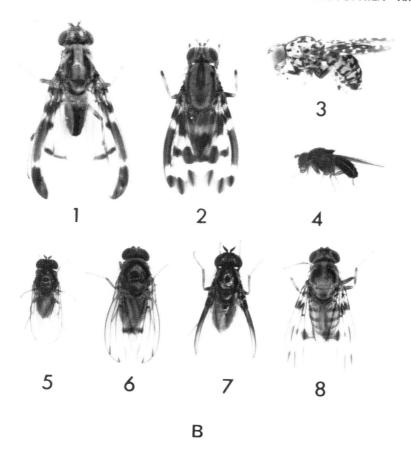

B

chromosomes were joined together; but, most unexpectedly, the centro-mere was repositioned near the terminus of this "double length" sub-metacentric chromosome (Yoon and Richardson, 1978a). Thus, we have suggested that the pseudochromocenter provides a mechanism for kary-otypic modifications in polytene chromosomes (Yoon and Richardson, 1976a; Carson and Yoon, 1982).

Earlier, we assumed that if there are ectopic joining configurations in the germ line, either premeiotically or meiotically, the pseudochromocen-ter-like configurations that we have seen in polytene cells are present also in dividing cells. Hence, one could postulate a model of genomic rear-rangements involving the heterochromatic sites associated with the pseu-dochromocenter. This assumption has been supported by the observation that inversion breakpoints differentiating closely related species, as well as polymorphic inversions, are identical to the sites involved in the pseu-dochromocenter. On the basis of several lines of cytogenetic evidence, we proposed a model involving the pseudochromocenter in the produc-tion of chromosome rearrangements (Yoon and Richardson, 1976a).

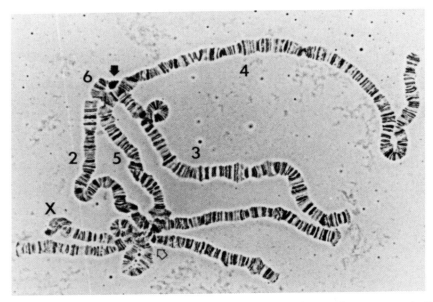

Figure 6.3. A pseudochromocenter of *D. biseriata*. Both true chromocenter (solid arrow) and pseudochromocenter (open arrow) are shown. Chromosomes X, 2, and 5 are involved in formation of the pseudochromocenter. Notice that the X chromosome's centromere with centromeric heterochromatin was detached from the true chromocenter while being squashed for slide preparation.

Our model is that, first, nonhomologous sites are joined in the pseudochromocenter; second, breaks may occur; and third, different chromosomal segments rejoin to complete the exchange. The ease of rejoining in new ways may be facilitated at the molecular level by the presence of highly repeated sequences of DNA located in the heterochromatin.

Whereas we proposed a *cytological* model associated with the pseudochromocenter, Lee and colleagues (Lee and Thomas, 1973; Lee, 1975) postulated a *molecular* model of chromosomal rearrangement between sites having high concentrations of highly repetitious DNA of the same nucleotide sequence. We also confirmed the expectation of these two models: The presence of highly repetitious DNA was found to correspond to heterochromatin at inversion breakpoints, and, further, highly repetitious DNA strands connected the sites that were ectopically joined. Recently, we also found that inversion breakpoints are clustered in a few heterochromatic sites along the chromosomes, even in species that seldom form a pseudochromocenter. We also have shown by DNA/DNA in situ hybridization that at least some of these heterochromatic sites contain highly repetitious DNA (unpublished, Fig. 6.4).

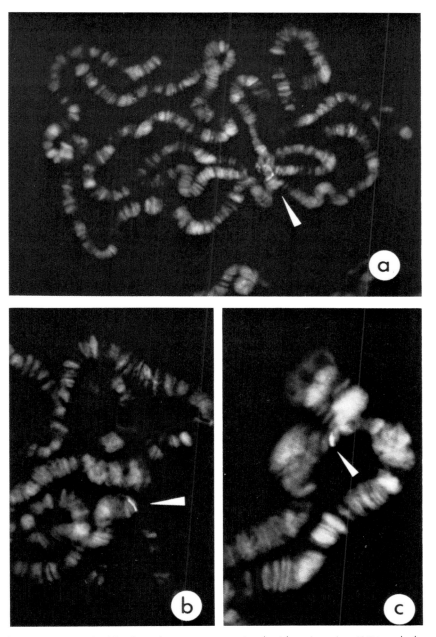

Figure 6.4. (**A**) *D. biseriata* chromosomes stained with quinacrine (1%) and photographed under UV epi-illumination reveal that the chromocenter is composed of AT-rich DNA. (**B,C**) The bright region (arrow) is one of the sites forming pseudo-chromocenters and is also one of the breakpoints for chromosomal rearrangements during speciation in the complex.

MECHANISM OF CHROMOSOME EVOLUTION

Clusters of breaks at certain intercalary heterochromatin sites that produce chromosomal rearrangements have been intensively studied in four species of Hawaiian *Drosophila: D. furvifacies, D. biseriata, D. hystricosa,* and *D. mitchelli* (Yoon, 1983). Data from DNA/DNA in situ hybridization and cytochemical studies show that these intercalary heterochromatin sites are AT-rich, and we have postulated that they probably originated from the centromeric heterochromatin. Furthermore, a study of staining with quinacrine reveals that the pseudochromocenter is composed of AT-rich DNA (Fig. 6.4). In addition to the laboratory strains of these four species, we observed other cases of pseudochromocenter-like configurations in other species complexes of Hawaiian *Drosophila* (unpublished).

We presented the pseudochromocenter model for the production of chromosomal rearrangements, a model that may be one of the major mechanisms for chromosome evolution. First, nonhomologous sites that are heterochromatic and contain similar sequences of highly repetitious DNA join in a chromocenter-like configuration. Second, chromatid exchanges by breakage and reunion occur at the ectopically joined sites. On the basis of this model, we expected many new chromosomal rearrangements, some of which have been observed and have been used to differentiate many species in Hawaiian *Drosophila.* One also can predict new chromosomal rearrangements among closely related species when a certain pseudochromocenter is observed (Yoon and Richardson, 1978a; Carson and Yoon, 1982).

These findings suggest, first, that the structural rearrangements are common under certain conditions in which heterochromatin (containing highly repetitious DNA) is similar at different chromosomal sites, and furthermore, that structural rearrangements with common breakpoints may be frequent. Second, phylogenies based on inversions being "unique" events are reliable because the substitution of an inversion into a gene pool is rare, but not because the occurrence of an inversion is rare. Substitution depends upon a rearrangement being highly fit, which, in turn, depends upon an extremely rare combination of interacting genes. This model therefore proposed one solution to the theoretical question of how an inversion-bound coadapted gene complex may arise. They would arise by chance, the rearrangement reoccurring many times until the proper genes were caught properly linked. Then the inversion would continue to evolve by conventional mechanisms involving two-strand double crossing-over, gene conversion, and mutation.

Recently, from some of the results of our work, we have estimated the rates of both chromosomal and molecular evolution in Hawaiian *Drosophila.* It appears that karyotypic and morphological evolution have been more rapid and their effect more profound than the evolution that has occurred through DNA base substitutions. We concluded, therefore, that chromosome (gene) rearrangements and the resulting effects on gene regulation may account for the most profound differences between species.

CHROMOSOME CHANGES AND SPECIATION

Many evolutionary geneticists have suggested that evolution at the organismic level (adaptive evolution) depends chiefly on mutations in genes that regulate other genes, and that gene arrangement on the separate chromosomes determines to a significant extent how they function as regulators. Therefore, rearranging the order of genes on a chromosome can be an important mechanism for achieving altered patterns of gene regulation. Regulatory mutations influence the expression of genes coding for proteins without affecting the structure of proteins. Thus a number of investigators have proposed that speciation of some animal groups may indeed proceed primarily by chromosomal rearrangements, and have suggested that shifts in the location of chromosomal segments create a new arrangement of genes which can become established in only a few generations with selective advantages (Yoon et al., 1972b; White, 1973, 1978; King and Wilson, 1975; Wilson et al., 1975; Bush et al., 1977). In addition, we have observed also that possible changes in gene regulation are associated with karyotypic changes of heterochromatin (Yoon and Richardson, 1978c). From the view of chromosomal theory of evolution, we and others have hypothesized that rearranged chromosomes both create a new species difference and provide the mechanism for maintaining it (Yoon et al., 1972a, b; Yoon and Richardson, 1978a; Myers and Shafer, 1979). Therefore, the chromosome rearrangements would alter the regulation of gene expression presumably by position effects (including those of heterochromatin), and would reduce the reproductive viability of hybrid offspring by interfering with the meiotic pairing process (Yoon and Richardson, 1978b; Myers and Shafer, 1979).

From the evidence of polytene chromosome analyses and other available data (see Carson and Yoon, 1982), it has been proposed, regarding the mode of evolution of new species, that a great deal of the explosive speciation of Hawaiian *Drosophila* has been due to the founder individuals transported from one island to another (Carson, 1971a). For example, the fact that two species from different islands are chromosomally polymorphic for certain inversions speaks in favor of the single founder event. Therefore, the model that the founder principle has played a role in colonization of Hawaiian *Drosophila* is acceptable on general principles.

Recent studies (Yoon and Richardson, 1976a,b; Richardson and Yoon, 1977; Yoon, 1984) in the Hawaiian Drosophilidae in many ways support the chromosomal models of speciation, indicating that chromosomal rearrangements play an important role in the process of speciation. However, in *Drosophila*, because of the existence of polytene chromosomes, we are able to study karyotypic changes in greater detail than is possible with vertebrates. Quantitative studies of speciation rates and chromosomal evolution in mammals (Bush et al., 1977) and in *Drosophila* (Yoon and Richardson, 1978b) have contributed to a better understanding of the mechanism for evolutionary change. Furthermore, our findings suggest mechanisms whereby rapid genome changes take place. These changes include both polytene chromosome banding (gene) sequences and metaphase modifications.

CYTOGENETICS OF "HOMOSEQUENTIAL" SPECIES

The term "homosequential" species has been used to refer to those species that have the same polytene chromosome karyotype (Carson et al., 1967). Even though the polytene chromosome banding patterns are the same, however, the amount and distribution of heterochromatin differ, and consequently, the metaphase karyotypes differ among such species. We designated these cases "anisohomosequential" species (Yoon et al., 1972b; Fig. 6.5). Pairs of species with no detectable cytological differences may be called "isohomosequential" species. Although true isohomosequential species may exist, our experience has shown that, as the species are studied in greater detail, heterochromatic differences become apparent (Yoon, 1983).

Recently, extensive cytogenetic analyses of homosequential species were carried out in four selected species of the Hawaiian *Drosophila: D. gymnobasis* and *D. silvarentis*, belonging to the picture-winged species group; and the *D. mimica* and *D. kambysellisi*, belonging to the modified mouthparts species group. The variable features of both polytene and metaphase chromosomes within these "homosequential" species are shown in Figure 6.6.

D. silvarentis from Hawaii and *D. gymnobasis* from Maui have the same polytene karyotypes (i.e., "homosequential"), as well as being morphologically very similar (Carson, 1971b). However, the F_1 hybrid males as well as females showed the existence of a sterility barrier, indicating that they are bona fide biological species. The metaphase configuration of each of the two species consists of five pairs of rods and one pair of dot chromosomes (5R1D). However, the length of the dot chromosome differs between the two species, that of *silvarentis* being twice the length of that of *gymnobasis* (see Clayton, 1971). It is also evident from the polytene chromosome data that the telomeric region (tip) of the chromosome does puff differently. In addition, the microchromosome (6) of *silvarentis* has an extra chromatin block in the tip of its polytene chromosome, as predicted from the metaphase chromosome. Therefore, these two species are "anisohomosequential" species (Yoon et al., 1972b).

D. mimica and *D. kambysellisi* are also anisohomosequential. Whereas

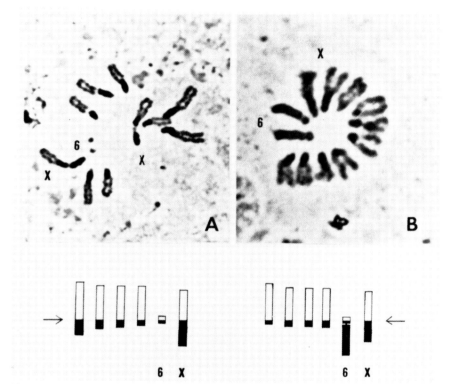

Figure 6.5. Photomicrographs of the metaphase karyotypes of two anisohomosequential species: (**A**) *D. kambysellisi* (5R1D); (**B**) *D. mimica* (6R); notice the size of chromosome 6.

D. kambysellisi has a metaphase configuration consisting of five pairs of rod chromosomes and one pair of microchromosomes (5R1D). *D. mimica* has a pair of long acrocentric chromosomes to go with the five pairs of rods (6R). In spite of differential distribution patterns, the total amount of heterochromatin appears to be the same in the two karyotypes (Fig. 6.5a, b).

In Yoon et al. (1975), we have discussed the phylogenetic changes of heterochromatic sites in polytene chromosomes of Hawaiian *Drosophila*, as well as in total heterochromatin associated with individual metaphase chromosomes. We also proposed that the distribution of heterochromatin is also related to speciation. Several phylogenies in the Hawaiian *Drosophila* exhibit an accumulation of heterochromatin on the "dot" (micro-) chromosome, thereby converting it into a rod. These large accumulations of heterochromatin on microchromosomes have so far been observed only in species that appear to have been recently derived from, but show no differences in gene sequence from their relatives (Yoon and Richardson, 1978b).

Upon more detailed analysis of the metaphase and polytene chromo-

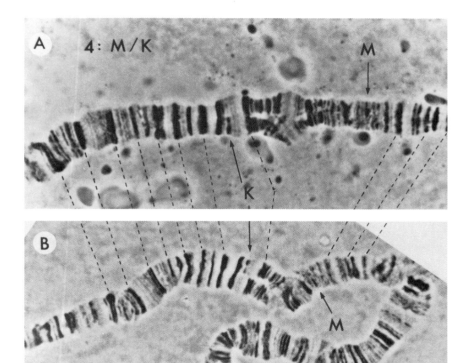

Figure 6.6. Chromosome 4 of a hybrid between anisohomosequential species: (**A**) *D. mimica* × *D. kambysellisi*; (**B**) *D. kambysellisi* × *D. mimica*. Band M belongs to *mimica*, and K to *kambysellisi*. Dashed lines between chromosomes indicate homologous band(s).

somes of 165 species (458 strains) of Hawaiian *Drosophila*, we have observed the following:

1. There are 14 species with large accumulations of heterochromatin on the microchromosome, converting it into a large acrocentric chromosome.
2. Each of these 14 species is a member of a different set of "homosequential" species. Therefore, species within each set have identical polytene chromosome karyotypes and thus have the same gene sequences (i.e., they contain no inversion differences).
3. Each of these 14 sets of homosequential species is situated at the terminal position of a phylogeny and therefore represents the most recently derived species of that lineage (Fig. 6.2). This provides a means of determining direction of evolution in several instances.

We have also found that localized accumulations of heterochromatin are correlated with absence of chromosome rearrangements. In other words, concentrations of heterochromatin in localized regions may signal reduced potential for rearrangements to occur elsewhere in the genome. This suggests that until the intercalary heterochromatic sites form, the accumulation of inversions may be delayed, thus contributing to the accumulation of homosequential species at the phylogenetic termini. In addition to the effect on structural changes, the location and/or amount of heterochromatin may serve as a major factor in differentiating newly formed species.

We are actively investigating strains of *D. mimica* (6R) that have variable heterochromatic lengths of rod and sex chromosomes (Yoon, 1983). This species is a member of an anisohomosequential species group containing a variety of heterochromatic variations, although the total heterochromatin content among species may be about the same. The major objective in the study of heterochromatin of these strains is to determine differences in heterochromatin among very closely related taxa, some of which are sympatric but reproductively isolated in nature (see Richardson and Johnston, 1975; Makela and Richardson, 1977).

In order to detect subtle differences in heterochromatin, we are using a variety of strains. G-bands are mapped largely for the *mimica* and *hystricosa* complexes (Yoon, unpublished), and more studies on *mimica* subspecies or cryptic species are in progress using various fluorescent dyes. Where possible, comparisons include the use of intertaxa (interspecific or intersubspecific) crosses so that chromosomes in the same cell include both types.

The existence of "apparent" homosequential species at least emphasizes the fact that speciation and evolutionary divergence can occur in the absence of drastic changes in chromosomal structure. However, the presence of heterozygous puffs or modifications of band(s) structures (Fig. 6.6A, B) and the degree of asynapsis of hybrids indicate that species have different regulatory mechanisms in their genomes, even though they have the same banding (gene) patterns. Furthermore, the effects of heterochromatin on gene regulation are well known, and are often associated with changes in regulation that are associated with changes in gene order.

ACKNOWLEDGMENTS

The research described in this chapter was supported by NSF Grant No. BSR-8400615.

LITERATURE CITED

Baker, W. K. 1968. Position–effect variegation. Adv. Genet. 14:133–169.
Bush, G. L., S. M. Case, A. C. Wilson, and J. L. Patton. 1977. Rapid speciation and chromosomal evolution in mammals. Proc. Natl. Acad. Sci. USA 74:3942–3946.

Carson, H. L. 1971a. Speciation and the founder principle. University of Missouri, Stadler Symp. 3:51–70.

Carson, H. L. 1971b. Polytene chromosome relationships in Hawaiian species of *Drosophila*. V. Addition to the chromosomal phylogeny of the picture-winged species. Univ. of Texas Publ. 7103:183–191.

Carson, H. L. 1981. Homosequential species of Hawaiian *Drosophila*. In M. D. Bennett, M. Bobrow, and G. M. Hewitt, (eds.), Chromosome Today, Vol. 7, pp. 150–164. George Allen & Unwin, London.

Carson, H. L., and K. Y. Kaneshiro. 1976. *Drosophila* of Hawaii: systematic and ecological genetics. Annu. Rev. Ecol. Systematics 7:311–346.

Carson, H. L. and J. S. Yoon. 1982. Genetics and evolution of Hawaiian *Drosophila*. In M. Ashburner, H. L. Carson, and J. N. Thompson, Jr. (eds.), The Genetics and Biology of *Drosophila*, Vol. 3b, pp. 298–343. Academic Press, New York.

Carson, H. L., F. E. Clayton, and H. D. Stalker. 1967. Karyotypic stability and speciation in Hawaiian *Drosophila*. Proc. Natl. Acad. Sci. USA 57:1280–1285.

Carson, H. L., D. E. Hardy, H. T. Spieth, and W. S. Stone. 1970. The evolutionary biology of the Hawaiian Drosophilidae. In M. K. Hecht and W. C. Steere (eds.), Essays in Evolution and Genetics in Honor of Th. Dobzhansky, pp. 437–543. Appleton-Century-Crofts, New York.

Clayton, F. E. 1966. Preliminary report on the karyotypes of Hawaiian Drosophilidae. Univ. Texas Publ. 6615:397–404.

Clayton, F. E. 1968. Metaphase configurations in species of the Hawaiian Drosophilidae. Univ. Texas Publ. 6818:263–278.

Clayton, F. E. 1971. Additional karyotypes of Hawaiian Drosophilidae. Univ. Texas Publ. 7103:171–181.

Clayton, F. E. 1976. Metaphase configuration in *Drosophila*. A comparison of endemic Hawaiian species and non-endemic species. Arkans. Acad. Sci. 30:32–35.

Clayton, F. E., and M. R. Wheeler. 1975. Catalog of *Drosophila* metaphase chromosome configurations. In R. C. King (ed.), Handbook of Genetics, Vol. 3, pp. 471–512. Plenum, New York.

Hannah, A. 1951. Localization and function of heterochromatin in *Drosophila melanogaster*. Adv. Genet. 4:87–127.

Hardy, D. E. 1965. Insects of Hawaii, Vol. 12: Diptera: Cyclorrhapha II, Series Schizophora, Sec. Acalypterae I. Family Drosophilidae. University of Hawaii Press, Honolulu.

Hardy, D. E. 1966. Descriptions and notes on Hawaiian Drosophilidae (Diptera). Univ. Texas Publ. 6615:195–244.

Hardy, D. E. 1974. Evolution in the Hawaiian Drosophilidae: introduction and background information. In M. J. D. White (ed.), Genetic Mechanisms of Speciation in Insects, pp. 71–80. Australia and New Zealand Book Co., Sydney.

Hardy, D. E., and K. Y. Kaneshiro. 1975a. Studies in Hawaiian *Drosophila*, modified mouthparts species No. 1: mitchelli subgroup. Proc. Hawaiian Entomol. Soc. 22:51–55.

Hardy, D. E., and K. Y. Kaneshiro. 1975b. Studies in Hawaiian *Drosophila*, miscellaneous new species, No. I. Proc. Hawaiian Entomol. Soc. 22:57–64.

Hardy, D. E., and K. Y. Kaneshiro. 1979. A review of the modified tarsus species group of Hawaiian *Drosophila*. I. "Split-tarsus" subgroup. Proc. Hawaiian Entomol. Soc. 23:71–90.

Heed, W. B. 1968. Ecology of the Hawaiian Drosophilidae. Univ. Texas Publ. 6818:387–419.

Heed, W. B. 1971. Host plant specificity and speciation in Hawaiian *Drosophila*. Taxon. 20(1):115–121.

Hsu, T. C. 1979. Human and Mammalian Cytogenetics. Springer-Verlag, New York.

John, B., and G. L. Miklos. 1979. Functional aspects of satellite DNA and heterochromatin. Int. Rev. Cytol. 58:1–115.

Kaneshiro, K. Y. 1974. Phylogenetic relationships of Hawaiian Drosophilidae based on morphology. In M. J. D. White (ed.), Genetic Mechanisms of Speciation in Insects, pp. 102–110. Australia and New Zealand Book Co., Sydney, Australia.

Kaneshiro, K. Y. 1976a. Ethological isolation and phylogeny in the planitibia subgroup of Hawaiian *Drosophila*. Evolution 30:740–745.

Kaneshiro, K. Y. 1976b. Evolutionary studies of two *Drosophila* species from Hawaii: a multidisciplinary approach. In Proc. First Conf. in Natural Sciences, Hawaii Volcanoes National Park, Cooperative National Park Resources Studies Unit, pp. 115–119. Department of Botany, University of Hawaii, Honolulu.

King, M. C., and A. C. Wilson. 1975. Evolution at two levels in human and chimpanzee. Science 188:107–116.

Lee, C. S. 1975. A possible role of repetitious DNA in recombinatory joining during chromosome rearrangement in *D. melanogaster*. Genetics 79:467–470.

Lee, C. S., and C. A. Thomas, Jr. 1973. Formation of rings from *Drosophila* DNA fragments. J. Mol. Biol. 77:25–55.

Makela, M. E., and R. H. Richardson. 1977. The detection of sympatric sibling species using genetic correlation analysis. I. Two loci, two gamedemes. Genetics 86:665–678.

Myers, R. H., and D. A. Shafer. 1979. Hybrid ape offspring of a mating of gibbon and siamang. Science 205:308–310.

Pathak, S., T. C. Hsu, and F. E. Arrighi. 1973. Chromosomes of *Peromyscus* (Rodentia, Cricetidae). IV. The role of heterochromatin in karyotype evolution. Cytogenet. Cell Genet. 12:315–326.

Richardson, R. H., and J. S. Johnston. 1975. Behavioral components of dispersal in *Drosophila mimica*. Oecologia 20:287–299.

Richardson, R. H., and J. S. Yoon. 1977. Chromosome and DNA divergence in Hawaiian *Drosophila*. Genetics 86:s51.

Spieth, H. T. 1966. Courtship behavior of Hawaiian Drosophilidae. Univ. Texas Publ. 6615:245–313.

Spieth, H. T. 1978. Courtship pattern and evolution of *Drosophila adiastola* and *plantibia* species subgroups. Evolution 32:435–451.

Throckmorton, L. H. 1966. The relationships of the endemic Hawaiian Drosophilidae. Univ. Texas Publ. 6615:335–396.

White, M. J. D. 1973. Animal Cytology and Evolution, 3rd ed. Cambridge University Press, Cambridge.

White, M. J. D. 1978. Modes of Speciation. W. H. Freeman, San Francisco.

Wilson, A. C., G. L. Bush, S. M. Case, and M. C. King. 1975. Proc. Natl. Acad. Sci. USA 72:5061.

Wilson, A. C., V. M. Sarich, and L. A. Maxon. 1974. The importance of gene rearrangement in evolution: evidence from studies on rates of chromosomal protein, and anatomical evolution. Proc. Natl. Acad. Sci USA 71:3028–3030.

Yoon, J. S. 1982. Chromosomal rearrangements and speciation in *D. hystricosa* complex. Genetics 100:s76.

Yoon, J. S. 1983. Cytogenetics of homosequential species of Hawaiian *Drosophila*. Genetics 104:s76.

Yoon, J. S. 1984. Chromosome evolution and speciation in Hawaiian *Drosophila*. In M. D. Bennett, A. Gropp, and U. Wolf (eds.), Chromosome Today, Vol. 8, p. 354. George Allen & Unwin, London.

Yoon, J. S., and H. L. Carson. 1973. Codification of polytene chromosome designations for Hawaiian Drosophilidae. Genetics 74:2302–2304 [abstract].

Yoon, J. S., and R. H. Richardson. 1974. Evolution in Hawaiian Drosophilidae: Karyotypic studies of the genus *Antopocerus*. Genetics 77:s72 [abstract].

Yoon, J. S., and R. H. Richardson. 1976a. A model for genome change: pseudochromocenter hypothesis. Genetics 83:s85.

Yoon, J. S., and R. H. Richardson. 1976b. Evolution of Hawaiian Drosophilidae II. Patterns and rates of chromosome evolution in an *Antopocerus* phylogeny. Genetics 83:827–843.

Yoon, J. S., and R. H. Richardson. 1977. Genome evolution and speciation in Hawaiian *Drosophila*. Genetics 86:s71–s72.

Yoon, J. S., and R. H. Richardson. 1978a. A mechanism of chromosomal rearrangements: the role of heterochromatin and ectopic joining. Genetics 88:305–316.

Yoon, J. S., and R. H. Richardson. 1978b. Rates and roles of chromosomal and molecular changes in speciation. In R. H. Richardson (ed.), The Scewworm Problem. Evolution of Resistance to Biological Control, pp. 129–143. Univ. of Texas Press, Austin.

Yoon, J. S., and R. H. Richardson. 1978c. Evolution of Hawaiian Drosophilidae III. The microchromosome and heterochromatin of *Drosophila*. Evolution 32:475–484.

Yoon, J. S., and R. H. Richardson. 1978d. Heterochromatin and chromosome evolution. Genetics 88:s114–115.

Yoon, J. S., and R. H. Richardson. 1978e. The role of heterochromatin as a mechanism of chromosomal rearrangements. XIV Int. Cong. Genet., Moscow [abstract 386].

Yoon, J. S., and M. R. Wheeler. 1973. Chromosome evolution: Repetitive euchromatic chromosomal segments in Hawaiian *Drosophila*. Can. J. Genet. Cytol. 15:171–175.

Yoon, J. S., K. Resch, and M. R. Wheeler. 1972a. Cytogenetic relationships in Hawaiian species of *Drosophila* I. The *D. hystricosa* subgroup of the "modified mouthparts" species group. Univ. Texas Publ. 7213:179–199.

Yoon, J. S., K. Resch, and M. R. Wheeler. 1972b. Cytogenetic relationships in Hawaiian species of *Drosophila* II. The *D. mimica* subgroup of the "modified mouthparts" species group. Univ. Texas Publ. 7213:201–212.

Yoon, J. S., K. Resch, and M. R. Wheeler. 1972c. Intergeneric chromosomal homology in the family Drosophilidae. Genetics 71:477–480.

Yoon, J. S., K. Resch, and M. R. Wheeler. 1972d. Cytogenetic evidence for "one ancestral" relationship within Hawaiian drosophilids. Genetics 71:s70–s71 [abstract].

Yoon, J. S., R. C. Richardson, and M. R. Wheeler. 1973a. Chromosomal reorganization with four allopatric sibling species of Hawaiian *Drosophila*. Genetics 74:s304 [abstract].

Yoon, J. S., R. H. Richardson, and M. R. Wheeler. 1973b. A technique for improving salivary chromosome preparations. Experientia 29:639–641.

Yoon, J. S., K. Resch, M. R. Wheeler, and R. H. Richardson. 1975. Evolution in Hawaiian Drosophilidae: chromosomal phylogeny of the *Drosophila crassifemur* complex. Evolution 29:249–256.

7

Is There a Role for Meiotic Drive in Karyotype Evolution?

TERRENCE W. LYTTLE

Much of population genetic theory depends to some degree on the assumptions that meiotic segregation of alleles is Mendelian and that natural selection proceeds to maximize sporophytic fitness. One area of evolutionary theory in which these assumptions figure prominently concerns the fixation in lineages of unusual chromosome types or rearrangements—that is, karyotype evolution. Because most chromosomal rearrangements are presumed to lead to abnormal meiotic segregation and a concomitant reduction in fecundity of heterozygous carriers, their evolutionary fate is thought to be best understood by invoking simple models of heterozygote disadvantage. In this case, there is a single internal unstable equilibrium frequency for an autosomal rearrangement, given by:

$$\hat{p} = t/(s + t)$$

where t and s are the selection coefficients favoring the normal and rearrangement homozygotes, respectively. Because $\hat{p}$ represents an unstable equilibrium, values of p below this will lead to deterministic loss of the rearrangement (see Fig. 7.1). On the other hand, when the rearrangement has relatively low homozygous fitness (i.e., s is small) and $\hat{p}$ approaches 1, then p must reach a very high level before deterministic fixation of the new karyotype occurs. In the absence of genetic mechanisms that act to increase the frequency of a new rearrangement past the unstable equilibrium, it is difficult to see how karyotype evolution can occur at all. In his classic work, *Modes of Speciation,* White (1978) suggested several mechanisms that might promote rearrangement fixation, among them genetic drift and meiotic drive. More recently, Lande (1979) and Hedrick (1981) have investigated these two forces and have reached somewhat different conclusions as to their relative importance, with Lande taking the position that meiotic drive plays a negligible evolutionary role, if any, in the fixation of chromosome rearrangements. However, both authors consid-

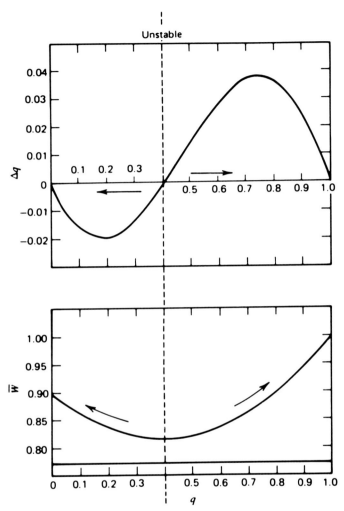

Figure 7.1. Rate of change in rearrangement frequency (Δq, upper graph) and mean fitness ($\bar{w}$, lower graph) when $s = 0.29$, $t = 0.43$, and there is no meiotic drive ($k^* = 0$) favoring the rearrangement in heterozygotes. Deterministic selection moves the frequency away from the unstable equilibrium value of 0.4 to cause fixation or loss of the rearrangement, thus increasing population fitness. (After Li, 1955, 1976)

ered only relatively simple models of drive. In the present discussion, I will try to take a broad view of the different classes of meiotic drive and their potential impact on karyotype evolution.

FIXATION OF CHROMOSOME REARRANGEMENTS: FIXATION MECHANISMS FOR DIFFERENT CLASSES

In fact, fixations of translocations in the majority of lineages *are* rare (see Lande, 1979, for a comprehensive review of observed mutation and fixation rates for rearrangements). Those that have occurred usually can be classified as Robertsonian fusions of two telocentric chromosomes, centromere dissociations to give telocentric derivatives, or at least whole arm exchanges. Inversion fixations are much more commonly observed. This is to be expected when one considers that fecundity should be lowered less by genetic exchange within a heterozygous inverted segment than from the improper segregation of unbalanced gametes from a translocation carrier. Moreover, when meiotic recombination is restricted to females, as in the case of many drosophiline species, paracentric inversion heterozygotes may show no reduction in fecundity, because the dicentric bridge products resulting from genetic exchange within the inversion loop generally lag at anaphase I and are subsequently excluded from the egg. As a consequence, we would expect negligible selection pressure opposing fixation of such an inversion within a population or species.

The situation is somewhat different when we consider the mechanisms by which unusual chromosome types become fixed in a lineage. It is common to see accumulation of extra blocks of heterochromatin in the genome, either as additions to one of the chromosomes making up the normal karyotype of a species (the A chromosomes) or as independent supernumerary chromosomes. Indeed, supernumerary B chromosomes are reported in both plant and animal species covering the whole range of the evolutionary spectrum (White, 1973, 1978; Hartl, 1977; Jones and Rees, 1982). Such largely heterochromatic chromosomes are relatively devoid of genes and can probably be more easily tolerated in odd dosages than would be the case for similarly sized pieces of gene-rich euchromatin; thus selection against their accumulation is probably weaker than for the rearrangements discussed above. Moreover, many of these chromosomes apparently persist in populations because they exhibit accumulation mechanisms, both meiotic and mitotic, that result in their preferential recovery in the gametes (see Jones and Rees, 1982, for a review). This is a type of *meiotic drive* which may help add extra genetic material to the karyotype.

NATURE OF MEIOTIC DRIVE AND RANGE OF POSSIBLE SYSTEMS

"Meiotic drive" is defined as occurring whenever, as a result of the mechanics of the meiotic divisions, one member of a pair of heterozygous alleles or heteromorphic chromosomes is recovered in excess in the gametes (Sandler and Novitski, 1957; Zimmering et al., 1970). Where meiotic drive occurs, the basic assumptions of Mendelian segregation are violated, and the conclusions of most simple population genetic models must be radically altered or even discarded. For example, the principle that populations always evolve to maximize their diploid fitness is no longer necessarily true (cf. Prout et al., 1973; Charlesworth and Hartl, 1978; Lyttle, 1979). In fact, meiotic drive may actually promote the fixation of alleles or chromosomes that should otherwise be lost as a result of their detrimental effect on overall viability or fecundity.

Cases of meiotic drive observed in nature have been numerous, and have been reported in both plant and animal species (see Lyttle, 1975; Hartl, 1977; Curtsinger, 1984a, for reviews). They generally fall into one of two main categories:

True Meiotic Drive

The original definition of meiotic drive requires that the meiotic divisions themselves be the cause of the nonrandom segregation; that is, a chromosome or allele is recovered in excess because it undergoes extra replications, or because it is recovered preferentially in those meiotic products destined to become gametes. Because replication and movement are properties of whole chromosomes, most cases of true meiotic drive must be, a fortiori, chromosomal rather than allelic. For example, in cases where a supernumerary chromosome shows drive, we might expect this to be due to a replication advantage. Conversely, structural rearrangements might have an orientation advantage that would lead to their excess recovery, particularly in females, where only one of the four products of meiosis is regularly destined to become a gamete.

Gamete Competition

Several of the best studied cases of meiotic drive (e.g., *Segregation distorter (SD)* in *Drosophila melanogaster,* *t*-allele drive in mouse, male drive (M^D) in mosquito) are actually a form of haploid selection arising from gametophyte competition. Often, as in the three cases cited, the driven element causes some lesion in its homologue during meiosis which ultimately leads to the dysfunction of gametes carrying that chromosome. Gamete competition may be of particular importance to plants, where the complex development of pollen provides the opportunity for significant selective competition (cf. the phenomenon of certation; Grant, 1975). If the gamete loss associated with this form of drive does not lead to a concomitant loss of fecundity, then gamete competition is mathematically indistinguishable from true meiotic drive. On the other hand, if there is a loss of fecundity strictly proportional to the loss of gametes, then there is effectively no drive at all. In brief, gamete competition will lead to effective meiotic drive for a chromosome or allele only if it results in a selective increase in the *absolute number* of progeny produced which carry that element. Consequently, meiotic drive arising from gamete competition is much more likely to occur in the males of monogamous species, where gamete wastage can be more easily tolerated.

Other Mechanisms

There are also a number of other mechanisms of non-Mendelian segregation that could be loosely collected under the umbrella term of meiotic drive, or at least exhibit similar population dynamics. An example is the phenomenon of gene conversion, where the DNA sequence of one member of a pair of heterozygous alleles is altered (or "converted") during meiosis to produce a copy of the other. In terms of the population genetics of a single such locus, asymmetric gene conversion can be considered to be identical to true meiotic drive (Nagylaki, 1983). However, an allele exhibiting true meiotic drive will also allow closely linked neighboring alleles to show associated secondary drive through "hitchhiking", whereas gene conversion extends only a few hundred base pairs, and is thus generally restricted to a single locus. To someone like myself who for some time has been interested in the evolutionary implications of meiotic drive, it seems redundant to recategorize this type of gene conversion as a form of molecular drive (Dover, 1982) and invoke it as an evolutionary force of major consequence. Because most of the current direct observation of what is called molecular drive appears to consist of either gene conversion or unequal crossing-over (Ohta, 1983; Ohta and Dover, 1984), it would seem more useful from both a theoretical and historical point of view to treat it simply as a subset of meiotic drive. For example, such a fusion of terms would make it much easier to consider the similarity between the meiotic drive of some B chromosomes (which might be thought of as extremely large pieces of selfish DNA, capable of quick change in number or type as a consequence of having their own centromere), and the apparently rapid intra- or interchromosomal fixation of identical copies for certain repetitive gene families.

Having considered the general range of possible systems of meiotic drive, let us now consider specific ways in which the selection caused by such mechanisms might interact with genetic drift to promote fixation of supernumerary chromosomes or structural rearrangements.

TRUE DRIVE FOR REARRANGEMENTS

Let us suppose that meiotic drive is superimposed on the fitness model described above for populations showing an autosomal rearrangement polymorphism, such that the heterozygotes transmit the rearrangement to a proportion k of their progeny. The modified model then becomes:

Genotype	RR	RN	NN
Frequency	p^2	$2pq$	q^2
Fitness	$1 + s$	1	$1 + t$
Proportion R-bearing gametes	1	k	0

Where R = rearranged, N = normal chromosome type. Note that if the two sexes show differential drive for the rearrangement, then k is simply the arithmetic average of the two values.

The value $\hat{p}$ can be obtained from considering that, at equilibrium:

$$0 = \Delta p = (1 + s)\hat{p}^2 + 2\hat{p}(1 - \hat{p})k - \hat{p}[1 + \hat{p}^2 s + (1 - \hat{p})^2 t] \quad (7.1)$$

which yields the solution

$$\hat{p} = (t + 1 - 2k)/(s + t)$$

or

$$= (t - 2k^*)/(s + t) \quad (7.2)$$

where $k^* = k - \frac{1}{2}$, the drive differential.

When there is symmetric diploid selection ($s = t$), this simplifies further to:

$$\hat{p} = (\frac{1}{2} - k^*)/t \quad (7.3)$$

Consideration of the direction of change in rearrangement frequency when the system is perturbed slightly away from this value show that $\hat{p}$ is a stable point whenever

$$k/(1 + t) > \frac{1}{2} \quad \text{and} \quad k < (1 - s)/2 \quad (7.4a)$$

and is unstable if

$$k/(1 + t) < \frac{1}{2} \quad \text{and} \quad k > (1 - s)/2 \quad (7.4b)$$

Since in our model $s > 0$, the first case can never arise. Consequently, an unstable equilibrium is the only attainable one, and even this will be realized only when drive is weak. For, if drive is of a strength that the absolute recovery of the rearrangement from heterozygotes (i.e., the product of k and the heterozygote relative fitness value $1/[1+t]$) exceeds the normal value of $\frac{1}{2}$, then the rearrangement will selectively increase to fixation. Hiraizumi, Sandler, and Crow (1960) and Hartl (1972) generated these same results, but were more interested in the case were $s < 0$, $t > 0$; that is, where diploid directional selection opposes meiotic drive. Hartl's work in particular is an elegant analysis of the case in which drive is the result of gamete competition, such that the fecundity is likely to be inversely related to the level of gamete death; consequently, he did not consider cases in which the drive homozygote was more fit than the heterozygote, or in which diploid genotypic fitness was independent of the strength of drive. On the other hand, both Bengtsson and Bodmer (1976) and Hedrick (1981) examined the case in which drive overrides negative heterosis. Hedrick carried out a quite comprehensive analysis for this simple situation, and it should be noted that the results described above essentially reproduce his.

Figure 7.2 shows the fate of a rearrangement with different values of k^* and t. Here I assume the simple case $s = t$. The solution for $\hat{p}$ is obtained from Eq. 7.3, with $\hat{p} = 0.5$ the expected value when meiotic drive is absent (i.e., $k^* = 0$). Three values of t are chosen as representative of reciprocal translocations (case A, $t = 1.0$), Robertsonian fusions or cen-

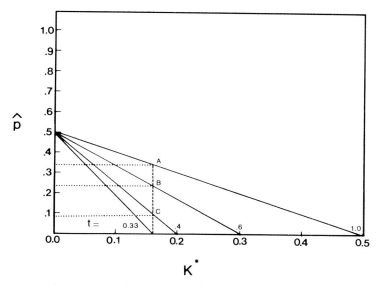

Figure 7.2. The relationship of equilibrium rearrangement frequency ($\hat{p}$) and meiotic drive differential (k^*). It is assumed that selection is symmetric ($s = t$) so that Eq. 7.3 applies. The vertical dotted line indicates a representative value of $k^* = 0.167$ ($k = 0.667$), typical of that observed for naturally occurring drive systems. When this line is reflected off the diagonal depicting the appropriate fitness advantage of the homozygote, it yields a reduced value for $\hat{p}$ below the value of 0.5, which is obtained for no drive. When $s = t < 0.33$, then $\hat{p} < 0$ and the driven rearrangement fixes unconditionally. Representative values of t are given by the marked lines A for translocations, B for Robertsonian fusions/dissociations, and C for inversions.

tromere dissociations (case B, $t = 0.6$), and pericentric inversions (case C, $t = 0.4$), in order of increasing heterozygote fitness. Paracentric inversions might be expected to give $t < 0.2$. For $k = 0.67$, a value comparable to some observed in nature (e.g., for knob *K10* in maize [Rhoades, 1952]; t alleles in mice [Braden, 1972]; *SD, RD*, and *SR* in *Drosophila;* and M^D in mosquito [Zimmering et al., 1970]), we obtain $k^* = 0.17$ on the abscissa of Figure 7.2. For paracentric inversions, and indeed any rearrangement giving $t < 0.33$, an intermediate level of drive such as this is capable of overriding the heterozygous fecundity disadvantage to allow for deterministic fixation of the mutation. Even when there is only sufficient drive to allow for an unstable equilibrium, the new value of $\hat{p}$ can be lowered considerably from the default value of $\hat{p} = 0.5$, perhaps low enough to allow some rearrangements in small populations to drift stochastically past the unstable equilibrium point and continue deterministically to fixation. Obviously, the closer $\hat{p}$ approaches zero, the more likely the establishment of the rearrangement. This produces the interesting result that drift on the one hand should *reduce* the fixation probability of rearrangements showing drive normally strong enough to cause their deterministic fixation (e.g., paracentric inversions), but *increases* the fixation probability for rearrangements with drive only strong enough to lower the value of $\hat{p}$ (e.g., translocations).

Lande (1979) also considered the effect of drive in promoting fixation of negatively heterotic rearrangements in concert with genetic drift. He concluded that because the overall rate of rearrangement fixation in lineages is explainable through the action of drift effects in small isolated populations alone, and because there is no evidence for currently segregating rearrangements showing drive, then the importance of drive in such fixation processes must be negligible. There are two reasons why I think that this assertion is stronger than is warranted from the analysis. First, if rearrangements showing meiotic drive occur, it is clear that their probability of fixation should be much higher than for similar mutations exhibiting normal Mendelian segregation (see appendix II of Lande, 1979). That is, although fixation of either type of rearrangement requires that conditions promoting drift persist long enough for the mutant frequency to reach $\hat{p}$, after which directional selection will generally ensure its further deterministic increase, the probability of this establishment should increase with increasing strength of drive. The reason is that drive causes a concomitant decrease in both $\hat{p}$ (see Fig. 7.2) and the strength of directional selection operating against the rearrangement. In effect, meiotic drive reduces the frequency (and generation) interval over which drift must act to ensure rearrangement establishment in the face of opposing selection, and increases the minimum population size necessary to generate this drift. Thus, when meiotic drive is aiding a rearrangement, not only does the fixation probability increase, but also we are allowed some relaxation in the constraints normally imposed, in the absence of drive, on the size and migration rates of populations.

This brings me to the second problem, which arises when one attempts to ascertain whether meiotic drive or the demographic conditions promoting fixation by drift are, a priori, more likely. Is meiotic drive common? Should we be able to identify easily rearrangements whose fixation is dependent on drive? To begin with, it is difficult (even in well-studied organisms like *Drosophila*) to be able to disprove statistically the existence of weak drive (cf. Curtsinger, 1984b). In addition, from the moment a meiotically driven rearrangement arises, there is immediately a strong selection for the accumulation of unlinked suppressors that reduce the drive (Lyttle, 1979, 1981), making it even harder to identify. Yet, as long as meiotic drive is active long enough to allow for its initial establishment, eventual fixation of a rearrangement may still be assured. It is simply necessary that the deterministic increase in p arising from selection (the right side of Fig. 7.1) is fast enough to keep the rearrangement frequency ahead of $\hat{p}$ (which is also increasing by virtue of the concomitant reduction in k^*, as can be seen in Fig. 7.2). In a similar manner, drift may act to create enough transient linkage disequilibrium between a rearrangement and an unrelated drive locus to allow the rearrangement to hitchhike past $\hat{p}$, and then proceed to fixation independently of meiotic drive. Thus, rearrangements may owe their establishment to an initial meiotic drive that is no longer measurable in normal genetic backgrounds. Moreover, even if it were demonstrable that rearrangements showing drive are a priori very rare, it is clear that in a Bayesian sense, their high fixation probabilities imply that they could have a larger role in affecting karyotype evolution than do rearrangements that rely on drift alone to carry them to fixation. Finally, rearrangements fixed by drift in small populations are required to spread through the species by colonizing areas vacated by population extinction, whereas those fixed by drive can more readily diffuse through larger populations by an essentially stasipatric mechanism (cf. White, 1978, pp. 212–213).

In short, it is somewhat problematic as to whether it is inherently more plausible to assume, as any model of drift-mediated fixation must, that populations remain small, and that interdemic migration rates remain much lower than local extinction rates over long periods of time (Lande, 1979), or simply to make the argument that when the fixation of rearrangements is unusual, only the unusual rearrangements (i.e., those showing segregational oddities) will fix. Even meiotic drive that is rare and weak enough in strength to escape current detection could be playing a preeminent role in karyotype evolution.

In summary, although rearrangements showing true meiotic drive probably arise at very low frequency, they require only a small drive advantage to ensure their deterministic fixation. When the effects of drift are superimposed, even weaker amounts of drive may still lead to fixation if stochastic frequency fluctuations carry the rearrangement above the unable equilibrium point, which is itself made lower by the existence of drive. As only a small proportion of spontaneously arising rearrangements are ever fixed, it is quite conceivable that many of these may owe their success to true meiotic drive.

GAMETIC COMPETITION FAVORING REARRANGEMENTS

Gametic competition leading to effective meiotic drive generally occurs in the male sex, as discussed earlier. Although a number of rearrangements involving the sex chromosomes of *D. melanogaster* have demonstrated an advantage in sperm competition (cf. Peacock and Miklos, 1973; McKee, 1984), it is more often the case that the excess recovery of a sperm class is due to the action of a specific genetic locus and its modifiers. Genic drive of this sort generally cannot deterministically promote the fixation of rearrangements, even when one of the involved breakpoints is quite near the drive locus itself (but see below), for in this case the rearrangement should still be at a disadvantage when compared to any structurally normal ancestral chromosome that also shows meiotic drive. Consequently, unless the chromosome rearrangement breakpoint itself is the cause of a gametic selective advantage, or the rearrangement and the drive element arrive linked together in a single immigrant, or the drive occurs secondarily in a preexisting rearrangement, gamete competition should not be an important force promoting fixation of simple autosomal rearrangements. However, it could conceivably play a role in the fixation of unusual rearrangements involving the sex chromosomes, in the accumulation of supernumerary chromosomes, and in the fixation of certain inversions acting to strengthen linkage of interacting drive elements.

Rearrangements Favored by Sex Chromosome Drive

The population genetic consequence of sex chromosome drive have been examined in some detail (Hamilton, 1967; Lyttle, 1977, 1979, 1982; Curtsinger and Feldman, 1980). In general, very strong drive favoring either of a pair of heteromorphic sex chromosomes should lead to fixation of the corresponding sex and the subsequent extinction of the affected population. It follows that any genetic mechanism that ameliorates such sex-ratio distortion will enjoy a large selective advantage. For example, laboratory populations of *Drosophila melanogaster* are able to cancel the effects of Y-chromosome drive by accumulating aneuploid XXY and XYY genotypes, whenever drive strength is above a threshold value dictated by aneuploid fitness and sex chromosome segregation parameters (Lyttle, 1981, 1982). Apparently the nondisjunctant XX-bearing eggs (see Fig. 7.3), produced in relatively high numbers (1–4%) by XXY mothers, become the main population sources of females, giving these aneuploid genotypes a significant selective advantage over their euploid XX counterparts. This should occur for cases in which the Y-drive arises either because a Y chromosome saves a gamete from dysfunction (abbreviated as YS) or because the presence of the X causes dysfunction (abbreviated as XE), although as Figure 7.4B illustrates, the drive threshold differs markedly for these two models. Representative threshold values for *Dro-*

Figure 7.3. Potential genotypes with sex chromosome aneuploids. The question marks refer to the fact that the survival and fertility of the *XXYY* genotype is questionable even in species in which other aneuploids have high fitness. We define l = proportion of X/YY segregations and m = proportion of XX/Y segregations in *XYY* and *XXY* genotypes, respectively. (Reproduced from Lyttle, 1982)

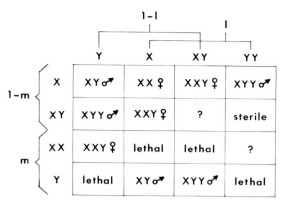

sophila are $k = 0.68$ ($z = 0.53$) and $k = 0.86$ ($z = 0.84$), for the YS and XE cases, respectively (Lyttle, 1982). Such counter selection for aneuploidy could conceivably maintain a stable polymorphism (as illustrated by the equilibrium values in Fig. 7.4) for the accessory Y chromosome long enough to allow for its full incorporation into the karyotype. Under certain conditions, natural selection might actually favor an increase in Y-drive strength concomitant with the fixation of XXYY females and XYY males as a way of overcoming sex-ratio distortion. In this case, all sperm would carry a saving Y, and the sex-ratio would return to 1 : 1 (see open triangles in Fig. 7.4A,B). In fact, the population would mimic an XO sex determination system with the Y chromosomes acting like an extra set of heterochromatic autosomes.

Other rearrangements might evolve similarly. For example, if the part of the Y saving a gamete from drive-induced dysfunction was moved directly to the X by translocation to become indepedent of the sex-determining portion, such a rearrangement could have the same gametic advantage as an XY sperm, while avoiding the problem of unbalanced sexual development. In fact, a spontaneously arising translocation was recovered from one of the experimental *D. melanogaster* populations mentioned above (Lyttle, 1981, and Fig. 7.5), which mimics just such an XY fusion. This rearrangement arose spontaneously and became established (with frequency >0.10) in less than 20 generations in a laboratory population of $N \simeq 1000$ segregating for a form of Y chromosome drive (Lyttle, 1981). Although the arrangement shows no meiotic drive of its own, its presence in the population is sufficient to eliminate most of the sex ratio distortion, and this provides enough of a selective advantage to keep the rearrangement segregating. *D. miranda* apparently exhibits a Y-autosome fusion, while an X-autosome fusion has recently arisen in *D. americana;* such events seem to be common in the evolution of *Drosophila* (White, 1973). A similar result would be obtained if a chromosome dissociation separated the parts of a sex chromosome involved in drive

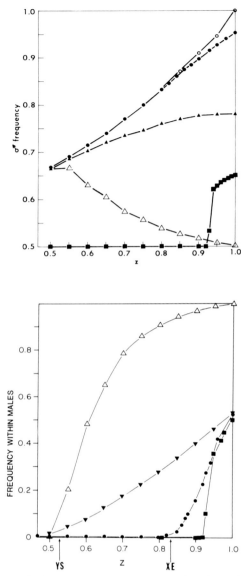

Figure 7.4. (**A**) Male frequency (sex ratio) as a function of z, the probability of gamete dysfunction ($z = (2k - 1)/k$), for various models of Y chromosome drive: (○) pure Y drive, no aneuploidy; (●) the XE case; (▲) the YS case with fitness zero for XXYY genotypes (see Fig. 7.3); (■) pseudo-Y drive with aneuploidy; and (△) the YS case with full XXYY fitness. Note that for the last case, increasing Y drive strength is actually associated with a return to the ideal 1:1 sex ratio, owing to the fixation of XXYY and XYY (see Fig. 7.4B) types. (**B**) Equilibrium frequency of aneuploid (XYY) genotypes among all males as a function of z. Notation is the same as for A. Arrows represent minimum values of z for the establishment of aneuploidy (see text). (A reproduced from Lyttle, 1982)

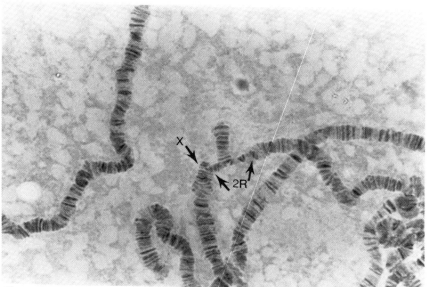

Figure 7.5. A spontaneous secondary translocation that cancels Y chromosome drive. Originally, the 2R tip of an *SD* chromosome was translocated to Y^L to create pseudo-Y chromosome drive. The above T(X;2) (shown in a heterozygous female) arose as the result of a subsequent secondary translocation event (occurring within 10–15 generations of population establishment) which eliminated the Y chromosome drive (Lyttle, 1981). The X breakpoint is in 1B, the 2R breakpoint in 59A, of the standard polytene map. The 2R base is capped by a Y^L telomere, while the translocated X tip is lost; thus, the translocation is lethal in hemizygous males.

and sex determination. There are a number of cases of unusual sex chromosome constitutions encountered in insects and lower vertebrates presumed to have arisen as a result of Y or X chromosome dissociation (White, 1978) whose origin might be explained by this kind of drive–aneuploidy interaction mechanism.

A further point to be noted is that such elements could accumulate as a secondary effect of meiotic drive elsewhere in the genome, without showing drive themselves. Again, the demonstration that a rearrangement itself lacks drive is not sufficient evidence to exclude meiotic drive as playing a role in its evolution.

The accumulation of supernumerary or B chromosomes could occur in an analogous manner, although there is currently no evidence that any of the known germinal mechanisms promoting B chromosome transmission are the result of gamete competition (Jones and Rees, 1982, pp. 46–68).

Rearrangements That Strengthen Linkage Among Drive Elements

Although under normal circumstances, a secondary rearrangement with breakpoints near a locus showing gametic drive should not be able to "hitchhike" to fixation (see above), there are instances in which such a variant might enjoy a selective advantage. For example, a rearrangement might strengthen a linkage between a locus showing drive and secondary loci segregating for drive enhancers. Such a variant would be selectively favored when it happened to create the right kind of linkage disequilibrium (i.e., linking enhancer alleles to the drive element), since the resulting rearrangement would on average enjoy stronger drive than would the corresponding unrearranged alternative (Prout et al., 1973). Because the advantage would be small, such linkage effects would probably be more important for the fixation of inversions (whose heterozygote fitness disadvantage would be correspondingly small) than for translocations or fusions. Note that this argument is analogous to that employed to explain the evolution of "supergene" complexes, in which loci in short chromosomal segments segregating for alleles showing positive epistasis for fitness are thought to become tied together recombinationally by virtue of their common inclusion within a paracentric inversion. In nature, multiple inversions or other similar chromosome aberrations that might owe their establishment to this type of interaction are found associated with the X chromosome drive system *sex ratio (SR)* of *Drosophila pseudoobscura* and *persimilis* (Wu and Beckenbach, 1984), as well as with cases of autosomal drive such as by *Segregation distorter* of *Drosophila*, and the *t* system of mouse (Braden et al., 1972; Hartl and Hiraizumi, 1976).

In summary, it seems clear that population genetic theory points to several ways in which true meiotic drive or gametic competition advantage might promote the fixation of initially rare chromosome rearrangements. Drive might be of sufficient strength to overcome the negative heterosis of a translocation or inversion and cause its deterministic fixation, or it might lower the unstable equilibrium point for such a rearrangement until even relatively transient periods of population size bottlenecks may allow the rearrangement to become established by drift. On the other hand, a rearrangement may be able to hitchhike to establishment by virtue of its effect in reducing recombination among loci having an advantage in gametic competition, or as a consequence of linkage disequilibrium between a drive locus and the rearrangement arising from drift or migration. Finally, some rearrangements, particularly those involving portions of Y or supernumerary chromosomes that can be tolerated in the genome in unusual doses, may be secondarily favored because they nullify the effect of drive operating elsewhere in the genome. The last two cases are of particular interest because they offer insights into ways that rearrangements that show no drive of their own may still owe their establishment to meiotic drive in other parts of the genome.

ACKNOWLEDGMENTS

This chapter is dedicated to Professor Hampton Carson, on the occasion of his 70th birthday, and in memory of Professor Michael J. D. White, whose ideas concerning chromosomal evolution stimulated my own. This work was supported by grant PCM-8207853 from the National Science Foundation.

LITERATURE CITED

Bengtsson, B. O., and W. F. Bodmer. 1976. On the increase of chromosome mutations under random mating. Theor. Popul. Biol. 9:260–281.

Braden, A. W. H. 1972. T-locus mice; segregation distortion and sterility in the male. In R. A. Beatty and S. Gluecksohn-Waelsch (eds.), Proceedings of the International Symposium: The Genetics of the Spermatozoon Bogtrykkeriet Forum, Copenhagen.

Braden, A. W. H., R. P. Erickson, S. Gluecksohn-Waelsch, D. L. Hartl, W. J. Peacock, and L. Sandler. 1972. A comparison of effects and properties of segregation distorting alleles in the mouse (t) and in Drosophila (SD). In R. A. Beatty and S. Gluecksohn-Waelsch (eds.), Proceedings of the International Symposium: The Genetics of the Spermatozoon, pp. 310–312. Bogtrykkeriet Forum, Copenhagen.

Charlesworth, B., and D. L. Hartl. 1978. Population dynamics of the segregation distorter polymorphism of Drosophila melanogaster. Genetics 89:171–192.

Curtsinger, J. W. 1984a. Evolutionary landscapes for complex selection. Evolution 38:359–367.

Curtsinger, J. W. 1984b. Components of selection in X chromosome lines of Drosophila melanogaster: sex ratio modification by meiotic drive and viability selection. Genetics 108:941–952.

Curtsinger, J. W., and M. W. Feldman. 1980. Experimental and theoretical analysis of the "sex ratio" polymorphism in Drosophila pseudoobscura. Genetics 94:445–466.

Dover, G. 1982. Molecular drive: a cohesive mode of species evolution. Nature 299:111–117.

Grant, V. 1975. Genetics of Flowering Plants. Columbia University Press, New York.

Hamilton, W. D. 1967. Extraordinary sex ratios. Science 156:477–488.

Hartl, D. L. 1972. Population dynamics of sperm and pollen killers. Theor. Appl. Genet. 42:81–88.

Hartl, D. 1977. Applications of meiotic drive in animal breeding and population control. In E. Pollack, O. Kempthorne, and T. B. Bailey, Jr. (eds.), Proceedings of the International Conference on Quantitative Genetics, pp. 63–88. Iowa State University Press, Ames.

Hartl, D. L., and Y. Hiraizumi. 1976. Segregation distortion after fifteen years. In E. Novitski and M. Ashburner (eds.), The Genetics and Biology of Drosophila, Vol. 1b. pp. 615–666 Academic Press, New York.

Hedrick, P. W. 1981. The establishment of chromosomal variants. Evolution 35:322–332.

Hiraizumi, Y., L. Sandler, and J. F. Crow. 1960. Meiotic drive in natural populations of *Drosophila melanogaster*. III. Population implications of the *Segregation-distorter* locus. Evolution 14:433–444.

Jones, R. N., and H. Rees. 1982. B Chromosomes. Academic Press, New York.

Lande, R. 1979. Effective deme sizes during long-term evolution estimated from rates of chromosomal rearrangement. Evolution 33:234–251.

Li, C. 1955. Population Genetics, 2nd ed. University of Chicago Press; First Course in Population Genetics. 1976. Boxwood Press, Pacific Grove, CA.

Lyttle, T. W. 1975. Population dynamics of pseudo-Y drive systems in Drosophila involving the Segregation distorter locus. Ph.D. thesis, University of Wisconsin, Madison.

Lyttle, T. W. 1977. Experimental population genetics of meiotic drive systems. I. Pseudo-Y chromosomal drive as a means of eliminating cage populations of *Drosophila melanogaster*. Genetics 86:413–445.

Lyttle, T. W. 1979. Experimental population genetics of meiotic drive systems. II. Accumulation of genetic modifiers of *Segregation distorter (SD)* in laboratory populations. Genetics 91:339–357.

Lyttle, T. W. 1981. Experimental population genetics of meiotic drive systems. III. Neutralization of sex-ratio distortion in *Drosophila* through sex chromosome aneuploidy. Genetics 98:317–334.

Lyttle, T. W. 1982. A theoretical analysis of the effects of sex chromosome aneuploidy on X and Y chromosome meiotic drive. Evolution 36:822–831.

McKee, B. 1984. Sex chromosome meiotic drive in *Drosophila melanogaster* males. Genetics 106:403–422.

Nagylaki, T. 1983. Evolution of a large population under gene conversion. Proc. Natl. Acad. Sci. USA 80:5941–5945.

Ohta, T. 1983. On the evolution of multigene families. Theor. Popul. Biol. 23:216–240.

Ohta, T., and G. Dover. 1984. The cohesive population genetics of molecular drive. Genetics 108:501–521.

Peacock, W. J., and G. L. G. Miklos. 1973. Meiotic drive in *Drosophila:* new interpretations of the *Segregation distorter* and sex chromosome systems. Adv. Genet. 17:361–407.

Prout, T., J. Bundgaard, and S. Bryant. 1973. Population genetics of modifiers of meiotic drive. I. The solution of a special case and some general implications. Theor. Popul. Biol. 4:446–465.

Rhoades, M. M. 1952. Preferential segregation in maize. In J. W. Gowern (ed.), Heterosis, pp. 66–80. Iowa State College Press, Ames.

Sandler, L., and E. Novitski. 1957. Meiotic drive as an evolutionary force. Am. Natur. 91:105–110.

White, M. J. D. 1973. Animal Cytology and Evolution, 3rd ed. Cambridge University Press, Cambridge.

White, M. J. D. 1978. Modes of Speciation. W. H. Freeman, San Francisco.

Wu, C-I., and A. Beckenbach. 1983. Evidence for extensive genetic differentiation between the sex-ratio and the standard arrangement of *Drosophila pseudoobscura* and *D. persimilis* and identification of hybrid sterility factors. Genetics 105:71–86.

Zimmering, S., L. Sandler, and B. Nicoletti. 1970. Mechanisms of meiotic drive. Annu. Rev. Genet. 4:409–436.

IV

FOUNDER EFFECTS IN MOLECULAR EVOLUTION AND DEVELOPMENTAL BIOLOGY

8

Genomic DNA Variation Within and Between Closely Related Species of Hawaiian *Drosophila*

JOHN A. HUNT, KATHLEEN A. HOUTCHENS,
LAURA BREZINSKY, FARIDEH SHADRAVAN,
AND JOHN G. BISHOP, III

We have chosen five species of Hawaiian *Drosophila* from the planitibia subgroup of the "picture wing" group for our studies. The species are *D. picticornis* from Kauai, *D. differens* from Molokai, *D. planitibia* from Maui, and *D. silvestris* and *D. heteroneura* from Hawaii. These species were chosen because they cover the evolutionary range of the Hawaiian Islands (4.4 to 0.4 million years; McDougal, 1979), and they were readily available as laboratory stocks. This made it possible to clone the genomic DNA from these species because the then existing NIH guidelines required the stocks to be grown for 10 generations, which is approximately three years for these species. More importantly this subgroup has been the subject of extensive analysis regarding the evolution of its behavioral, morphological, and enzyme polymorphism relationships (Carson and Kaneshiro, 1976; Carson and Yoon, 1982). The evolutionary relationships so derived are highly correlated with the age of the island on which the species is found (Carson, 1976).

It appeared, then, that analysis of the DNA of these species could add considerably to the knowledge of the evolution of these species and also might suggest new ways to determine the relatedness of each of the species as well as the relationships within species.

METHODS OF COMPARING GENOMIC DNA SEQUENCES

Three methods of comparison of genomic DNA sequences from these species were used. The first method involved hybridizing radiolabeled single-copy tracer DNA with unlabeled (driver) DNA from the various species, and measuring the lowering of the melting point (T_m) of the hybrid by comparing it to the melting point of the tracer hybridized with its own driver DNA (Britten et al., 1978). It is assumed that a mismatch of 1% causes a lowering of melting temperature by 1°C. This method has the advantage that it measures the similarity of the DNA that hybridizes, which is usually a large proportion of the genome. However, the amount of DNA that hybridizes is reduced as the lowering of the melting point increases, and this reduction is not consistent between pairs of species of different taxa that show the same lowering of melting points (Hunt et al., 1981). The closest distances that have been measured are ΔT_m of 0.5°C. These have a standard deviation of 0.2°C, provided that reciprocal measurements are made using the radiolabeled tracer from each of the species being compared.

The second method used was genomic DNA restriction enzyme analysis around the alcohol dehydrogenase *(Adh)* gene. This gene was chosen because it had been isolated from *D. melanogaster* (Goldberg, 1980) and had sufficient homology with the DNA of *D. heteroneura* to make clonal isolation of the gene from *D. heteroneura* possible. The *Adh* gene has been extensively studied in sibling species of *D. melanogaster* (Langley et al., 1981), both by restriction mapping and by DNA sequence analysis (Bodmer and Ashburner, 1984), as well as in several lines within the species (Kreitman, 1983). The technique involves digestion of genomic DNA isolated from mass cultures of isofemale lines and the transfer of the electrophoretically fractionated digest by blotting onto nitrocellulose filters. The size of the homologous DNA restriction fragments is determined by probing with radiolabeled DNA from specific (cloned) sequences isolated from around the *Adh* gene. Because restriction enzymes are highly specific, single-nucleotide substitutions are sampled by the gain or loss of restriction sites. An exception is the case in which small deletions and insertions of DNA mimic the effect of single base changes by causing loss of sites; the reason is that it is impossible to detect 10–20 base-pair changes in the size of most restriction fragments that are detectable by the blotting techniques. The presence of large deletions and insertions caused by transposable elements also complicates the construction of restriction maps. It is usually assumed that the majority of changes that affect restriction sites are caused by single base changes, and therefore, distances can be calculated from the differences between species (Nei and Li, 1979). In addition, the assumption that a single change in a restriction site is due to a single base change allows character-state changes to be assigned to the differences, and these can be used for phylogenetic analysis (Templeton, 1983). An additional advantage in the use of restriction sites is that there is a higher probability of losing a site than of gaining one by a single base change at a specific locus, and thus weighting can be used in assessing the probability of convergent gain or loss of sites in phylogenetic analysis.

Because this method uses radioactive probes to detect the sequences in genomic DNA, it is possible, under optimal conditions, to perform restriction digests on the DNA from an individual fly. However, because of the small amount of DNA available, it is not possible to do more than one or two digestions, and maps must be constructed using DNA from several flies. This is not the case for restriction enzyme analysis of mitochondrial DNA because of its higher multiplicity in the cell, and because, even when enough DNA cannot be obtained from an individual, the fact of its maternal inheritance allows the progeny of an isofemale line to provide DNA from a single molecular species of mitochondrion.

It is sometimes quite difficult to compare restriction maps made in different laboratories, mostly because different sets of enzymes are generally used, but also because the differences in techniques can produce somewhat different estimates of sizes of restriction fragments, making it difficult to detect the presence of small deletions and insertions.

The third method that we have used, and that is now our major technique for DNA comparison, is that of DNA sequencing of the *Adh* gene region. We have chosen a 1600-base pair (bp) region that covers all of the coding region and most of the 5' untranslated DNA, but not the two promoter regions for the adult and larval forms of *Adh* mRNA (Benyajati et al. 1983). The advantages of DNA sequencing are that (1) it allows a larger region of DNA to be sampled than is usually the case with restriction mapping, and (2) comparison of the sequences obtained by others for different species is immediate, and is limited only by the algorithms used for the alignment of the sequences or the imagination of the comparer. The disadvantage is that because of the time taken to determine a DNA sequence as compared with the time required to perform restriction enzyme analysis, it is not practical to use DNA sequence analysis as a method for comparing populations (see Kreitman, 1983). It is possible to isolate the genes from different chromosomes in a single genomic library of DNA, provided that polymorphic markers (usually restriction sites) are present. In this case it is possible to assess the level of polymorphism within the population from the differences in sequences of genes from two chromosomes.

The previous discussion, although specific for the work we have done with the Hawaiian *Drosophila*, is general for the methods themselves. These methods are relevant to the discussion of DNA differences between any group of species. In principle, the same methods can be used for comparison of individuals within species, and hence changes in the DNA that may be important in the speciation process may be measured. The three methods of comparison of DNA sequences differ in their usefulness for this type of comparison: For example, it is possible to estimate the heterozygosity of individuals by using the DNA hybridization method (Britten et al., 1978), but it is not possible to isolate sufficient DNA from individual Hawaiian *Drosophila*, nor is it clear that the melting point differences would be sufficiently large to make this method a useful one for the Hawaiian *Drosophila*. Genomic restriction enzyme analysis on DNA from individuals is still difficult to do for more than one or two enzymes so that unless a specific polymorphism is being studied, this method is of limited usefulness for the comparison of individuals. This is not true for restriction enzyme analysis using mitochondrial DNA (DeSalle et al., 1986a,b). Because it is not practical to do population surveys by DNA sequencing, analysis of the sequence differences between two or a few individuals is feasible only by comparing long sequences from a few clonal isolates of the same gene either from the same library or from different libraries. Alternatively, if the cloning is done by using cosmids, genomic regions of the order of 50 kilobase pairs (kb) can be compared by restriction enzyme analysis on several clones from one or several genomic libraries.

RESULTS AND DISCUSSION

Distance Measurements

DNA Hybridization

The various distance measurements that we have obtained thus far are shown in Table 8.1. From the DNA hybridization data the smallest distances are those obtained for the comparison of either *D. heteroneura* and *D. silvestris* (Hunt et al., 1981) or *D. differens* and *D. planitibia* (Hunt and Carson, 1983). There are progressively larger distances between these

Table 8.1. Distance measurements for the five species of Hawaiian *Drosophila*

					Adh region	
		Protein	Total DNA ΔTm	Restriction[a]	Sequence (%) with intron 2	Sequence (%) without intron 2
Group 1	*silv./het.*	0.08	0.63	1.04		
	diff./plan.	0.11	0.65	1.15		
	Mean	0.10	0.64	1.10	0.4	0
Group 2	*silv./plan.*	0.25	0.99	2.35	3.8	2.8
	het./plan.	0.21	0.96	2.50	3.8	2.8
	silv./diff.	0.26	1.12	2.02	4.1	2.8
	het./diff.	0.23	1.17	2.17	3.4	2.8
	Mean	0.24	1.06	2.26	3.8	2.8
Group 3	*silv./pict.*	1.10	2.25	5.48	10.3	7.4
	het./pict.	1.10	2.02	5.52	10.0	7.4
	plan./pict.	1.60	—	5.02	9.3	6.4
	diff./pict.	1.40	—	4.65	9.0	6.4
	Mean	1.30	2.13	5.17	9.7	6.9
Ratio group 3/group 2		5.4	2.0	2.3	2.6	2.5

[a]From Nei and Li (1979).

two groups and between the *D. heteroneura*/*D. silvestris* pair and *D. picticornis*. Because the reference measurement for DNA hybridization is from self-hybridization and the DNA itself is from a mass culture, it is assumed that heterozygosity of the tracer DNA is effectively removed by using this melting point rather than the melting point of "pure" DNA (Hunt et al., 1981). It should be noted that errors may be generated unless reciprocal measurements are made if the degree of heterozygosity of the DNA from the two species being compared is different in the two samples. Because the level of heterozygosity can be as much as half of the between-species difference (Britten et al., 1978), the error could be significant. The data obtained are correlated with the previous measurements obtained by protein isozyme studies with the same species (Hunt and Carson, 1983). However, the data are still low in statistical significance, and it was decided to shift our attention to a more comprehensive comparison of smaller regions of the genome.

Restriction Mapping

Our first analysis used two bacteriophage λ clones covering a region of 20 kb of DNA around the *Adh* gene region, isolated from a genomic library of *D. heteroneura*. The restriction maps of these two clones are shown in Figure 8.1. Subclones from these two clones were made in the plasmid pBR322 and are shown in Figure 8.1 below the *D. heteroneura* clones. The Hind III and Sal I subclones were used as the probes for detection of the homologous regions in blots of restriction enzyme-digested genomic DNA from the five species. Restriction maps were constructed by using single and double enzyme digests; these are shown in Figure 8.2. Unfortunately, a region of repetitive DNA was found in the cloned *D. heteroneura* DNA, which is located in the overlapping region of the two clones p3.5H and p3.9S and which covers a region of 1–2 kb (Hunt et al., 1984). The repetitive DNA in these two plasmids did not allow them to be used for restriction mapping. It is also clear from the restriction maps that the area covered by the repetitive element is not present in the other species. Because of the confusion caused by the deletion of 1–2 kb of DNA in this area, it was not possible to align the maps properly on the 3' side of the insertion. As a result, only sites that appear between the two arrows at −12 kb and +2 kb shown in Figure 8.2 could be used for the comparison of the species. It is interesting to note that similar insertions are found on the 3' region of the *D. melanogaster Adh* gene (Langley et al., 1981). We have used both the method of Nei and Li (1979) and that of Engels (1981) and Ewens et al. (1981) to generate distances from these data. There is essentially no difference between the two sets of distances obtained, even though the standard deviations measure different errors. These data correlate well with the other two distance measurements given in Table 8.1. Again, partly because of the lack of restriction sites and partly because of the problem with the repetitive element, the accuracy of these determinations is not sufficiently high to mea-

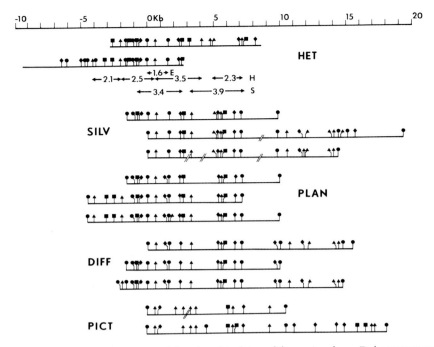

Figure 8.1. Restriction maps of the clonal isolates of the region from *D. heteroneura* (HET), *D. silvestris* (SILV), *D. planitibia* (PLAN), *D. differens* (DIFF), and *D. picticornis* (PICT) genomic DNA. The restriction enzymes are shown by EcoR I(●), Sal I (■), Hind III (▲), and Pst I (◆). The double slashes indicate that the regions marked are shorter by the amount shown (about 200 bp) than the corresponding regions in the other clones. The numbered regions below the *D. heteroneura* clones are as follows: 1.6 E is the EcoR I fragment that hybridizes with the *D. melanogaster Adh* clone; 2.1, 2.5, 3.5, and 2.3 H are Hind III fragments from the *D. heteroneura* clone which were subcloned into pBR322; and similarly the fragments 3.4 and 3.9 S are Sal I fragments from *D. heteroneura* which were subcloned into pBR322.

sure significant differences between the distances of the closest species. What can be seen in Table 8.1 is that the rank order of species comparisons generated by all of the three distance measurements is the same if the group comparisons between the *D. heteroneura/D. silvestris* or *D. differens/D. planitibia* pairs (group 2, Table 8.1), or the comparison of these four species with *D. picticornis* (group 3) is considered. Within these groups there is some heterogeneity, but this is all within the variance of the measurements. It would be possible to generate more data by use of restriction enzymes that are specific for 4-bp sequences rather than the 6-bp sequences of the enzymes used; however, these enzymes would generate smaller fragments, many of which would not transfer to the nitrocellulose "blot," and make it much more difficult to generate unambiguous maps.

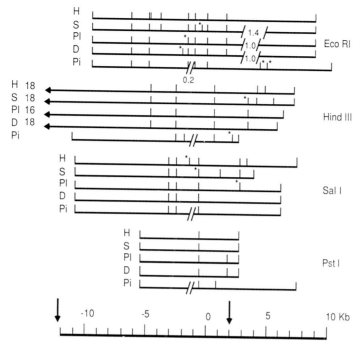

Figure 8.2.   Restriction maps of genomic DNA digested with four restriction enzymes from the five species: *D. heteroneura* (H), *D. heteroneura* (S), *D. planitibia* (Pl), *D. differens* (D), and *D. picticornis* (Pi). Polymorphic loci are marked by an asterisk, and the site and size of presumed deletions are shown where they could be inferred. The two arrows mark the region used for comparison of the species.

Several polymorphic sites are found in the DNA of the different species. In general, the presence of each polymorphism has been confirmed by the clonal isolation of the *Adh* region from the five species. The restriction maps of several of the clones are shown in Figure 8.1. If we assume that each site that is polymorphic is present as 50% of the DNA in the species, the upper limit of the heterozygosity of each species over this 15-kb region is $0.6 \pm 0.2\%$ (range, 0.4–0.9%). No apparent difference between any of the species can be shown by the measurements of polymorphism. However, when we examine the region on the 3′ side of the *Adh* gene in *D. picticornis*, the level of heterozygosity in restriction sites is 3%, which is not matched by any of the other species. The high value of the heterozygosity found in this region of *D. picticornis* is reflective of a general variability of the region which is manifested in several of the other species. In *D. heteroneura* there is the repetitive element, and in *D. silvestris* there are a number of small (50–200 bp) deletions and insertions that are detected in the cloned DNA and would probably be undetected in the blots. This kind of variability presents a challenge in terms of the evolutionary analysis of the species because it severely limits the regions of DNA that can be compared by restriction enzyme analysis or by DNA sequence analysis, but they cannot be ignored in terms of the variability of the species. We also find that this high degree of variability is present in other species of Hawaiian *Drosophila*. The restriction map of the *Adh* region of *D. affinidisjuncta* (Brennan et al., 1984) can be aligned and compared with the *D. heteroneura* map if the area compared is restricted to the same region used in the genomic DNA comparisons of the planitibia subgroup species. However, it is not possible to compare the restriction map of the 3′ side of the *D. affinidisjuncta Adh* region with the same region from any of the planitibia subgroup species. The distance measured between *D. heteroneura* and *D. affinidisjuncta* was 7.6%. In the case of *D. silvestris*, *D. differens*, and *D. planitibia*, the cloned fragment comparison could be made only over a smaller region, and *D. picticornis* could not be compared at all.

DNA Sequences

The complete sequence of the 1.6-kb EcoR I region of the five species containing the *Adh* region has not yet been completed. However, a region covering almost 300 bp from the five species has been sequenced and is shown in Figure 8.3. This region encompasses part of the 5′ untranslated region, exon 1, intron 2, and part of exon 2 of the *Adh* gene. Although, in terms of sequence analysis, this is not a very large amount of information, it is equivalent to a comparison of 50 restriction sites generated by 6-bp-specific restriction enzymes.

First, it is noted that *D. heteroneura* and *D. silvestris* differ by a single base pair in the first intron, as do the pair *D. differens/D. planitibia*. However there are several differences between the two pairs. There are three changes in the 5′ untranslated region, one synonymous change (i.e., a change that does not alter the amino acid sequence) in the first exon, 4- to 6-bp changes in the second intron, and one synonymous and one non-synonymous (isoleucine to valine) base change in the second exon. In addition, there are two single base-pair deletions and a 10-bp deletion in the second intron of the *D. differens/D. planitibia* pair. The overall rate of substitution between the two pairs is 3.6%, as compared to the within-pair rate of 0.4%. If the second intron is omitted, the value for the between-pair comparison falls to 3.2%. The low value for the within-pair comparison, where the changes are limited to the second intron, could be due to sampling bias; indeed, in more recent sequence analysis a substitution rate of more than 1% is found for the 3′ nontranscribed region of the *D. heteroneura/D. silvestris* pair. The overall substitution rate between *D. silvestris* and *D. picticornis* is 10.3%, and the substitution rate between *D. planitibia* and *D. picticornis* is 9.3%. In the case of the *D. planitibia/D. picticornis* comparison, the deletions in the second intron were ignored. It has now become apparent through examination of the introns of these species and of several other closely related *Drosophila* species (Hunt and Jeffs, unpublished) that many of the changes that are

```
              10        20        30        40        50        60        70        80        90        100       110       120
SILV  GAATTCTCTGCCCAAACAGGTGAACAGAGTTGAGCGTAGCAGTGAAAAAATGGTTATCGCTAACAGTAACATCATCTTTGTGCTGGTGGTCTGGCTGGCATTGGCCTTGGACACCAGTCGCGA
HET   ------------------------------------------------------------------------------------------------------------------------
DIFF  ------T-G-----------------------------------------------------------------------------------------------------------------
PLAN  ------T-G-----------------------------------------------------------------------------------------------------------------
PICT  ------T-G----------------------------------G--T-GC----------------C--C-------T--------------------------------------------

              130       140       150       160       170       180       190       200       210       220       230       240
SILV  GATTGTCAAGAGCGGCGCCCAAGGTAGGTTATAGCCTGTAAATCTCAGGGAAAAGTAGTGTACAAATAAGAAATAAATTATTATTATTTGTTCGTTTAGAACTTGGTGGTGC
HET   --------------------------------------------------A------------------------------------------------------
DIFF  -----T------A-G---------G-------------A----------A--------C------
PLAN  -----T------A-G-----------------------A----------A--------C------
PICT  ------T------------A----------AG-C---T-------G------A---T---T--T-C---C----T---

              250       260       270       280       290       300
SILV  TTGATCGCATTGACAACCCCGCTGCCATTGCCGAGTTGAAAGCCCTTAA
HET   -----G------------------------A---
DIFF  -----G------------------------A---
PLAN  -----G-----------------------------------TCCCAAGGTGACCGTTA
PICT  -------C---
```

Figure 8.3. Comparison of the sequence of the Adh region from the species *D. heteroneura* (HET), *D. silvestris* (SILV), *D. differens* (DIFF), *D. planitibia* (PLAN), and *D. picticornis* (PICT) starting at the EcoR I site of the 1.6-kb region. The first exon (coding sequence) starts at residue 50, the second intron starts at residue 143 and ends at residue 227. The exons are underscored. The major sequence is shown for *D. silvestris*. The dashes indicate the same sequence, and gaps denote deletions.

found are due not to single base-pair changes but rather to duplication of small regions of the sequences either by the action of transposable elements or by other mechanisms. If we ignore the changes found in the introns, then there are no differences between the two pairs of species: The difference between the pairs is 2.8%, and that between each of the pairs and *D. picticornis* is 7.4% and 6.4%, respectively. The amount of information available is not yet sufficient to make comparisons of synonymous and nonsynonymous substitution rates between these species. The mean distances for the group 2 and group 3 comparisons are quite similar for the restriction enzyme distances and the DNA distances measured without intron 2.

Phylogeny and the Founder Effect

How do the nuclear DNA comparisons fit with phylogenies that have been proposed for this subgroup, and can we make any inferences about the founder effect? The distance measurements from DNA hybridization and restriction mapping as well as the protein polymorphism data produce a phenogram that is consistent with the phylogeny shown in Figure 8.4A. However, this is at odds with the phylogeny found both from distance measurements and from character state analysis in the mitochondrial DNA of the species *D. heteroneura*, *D. silvestris*, *D. differens*, and *D. planitibia* (Fig. 8.4B; DeSalle and Giddings, 1986). Character-state analysis of the protein polymorphism and restriction enzyme data shown here are inconclusive for distinguishing between the two phylogenies (Bishop and Hunt, 1988); however, the character states found in the DNA sequences analysis are strongly in favor of the phylogeny depicted in Figure 8.4A, by a score of 17-bp changes versus 24. The inclusion of the 10-bp deletion would require two identical events in phylogeny 4B over the single event required for phylogeny 4A and would be considered highly unlikely.

The discrepancy between the mitochondrial and nuclear DNA phylogenies is intriguing. There are two alternative hypotheses that could account for the differences, both of which require hybridization between the two

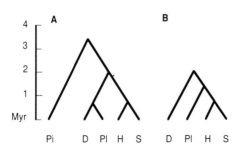

Figure 8.4. Phylogenies of the planitibia group of Hawaiian *Drosophila*: (**A**) This phylogeny is suggested by the nuclear DNA measurements. (**B**) This phylogeny is derived from the DeSalle and Giddings data from mitochondrial DNA restriction maps. The species shown are *D. heteroneura* (H), *D. silvestris* (S), *D. planitibia* (Pl), *D. differens* (D), and *D. picticornis* (Pi). my = millions of years.

species some time after their divergence. If we accept the nuclear phylogeny, then introgression would require the fixation of the ancestral *D. heteroneura/D. silvestris* mitochondrial type in *D. planitibia*, some time after the split between *D. differens* and *D. planitibia*. This would require a female preference of the ancestral *D. heteroneura/D. silvestris* species over *D. planitibia*. The female-preference studies on the present-day species indicate that *D. planitibia* females will not readily accept *D. heteroneura* or *D. silvestris* males, but that *D. heteroneura* or *D. silvestris* females will accept *D. planitibia* males as readily as their own males (Kaneshiro, 1976). This would allow mitochondrial transfer and fixation in the right direction, provided that the hybrid females in the *D. planitibia* population can also mate with *D. planitibia* males. The alternate hypothesis is that the mitochondrial phylogeny is correct and that nuclear genes have flowed from *D. differens* to *D. planitibia* without concomitant mitochondrial flow. This would require introgression of *D. differens* males, which is again in the acceptable direction. It might, however, be expected that the nuclear DNA changes could have occurred by gene conversion, in which case the sequences of other genes would not be expected to show the same close relationship. The major difference between these two models is that in the former case the hybridization must occur before the migration of the ancestor of the *D. heteroneura/D. silvestris* pair to Hawaii and before it became extinct on Maui, and before the divergence of *D. planitibia* from *D. differens*. In the latter case the hybridization could have occurred any time after the divergence of *D. planitibia* from *D. differens*.

One of the expected effects of the founder hypothesis is that the divergence of the species can occur only after a new island is formed, and therefore, there should be a correlation between the DNA distances and the age of the island on which the species are found. That is, we would expect the divergence of *D. picticornis* from the result of the planitibia subgroup to have occurred at the time of the formation of the island of Oahu (3.5 million years ago); and the divergence of ancestors of the two pairs *D. heteroneura/D. silvestris* and *D. differens/D. planitibia* to have occurred at the time of the formation of Molokai (1.9 million years ago) or, more likely, Maui (1.3 million years ago); and the divergence of *D. heteroneura* from *D. silvestris* after the formation of Hawaii (0.4 million years ago). The regression line obtained by using these dates with the restriction enzyme polymorphism distances produces a rate of DNA change of 0.6% per million years (Hunt and Bishop, unpublished). The DNA sequence data produce a similar rate. When the DNA hybridization distances are used, the rate is approximately half. It is not clear why there is this discrepancy between the two measurements. It may be that equating 1% mismatch with 1°C lowering of the melting temperature is an overestimate; it could also be that the region around the *Adh* gene does not properly reflect the rate of change in the DNA of the whole genome. Preliminary comparisons of the coding region of *D. heteroneura* and *D. sil-*

vestris with that of *D. affinidisjuncta* (Hunt and Rowan, unpublished) enable a maximal rate of nucleotide change of 1.3% per million years to be calculated from synonymous base-pair changes.

SUMMARY

Much work must yet be done to obtain enough data from DNA sequencing to answer all the questions that have been posed. It is clear that the kind of variation that is present between the species is not simple and is certainly not confined to the single base-pair substitutions that have been the basis of most of the methods of comparison used thus far. The presence of numerous deletions and insertions in the DNA sequences that affect the different ways of measuring sequence variation make it almost mandatory to use DNA sequencing to get the most out of these comparisons, especially since there are gross state changes that can add more to phylogeny construction than can base changes alone. However, it is clear that the difficulty of doing sequence analysis on genes isolated from populations makes such methods as restriction enzyme analysis of the mitochondrial DNA very important in the analysis of populations, especially those in which incipient speciation events are suspected. It is also clear from the level of heterozygosity found in the DNA sequences within a species that finding fixed differences between close species is going to be very difficult, but that such differences, when found, will be of importance in adding to our knowledge of the founder effect and speciation.

ACKNOWLEDGMENTS

We thank Dr. H. L. Carson for his constant advice and encouragement during the course of this work. Without his pioneering work and enthusiasm for the project, none of this would have been possible. We thank D. Aratani, A. Kennedy, K. Koo, F. Yost, and D. Sato, who have assisted in the isolation and restriction mapping of various clones. This work was supported by grants BSR82-14338 and DEB79-22820 from the National Science Foundation and a grant from the Hawaiian Evolutionary Biology Program.

LITERATURE CITED

Benyajati, C., N. Spoerel, H. Haymerle, and M. Ashburner. 1981. The messenger RNA for alcohol dehydrogenase in *Drosophila melanogaster* differs in its 5' ends in different developmental stages. Cell 33:125–133.
Bishop, J. G. III, and J. A. Hunt, 1988. DNA divergence in and around the alcohol dehydrogenase locus in five closely related species of Hawaiian *Drosophila*. Mol. Biol. Evol. 5:415–431.

Bodmer, M., and M. Ashburner. 1984. Conservation and change in the DNA sequences coding for alcohol dehydrogenase in sibling species of *Drosophila*. Nature 309:425–430.

Brennan, M. D., R. G. Rowan, L. Rabinow, and W. J. Dickinson. 1984. Isolation and characterization of the alcohol dehydrogenase gene from *Drosophila affinidisjuncta*. J. Mol. Appl. Genet. 2:436–446.

Britten, R. J., A. Cetta, and E. H. Davidson 1978. The single-copy DNA sequence polymorphism of the sea urchin *strongylocentrotus purpuratus*. Cell 14:1175–1186.

Carson, H. L. 1976. Inference of the time of divergence of some *Drosophila* species. Nature 259:395–396.

Carson, H. L., and K. Y. Kaneshiro, 1976. *Drosophila* of Hawaii: systematics and ecological genetics. Annu. Rev. Ecol. Systematics 7:311–346.

Carson, H. L., and J. S. Yoon. 1982. Genetics and evolution of Hawaiian Drosophila. In M. J. Ashburner and H. L. Carson (eds.), The Genetics and Biology of *Drosophila*, Vol. 3b, pp. 298–344. Academic Press, New York.

DeSalle, R. and L. V. Giddings, 1986. Discordance of nuclear and mitochondrial DNA phylogenies in Hawaiian *Drosophila*. Proc. Natl. Acad. Sci. USA 83:6902–6906.

DeSalle, R., L. V. Giddings, and K. Y. Kaneshiro. 1986a. Mitochondrial DNA variability in natural populations of Hawaiian *Drosophila*. II. Genetic and phylogenetic relationships of natural populations of *D. silvestris* and *D. heteroneura*. Heredity 56:87–96.

DeSalle, R., L. V. Giddings, and A. R. Templeton. 1986b. Mitochondrial DNA variability in natural populations of Hawaiian *Drosophila*. I. Methods and levels of variability in *D. silvestris* and *D. heteroneura* populations. Heredity 56:75–85.

Engels, W. R. 1981. Estimating genetic divergence and genetic variability with restriction endonucleases. Proc. Natl. Acad. Sci. USA 78:6329–6333.

Ewens, W. J., R. S. Spielman, and H. Harris. 1981. Estimation of genetic variation at the DNA level from restriction endonuclease data. Proc. Natl. Acad. Sci. USA 78:3748–3650.

Goldberg, D. A. 1980. Isolation and characterization of the *Drosophila* alcohol dehydrogenase gene. Proc. Natl. Acad. Sci. USA 77:5794–5798.

Hunt, J. A., and H. L. Carson. 1983. Evolutionary relationships of four species of *Hawaiian Drosophila* as measured by DNA reassociation. Genetics 104:353–364.

Hunt, J. A., T. J. Hall, and R. J. Britten. 1981. Evolutionary distances in Hawaiian *Drosophila* measured by DNA reassociation. J. Mol. Evol. 17:361–367.

Hunt, J. A., J. G. Bishop, III, and H. L. Carson. 1984. Chromosomal mapping of a middle repetitive DNA sequence in a cluster of five species of Hawaiian *Drosophila*. Proc. Natl. Acad. Sci. USA 81:7146–7150.

Kaneshiro, K. Y. 1976. Ethological isolation and phylogeny in the planitibia subgroup of Hawaiian *Drosophila*. Evolution 30:740–745.

Kreitman, M. 1983. Nucleotide polymorphism at the alcohol dehydrogenase locus of *Drosophila melanogaster*. Nature 304:412–417.

Langley, C. H., E. Montgomery, and W. Quattlebaum. 1981. Restriction map variation in the Adh region of *Drosophila*. Proc. Natl. Acad. Sci. USA 79:5631–5635.

McDougall, I. 1979. Age of the shield-building volcanism of Kauai and linear migration of volcanism on the Hawaiian Island chain. Earth Planet. Sci. Lett. 46:31–42.

Nei, M., and W.-H. Li. 1979. Mathematical model of studying genetic variation in terms of restriction endonucleases. Proc. Natl. Acad. Sci. USA 76:5269–5273.

Templeton, A. R. 1983. Phylogenetic inference from restriction endonuclease cleavage site maps with particular reference to the evolution of humans and apes. Evolution 37:221–244.

9

Gene Regulation and Evolution

W. JOSEPH DICKINSON

More than 20 years ago, it was suggested that changes in gene regulation are an important component of evolutionary change—perhaps more important than changes in primary structure of proteins (e.g., Wallace, 1963; Zuckerkandl, 1963). In recent years, the role of regulatory genes in evolutionary change has received increasing attention (e.g., King and Wilson, 1975; Valentine and Campbell, 1975; Wilson, 1976; Hedrick and McDonald, 1980; MacIntyre, 1982). In particular, Allan Wilson and his colleagues have argued persuasively for the potential significance of regulatory change. However, much of the evidence is indirect. My initial work on Hawaiian *Drosophila* (and my work with Hamp Carson in 1978–9) was undertaken in hopes of obtaining more direct evidence.

The remarkable diversification of the Hawaiian *Drosophila*, together with the extensive background on phylogeny and ecology developed by Carson and his colleagues, makes this an ideal system for a variety of evolutionary studies (Carson and Kaneshiro, 1976; Carson and Yoon, 1982). Work undertaken while I was spending a sabbatical year with Carson in 1978–9 certainly did not resolve the relative importance of structural and regulatory change, but it did demonstrate that remarkable diversity in patterns of regulation of specific genes has arisen during the evolution of the picture-winged group (Dickinson and Carson, 1979; Dickinson, 1980a–c). I take this opportunity to present a somewhat speculative assessment of the possible significance of this regulatory diversity and to record some thoughts on the limitations of existing data and on important directions for future research. Although less directly concerned with founder effects and speciation than some other contributors to this volume, I will have occasion to note possible relationships between alterations in gene regulation and the speciation events that have been so common in the evolution of the Hawaiian *Drosophila*.

INTERSPECIFIC REGULATORY DIFFERENCES

There are still relatively few studies in which regulatory patterns of specific genes have been compared across a large group of related species. Although completed several years ago, our Hawaiian *Drosophila* study still provides some of the best examples of dramatic and complex differences between closely related species (Dickinson, 1980a). Electrophoretic analysis of the activities of six enzymes in 14 larval and adult organs revealed that every enzyme showed interspecific regulatory differences, that it was rare for two species to have identical regulatory patterns, and that at least 30% of the individual traits (one enzyme in one tissue) showed significant interspecific variation. These are conservative estimates in at least two respects: It is now clear that many two- to fourfold differences in detected enzyme activity, not considered significant in our initial tabulations, are real and reproducible. In addition, analysis of extracts of whole organs obscures finer-scale pattern differences. Histochemical analyses indicate that there is much additional variation at this level. Figure 9.1 gives the reader some feeling for the detailed differences that are observed (Dickinson, unpublished). Consult our earlier publications (Dickinson 1980b, c, 1983) for other examples. It should be noted that almost all of these differences are observed in flies reared in the laboratory under standard conditions, so they do not reflect responses to different environments. Two major points may be derived from these data: (1) Novel regulatory patterns can arise and become established rapidly on an evolutionary time scale, and (2) the organization of regulatory systems permits a remarkable degree of pattern diversity.

Comparisons to other species groups might provide some indication as to whether this observed diversity is related to the rapid evolution of the Hawaiian *Drosophila*. The most extensive comparable studies are probably those of Whitt and his colleagues on fish (reviewed by Whitt, 1983). Again, qualitative regulatory differences have been found. For example, the lactate dehydrogenase *c* (LDHc) locus is expressed predominantly in the retina in most fish groups, but in the liver in a few groups (Markert et al., 1975). However, these differences distinguish families or orders rather than species. Moreover, most major qualitative differences involve divergence in regulation of one member of a family of duplicated loci (Ferris and Whitt, 1979), a circumstance in which relaxed selection might be important. In some cases, there are differences in timing of expression during embryogenesis that are also taken as indications of altered regulation (Parker et al., 1985b), but again differences are found primarily for taxonomic groups above the genus.

Greater diversity among closely related species was found in a recent survey of the major urinary proteins (MUPS) in several species of mice (Sampsell and Held, 1985). As the name suggests, these proteins are secreted in the urine of laboratory mice. There are dramatic sex differences,

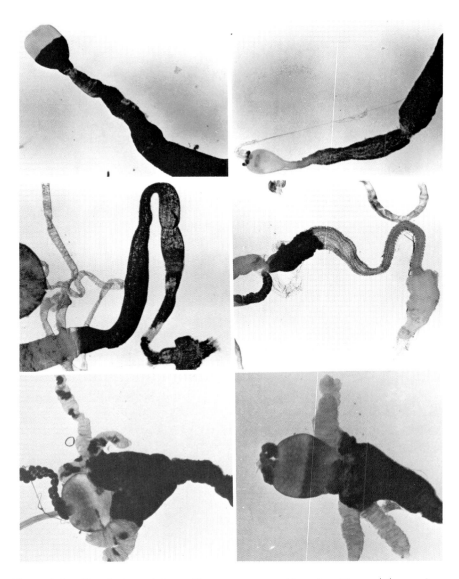

Figure 9.1.   Detailed regulatory differences among Hawaiian *Drosophila* species. Tissues have been stained for aldehyde oxidase (AO) activity and arranged, in each case, with anterior to the left. **Top:** *D. grimshawi* (left) and *D. adiastola* (right) have comparable total AO-2 activity in adult midgut, but the anterior boundary extends onto the bulb-shaped cardia in the former case but not in the latter. **Middle:** *D. heteroneura* (left) and *D. adiastola* (right) both have AO-1 in adult hindgut, but it is confined to the anterior section in the latter. **Bottom:** In feeding larvae of most species, AO-2 is found in the enlarged ventriculus region of the midgut but not in the four gastric ceca that project anteriorly and laterally from this region. In the very late third instar of some species including *D. gradata* (left), scattered cells in the ceca have the enzyme. In others, including *D. hawaiiensis* (right), this is never seen.

with males secreting much higher levels. In a series of related wild species, these sex differences are highly variable, with the male/female ratio ranging from 1 to 500. Relative levels of the proteins in different organs also vary between species. Other examples not derived from such systematic studies nevertheless illustrate the range of possible regulatory effects and suggest that regulatory changes may be common during evolution. Wilson et al. (1977) have listed several cases of organ-specific differences in enzyme levels among mammals. Of particular interest (see below) is the elevated level of lysozyme in the stomach of ruminants and certain monkeys (Dobson et al., 1984). Also notable is the alkaline phosphatase present in placenta of humans and great apes but not of other primates (Goldstein et al., 1982). A similar enzyme is found in lung of old-world monkeys, suggesting a recent regulatory change. In other groups of *Drosophila*, Powell (1980; Powell et al., 1980) reported substantial variability in the distribution of amylase along the gut within species of the obscura willistoni groups, but not fixed differences between species. Korochkin (1980) has recognized significant differences in esterase patterns in species of the virilis group. MacIntyre (1982) has reviewed additional cases.

These examples of regulatory differences in other groups of species not withstanding, the Hawaiian *Drosophila* stand out in terms of the frequency of qualitative differences between closely related species. It is tempting to speculate that this is related to the rapid evolution of this group. However, several important methodological differences make this conclusion premature. As indicated, many of the reported examples are not derived from systematic surveys and could represent only the tip of the iceberg. The more extensive surveys often have not examined regulatory patterns in much detail (e.g., they look at total enzyme activity in whole organisms at different ages) and may have missed many qualitative differences. Finally, one may need to consider the nature of the gene products examined. A number of studies suggest that some structural genes are generally more polymorphic than others (Kojima et al., 1970; Gillespie and Langley, 1974; Johnson, 1974; Powell, 1975; Selander 1976). Various explanations have been offered. For example, Kojima et al. (1970) and Gillespie and Langley (1974) have suggested that enzymes whose substrates are diverse molecules of external origin are more likely to be polymorphic than are enzymes whose substrates are specific molecules generated by internal metabolism. All of the enzymes used in the survey of Hawaiian *Drosophila* fit the former description and are polymorphic in most species of *Drosophila* (Johnson, 1974; Powell, 1975; Selander, 1976). It is at least plausible that factors that permit or favor electrophoretic polymorphism also permit regulatory diversity. In addition, an important component of the radiation of picture-winged *Drosophila* has been adaptation to new host plant species. Thus, it is likely that even closely related species encounter quite different arrays of plant secondary compounds and have been subjected to selective pressures affecting enzymes that metabolize these substances. The enzymes used in other extensive surveys of regulatory patterns include many more that function in essential steps of internal metabolism. To reach conclusions on the relationship of rapid evolution to regulatory change, we need comparative studies that are methodologically similar and with potential bias controlled as carefully as possible. Ideally, one would like surveys of rapidly evolving and conservative groups in which orthologous gene products have been examined by means of identical methods. With these considerations in mind, we have begun a survey of a "conservative" group of *Drosophila*—the virilis group—using the same methods and enzymes employed with the Hawaiian *Drosophila*. Preliminary results suggest that there is much less regulatory diversity in this group. We also plan studies examining a wider range of gene products. These results will be more meaningful if a reasonably secure time scale is available. Where, as with *Drosophila*, dates based on a fossil record are hard to obtain, comparative measurements of DNA sequence divergence might be useful.

INTRASPECIFIC REGULATORY VARIANTS

A complete understanding of the significance of interspecific regulatory differences will depend on clarification of the processes by which they are established, including an adequate description of the intraspecific variants that might serve as raw material for evolution of fixed differences. Studies aimed at uncovering and measuring regulatory diversity in natural populations are still scarce and are not directly comparable to electrophoretic surveys of structural gene polymorphism. Nevertheless, enough information exists to suggest that variants reasonably termed "regulatory" are common.

Probably the most extensive studies are those of Laurie-Ahlberg and her colleagues (Laurie-Ahlberg et al., 1980, 1982; Laurie-Ahlberg, 1985) on chromosome substitution lines in which chromosomes sampled from wild populations of *D. melanogaster* were made homozygous in a genetic background otherwise derived from an inbred laboratory stock. Activities of over 20 enzymes have been measured on samples from 100 lines. Most of the enzymes show significant quantitative variation between lines. In most cases, there are contributions of factors not linked to the structural gene for the affected enzyme and, hence, presumed to be due to regulatory influences. A two- to three-fold range in total activity is common. In a few cases, quantitative variants have subsequently been associated with tissue- or stage-specific differences in enzyme expression (Bewley 1981; Maroni and Laurie-Ahlberg, 1983; Clark et al., 1984). However, most systematic surveys have not looked for qualitative differences like those found in the Hawaiian *Drosophila*. Reconstruction experiments with some of our interspecific variants suggests that they would escape detection in such surveys. Systematic searches designed to detect *qualitative* variants are much to be desired.

A number of qualitative intraspecific variants have been described, mostly from work focusing on mechanistic questions. Although these studies were not designed to yield information on the frequency of the variants in natural populations, the relative ease with which variants are found suggests that they are common. For example, in *D. melanogaster,* variants affecting timing and tissue specificity of aldehyde oxidase (AO) were found in surveys of 50–60 laboratory stocks or isogenic lines derived from wild populations (Dickinson, 1975, 1978), with the less common regulatory pattern found in several stocks in each case. A survey of the distribution of AO along the gut in larvae from 54 isogenic lines derived from wild populations revealed 17 stocks with patterns significantly different from a standard (unpublished results of Kathy Black working in my laboratory). These fell into seven different classes with respect to the set of features affected. Later extension of this survey to over 100 stocks yielded additional variant strains at about the same frequency, but no new patterns. Adults were somewhat less variable, but four patterns were recognized. Examples illustrated in Figure 9.2 show that these intraspecific variants are similar to certain interspecific differences (compare Fig. 9.1). Abraham and Doane (1978) recognized three patterns of amylase distribution in adult guts in a survey of 11 laboratory stocks. Variants characterized by both abnormally high and abnormally low levels of xanthine dehydrogenase (XDH) were included in a set of 12 stocks carrying wild-type isoalleles at the *rosy* locus (Chovnick et al., 1976, 1977), although it is not clear how many stocks were surveyed to assemble that set. At least one of these variants is now known to have tissue-specific effects (Clark et al., 1984). We have also reported polymorphisms for regulatory traits within species of Hawaiian *Drosophila* (Dickinson, 1980a, 1983). These were found by examining only 5–10 stocks of each species. Surveys of modest size have also revealed qualitative regulatory variants in a variety of other species. For reviews see Paigen (1979a, b) and MacIntyre (1982).

The few reported systematic surveys specifically designed to sample qualitative variants in natural populations also detect much apparent regulatory variation. For example, Powell (1979) recognized 13 different patterns of amylase distribution in guts of *D. pseudoobscura* adults in a survey of 10 wild populations. Most of the populations were polymorphic for 8–10 of the patterns. The average frequency of the most common pattern was only 0.39, whereas the next most common pattern occurred at an average frequency of 0.22.

These results suggest that variability in regulatory genes may be about as common and extensive as is structural gene variability. However, it will be difficult to obtain estimates directly comparable to those derived from studies on electrophoretic polymorphisms (e.g., number and relative frequency of alleles, fraction of loci that are polymorphic, average heterozygosity, etc.—see Lewontin, 1974), even given much more extensive data. To begin with, despite the technical difficulty of establishing the identity of two independently sampled copies of a structural gene, there

Figure 9.2. Aldehyde oxidase pattern variants in *D. melanogaster.* Each pair of panels compares a specific features in two inbred strains. Anterior is to the left in each panel. The strains in each pair have roughly equal total enzyme activity. **Top:** anterior larval midgut. Note the difference in the bulb-shaped proventriculus. **Middle:** posterior larval midgut. Note the difference just anterior to the attachment of the Malpighian tubules. **Bottom:** anterior adult midgut. Note the different anterior limit of activity. Compare this to the interspecific difference shown at the top of Figure 9.1.

is little difficulty in at least specifying appropriate criteria—for example, identity of amino acid sequence or nucleotide sequence of the structural gene. In contrast, identity of regulatory patterns can be established only by evaluation of multidimensional patterns for which we probably cannot even name all of the relevant dimensions. A regulatory gene could affect enzyme activity in any of hundreds of cell types, at any stage of development and in response to any environmental cue. Known examples clearly indicate that complex and highly specific regulatory effects are possible (Dickinson, 1980a–c). The bottom pair in Figure 9.1 dramatizes this point.

The situation is further complicated by the probability that multiple, interacting genes contribute to any observed regulatory phenotype. In Powell's (1979) study, the amylase pattern frequencies are not equivalent to allele frequencies at any specific locus since the pattern differences are based on more complex genetic differences (Powell and Lichtenfels, 1979). I have also documented cases where different genes contribute to variation of different aspects of the overall regulatory pattern of a specific enzyme (Dickinson, 1980b). Similarly, in the unpublished study by Black on AO gut patterns, four of the seven variant phenotypes yielded results consistent with control by a single major gene (roughly 1 : 1 segregation in backcrosses), but three cases were more complex. Environmental contributions to realized enzyme activity were also noted in Powell and Lichtenfels's study and in others (Clarke et al., 1979; Doane 1969). Nonspecific effects must also be considered. Laurie-Ahlberg et al. (1980) evaluated the possibility that differences in total enzyme activities reflected differences in size of the flies in different stocks. Analysis of variance showed that size variation could account for part, but not all, of the observed differences in the enzymes. More subtle factors, such as variations in the size of specific organs or in the number of cells of a specific type, would be very much harder to detect. Laurie-Ahlberg et al. (1980), McDonald and Ayala (1978), and Powell (1979) all used comparisons between different enzymes as a test for the specificity of the effects they were looking at. However, one would need rather precise knowledge of the distributions of the enzymes chosen as controls (or remarkably good luck) for that control to be adequate.

We have analyzed a case that illustrates the importance of such nonspecific factors (Dickinson and Fey, unpublished). We thought we had detected a sexual dimorphism in the regulation of ADH. Males of *D. grimshawi* and some related species have readily detectable ADH activity in extracts of hindgut, whereas females do not. Examination by histochemical techniques reveals that the sexual dimorphism is actually structural (Fig. 9.3). In males of these species, the region of the hindgut immediately posterior to the Malpighian tubules is greatly enlarged. The ADH activity is confined precisely to this region. The corresponding region in the females of the same species is not enlarged, but there is a narrow band of cells containing ADH in the same position. The region is simply too small for the activity to be detected easily in extracts of the entire hindgut. Note that use of a second enzyme as a control would have been useful only if it happened to be concentrated in that same region. Clarke et al. (1979) and Dickinson et al. (1984) have reported other cases in which determination of activity in whole organisms reveals differences possibly due to nonspecific factors.

Finally, consider the difficulty of assigning even well-characterized regulatory variants to specific genetic loci. The phenotypes are frequently

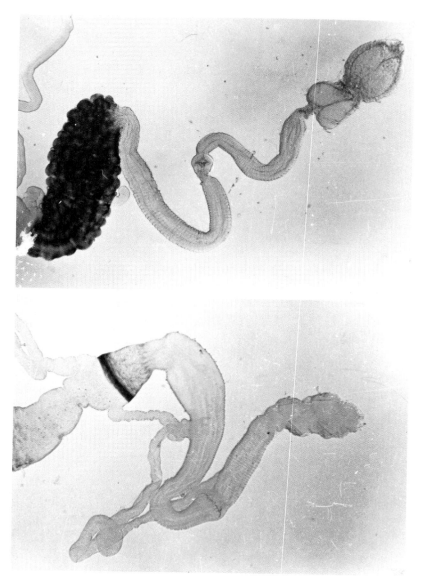

Figure 9.3. An enzyme activity difference due to morphological factors. Alcohol dehydrogenase is easily detected in extracts of hindgut from adult males, but not females, of *D. grimshawi*. Histochemical staining reveals ADH activity in a morphologically specialized enlargement posterior to the Malpighian tubules in males (**top**). Females lack the enlarged region, but cells in a corresponding position stain for ADH (**bottom**).

not easy to score, and therefore, large-scale mapping experiments are not practical. Complementation tests are not a straightforward alternative when one is confronted with a variety of regulatory phenotypes. What phenotype constitutes noncomplementation in a cross between a line that has elevated enzyme activity in larval salivary glands and another that has reduced activity in adult fat body?

Direct DNA sequence comparisons of regulatory regions eventually will be feasible (once we learn to recognize such regions) but may not be very meaningful until we understand the function of such regions well enough to recognize changes that are functionally significant (which brings us back to the *n*-dimensional problem raised above). Nor does this approach deal with the problem of identifying the complete set of regulatory genes relevant to the expression of any given structural gene.

As we have previously pointed out (Dickinson, 1980c, 1983), prospects for a meaningful analysis of the population genetics of regulatory diversity are much better if one considers only cis-acting regulatory sites. In that case, tissue- or stage-specific variation in the *relative* expression of different electrophoretic alleles in heterozygotes can reveal even subtle regulatory differences. However, cis-acting regulatory variants can be difficult to distinguish from structural gene variants (Dickinson, 1975, 1978; MacIntyre, 1982).

ADAPTIVE SIGNIFICANCE

From the foregoing discussion, it is evident that significant regulatory diversity exists in natural populations and that related species can differ dramatically in the regulation of specific genes. However, the starting point of our inquiry was the suggestion that regulatory differences are the real stuff of *adaptive* evolution. In a volume devoting considerable attention to the founder principle and related ideas on speciation, it need hardly be said that existence of extensive regulatory diversification within a rapidly evolving species group does not establish an adaptive role of those differences. For example, speciation in small populations may lead to random fixation of alternative alleles in polymorphic systems, or the new patterns may be reflections of the sort of "genetic revolution" discussed elsewhere in this volume. Indeed, incompatible regulatory programs are likely to be at least partially responsible for the developmental failures that characterize many interspecific hybrids (Parker et al., 1985a, b). This could be true even if expressed regulatory *differences* have no direct adaptive value. Thus, it will be important to seek direct evidence on the adaptive significance of specific regulatory differences. This is not a trivial problem.

In a few instances where prior knowledge has told us where to look, plausible cases for adaptive significance have been made. We have described a difference in regulation of alcohol dehydrogenase (ADH) between *D. melanogaster* and *D. simulans* that seems to relate to significant differences in ethanol tolerance which, in turn, relate to well-known differences in microhabitat selection (Dickinson et al., 1984). Dobson et al. (1984) have related the high level of stomach lysozyme in ruminants to the distinctive digestive system of that group. Their case is strengthened by their finding of a similar adaptation in leaf eating monkeys. Allendorf et al. (1983) have reported growth rate differences associated with a variant affecting phosphoglucomutase expression in liver of rainbow trout and have argued that these differences have adaptive significance. However, this case is not conclusive because of possible confounding effects of genes in linkage disequilibrium and because of the lack of a clear relationship of the alternative alleles to *different* adaptations. It is worth noting that the advent of germline transformation (Rubin and Spradling, 1982) offers an opportunity to test some hypotheses. For example, one might link the *D. melanogaster* ADH structural gene to the *D. simulans* regulatory region and transform it into a *D. melanogaster* ADH-null stock. If transformants showed the *D. simulans* pattern of ethanol sensitivity, the importance of the regulatory difference would be convincingly established. Note, however, that a negative result would not be conclusive. Many adaptively important regulatory differences might function only as part of an integrated set of differences.

In general, we can expect it to be difficult to establish whether or not a given regulatory difference has any adaptive significance. In particular, negative results may never be very significant; one simply may not have discovered the relevant environmental parameter(s). Nevertheless, some observations *suggest* that many such variants are *not* adaptive. Despite the variability in patterns of regulation in Hawaiian *Drosophila*, one can detect consistently conserved features, including the presence of ADH in larval fat body, octanol dehydrogenase (ODH) and AO-1 in ovary, and AO-2 and xanthine dehydrogenase (XDH) in midgut (Dickinson, 1980a). There are even indications that some features are conserved throughout the genus *Drosophila* (Dickinson and Gaughan, 1981). The tissues in which expression is conserved are almost always also those with the highest level of expression in most species. This has led me to consider the possibility that only the conserved features "matter" to the development and/or physiology of the organism and that interspecific variability represents evolutionary "noise" in the sense that the enzyme in other tissues has no significant function. For example, perhaps substrates for ADH accumulate only in fat body or a key product of ODH is needed only in oocytes. This line of thought has implications for developmental mechanisms that are considered in the following section.

DEVELOPMENTAL IMPLICATIONS

Could it be the case for a *variety* of gene products that some aspects of the normal pattern of expression are not functionally important? One can identify both empirical and theoretical considerations consistent with this possibility. Waddington (1942, 1962) pioneered the concept of developmental canalization, in part as an explanation for the dominance of wild-type alleles over deficiencies for many essential genes. He suggested that evolution should select for developmental mechanisms that produce similar end results despite substantial perturbation. Canalization could account for the tolerance of natural quantitative variation of the sort discussed in preceding sections. That such variation is tolerated for gene products other than the commonly studied enzymes is suggested, for example, by the ability of *D. melanogaster* to develop normally with from one to three doses of most chromosome segments (and a corresponding threefold range of concentration of gene products).

I would like to suggest also that canalization could account for essentially equivalent development in the face of *qualitative* differences in gene regulation like those seen in Hawaiian *Drosophila*. This might be true even for genes thought to control fundamental developmental decisions. For example, the function of many developmental "switches" may depend on satisfaction of specific prior conditions. Expression outside this context could be irrelevant. Note that this is not in conflict with the notion that regulatory incompatibilities arising, for example, during a "genetic revolution" can be a key event in speciation. It can simultaneously be true that some features of regulation are developmentally neutral and that those features that are essential are achieved by different mechanisms in different species. In this context, I believe it is important to know whether gene products with a range of different physiological and developmental functions show the substantial qualitative regulatory variability we have found for the limited set of enzymes so far examined. Genes implicated in the control of important developmental decisions in *D. melanogaster* will be of particular interest.

Nonadaptive features of regulatory patterns are particularly likely if regulatory networks are interconnected in ways that make alterations in one component of a system likely to affect several aspects of the overall pattern. The diversity of patterns controlled by cis-acting factors in the cases studied in the Hawaiian *Drosophila* led me to suggest that cis-acting regulatory regions are complex in the sense that they contain subcomponents involved in different aspects of the developmental regulation of the associated structural gene (Dickinson, 1980c, 1983). A rapidly increasing number of molecular analyses support the same conclusion (e.g., Garabedian et al., 1985; Hayashi et al., 1985; Hiromi et al., 1985; Levis et al., 1985; Heberlein et al., 1985; Wu et al., 1986). This does not preclude interactions. For example, Yamamoto (1986) suggests a combinatorial model in which interactions between a small number of cis-acting DNA sequences and trans-acting transcription factors could produce a wide range of regulatory specificities. A model along those lines could account for the diverse patterns of regulation we see for some enzymes in the Hawaiian *Drosophila* and logically leads to the expectation that some aspects of normal regulatory patterns are nonadaptive. Specifically, a combinatorial model implies that any given regulatory element, in combination with various sets of other elements, is likely to influence expression in several developmental contexts. In that case, selection operating to produce favorable levels, timing, etc. of gene expression in one tissue might lead to associated changes in expression elsewhere, including expression in tissues where no useful function is performed by the gene product (Dickinson, 1988).

Gould and Lewontin (1979) have criticized the assumption that every aspect of morphology must have an adaptive explanation. In part, they argue that poorly understand developmental mechanisms may constrain the range of possible body plans and that developmental interrelationships may cause selection for one trait to produce nonadaptive features in some other part of the organism. In essence, what I am suggesting here is that interconnected regulatory networks could represent a molecular corollary of the same idea. I think this is particularly likely if combinatorial mechanisms along the lines proposed by Yamamoto are important. This view also could contribute to our understanding of the concept of a coadapted gene complex which, in turn, is important for understanding the kind of "genetic revolution" that may follow founder events as discussed elsewhere in this volume.

These ideas clearly are important for those interested in basic developmental mechanisms. It is frequently assumed that knowledge of the time and place of expression of products of developmentally important genes will provide clues to their functions. In some cases, patterns of expression do make sense—as with the segmental distribution of transcripts of genes like *fushi tarazu* and *engrailed*, mutations of which disturb segmentation (Hafen et al., 1984; Kornberg et al., 1985). In other cases, the relationship of transcript distribution to genetically defined functions is less obvious (Levine et al., 1983). For example, Bruce Baker (personal communication) reports that the transcript of *double sex*, a gene involved in controlling somatic sexual dimorphism in *Drosophila*, is most prevalent in a region of the midgut that displays no obvious sexual dimorphism. Determination of evolutionarily *conserved* features may be an important aid in recognizing the functionally significant aspects of expression of a given gene.

PHYLOGENETIC SIGNIFICANCE

A number of individuals have suggested that our data on gene regulation might be useful for assessing evolutionary relationships. There are a few precedents. As mentioned above, some patterns of regulation in fish correspond to major taxonomic groups, and there has been some work on use of these traits in phylogenetic analysis (Whitt et al., 1977; Ferris and Whitt, 1978; Fisher and Whitt, 1978; Shaklee and Whitt, 1981). However, there has been little serious evaluation of the general utility of regulatory differences in this context.

A well-established phylogeny like that available for the picture-winged

Drosophila offers the opportunity to test the possibilities. Simple inspection of the distribution of a number of regulatory phenotypes across the phylogeny initially suggested independent origin of similar phenotypes in a number of cases (Dickinson, 1980a), an observation that discouraged me from serious attempts at phylogenetic reconstruction based on regulatory phenotypes. Nevertheless, for this occasion we have undertaken a more systematic evaluation of this issue. Two approaches are possible: (1) One could determine how well regulatory phenotypes conform to the assumptions implicit in standard methods for inferring phylogenies, or (2) one could construct phylogenies based on regulatory phenotypes and compare them to the "standard" phylogeny. We have used our basic data set, together with the latest chromosome phylogeny (Carson and Yoon, 1982), to conduct some preliminary analyses of both types.

Felsenstein (1982, and references therein) has discussed the assumptions underlying a number of methods for phylogenetic inference. Some of the important ones are summarized here together with a few comments on how well the Hawaiian *Drosophila* data conform. Methods that construct rooted trees generally assume that the ancestral state is known. There is no real possibility of firmly establishing ancestral states in our case. Patterns of enzyme regulation are not preserved in the fossil record. We have noted that some features of the enzyme regulatory patterns are conserved throughout the picture-winged phylogeny (Dickinson, 1980a) or even widely in the genus *Drosophila* (Dickinson and Gaughan, 1981). We might assume that these conserved features represent an ancestral program and that all other expression reflects a gain of function. Alternatively, we might treat the "standard" phenotype as previously defined (Dickinson, 1980a) as ancestral. This standard was constructed by taking the *most common* state for each separate component of the total pattern. Equating "most common" to "ancestral" would imply that, for each trait, changes from the ancestral condition occurred in a minority of lineages. Neither assumption can be particularly well justified.

Most methods assume that various traits evolve independently. Failure to meet this condition results, in essence, in inappropriate weight being given to certain traits because each aspect of a pleiotropic change is counted separately. Because we have treated expression of a given enzyme in each tissue at each stage as a distinct aspect of the phenotype, validity of this assumption depends heavily on poorly understood mechanisms of gene regulation. Our results make it clear that at least *some* aspects of the developmental program of an enzyme *can* change independently of others (e.g., we find cases in which only one aspect of a program differs between closely related species) but it is far from clear that each aspect of the program can vary independently. Some combinations of traits have not been found. For example, we have seen ADH activity in both larval carcass and midgut, in neither, and in carcass alone, but never in midgut alone. The genetic analysis suggests that many differences are due to changes in complex, cis-acting regulatory regions (Dickinson,

1980c), and it appears likely that a single-step event (e.g., insertion of a transposable element, or a local sequence rearrangement) can change several aspects of the phenotype. Ongoing molecular analyses should clarify this situation. The preceding discussion of possible interconnections in regulatory networks is obviously relevant here.

Various methods make different assumptions about relative ease of gain versus loss of traits. Because we lack knowledge of ancestral states and of detailed mechanisms of regulatory change, these relative probabilities are hard to judge. Using the aforementioned "standard" phenotype as ancestral, we find that 13 of the discrete changes previously summarized (Dickinson, 1980a) are gains and 7 are losses. If the standard phylogeny is accepted, there is clear evidence for both independent gain and independent loss of the same function in different branches (see Fig. 9.4).

A method known as "polymorphism parsimony" allows one to assume that species can remain polymorphic for an innovation for some time and attempts to account for apparent independent origins of a trait on this basis. The standard phylogeny based on chromosome inversions makes use of this assumption. In a previous section, we summarized evidence that populations can be polymorphic for regulatory alleles, even those causing dramatic differences. We do not know how long such polymorphisms can persist in evolutionary time. We would like to examine larger samples of species related to some in which polymorphisms have been identified to see if we can find shared polymorphisms.

Maximum likelihood methods make it possible to use quantitative traits (as opposed to discrete character differences) in constructing phylogenies. In practice, such procedures generally assume that quantitative changes accumulate gradually by a path that can be modeled as Brownian motion. We have clear evidence that major *qualitative* shifts in regulation are controlled by cis-acting factors that segregate as single genes linked to the affected structural gene (Dickinson, 1980c, 1983). These cases seem unlikely to be well suited to treatment as continuously variable characters. We also know of trans-acting regulatory influences (Dickinson, 1980b). These have not been analyzed genetically and could reflect cumulative effects of small changes. Some data on intraspecific variation in enzyme activity are consistent with this (e.g., McDonald and Ayala, 1978; Laurie-Ahlberg et al., 1980).

Perhaps the most fundamental assumption of all methods is that establishment of any new trait is rare, and hence, that independent origin of a trait in different lineages is unlikely. However, the very extensive diversity found among the Hawaiian *Drosophila* and the indications of independent origin noted above suggest that even this assumption may not be safe.

The above considerations do not lead to a definitive choice of methods for employing regulatory data to make phylogenetic reconstructions, but do point toward unrooted "Wagner Parsimony" as appropriate. This method handles discrete character differences, makes no assumptions

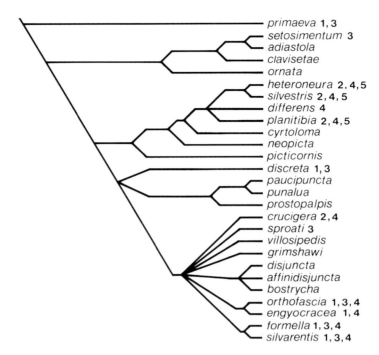

Figure 9.4. Phylogenetic relationships and regulatory phenotypes. This tree depicts relationships between species considered in our study. The relationships are inferred from data on chromosome inversions. It is based on the more extensive phylogeny given by Carson and Yoon (1982). Only the branch order is significant. Relative distances along branches do not represent relative times. Multiple branch points are not resolved by the existing inversion data. The bold numbers following some species' names identify sets of species related as follows: (**1,2**) Two sets of species identified as monophyletic groups on the basis of regulatory phenotypes. (**3,4**) Species with regulatory traits likely to represent convergence. Species marked "3" lack ADH in larval midgut, whereas those marked "4" have AO-1 present in larval salivary glands. (**5**) Species sharing a trait that may lead to resolution of a trifurcation not resolved in the chromosome phylogeny. Three species in the planitibia subgroup have AO-1 in larval hindgut, whereas *D. differens* does not. These marked examples are illustrative, not exhaustive.

about ancestral states, and allows both gain and loss of traits. The relevant algorithms attempt to find a tree with a minimum number of steps, where a "step" is gain or loss of a discrete character state. We have used the Wagner Parsimony Program contained in Felsenstein's Phylogenic Inference Package to conduct a number of preliminary tests using our data on regulatory patterns in picture-winged *Drosophila*. The data are those summarized in table 4 of Dickinson (1980a), treated as character-state data. Each species was scored as either displaying the variant character state (1) or the standard state (0). The topography of the standard tree for the relevant species is shown in Figure 9.4.

In general, trees generated with data on regulatory patterns are not consistent with the standard phylogeny. Moreover, multiple runs with different orders of data presentation (important with existing algorithms) produce different trees of similar or identical length. Nevertheless, a few features, some "correct" and some not, are found rather consistently. For example, *D. planitibia*, *D. heteroneura*, and *D. silvestris* are correctly placed in a near-terminal monophyletic group, but *D. crucigera* is incorrectly placed with them (cf. Fig. 9.4). Similarly, *D. formella* and *D. silvarentis* are always placed close together, but *D. orthofascia*, *D. engyocracia*, *D. primaeva*, and sometimes *D. discreta* are grouped with them. Many of these features also are generated by algorithms other than Wagner Parsimony.

An alternative way of evaluating the fit of regulatory traits to the chromosome phylogeny is to *specify* the chromosomal phylogeny and ask the computer to find the minimum number of steps necessary to superimpose the regulatory traits on this tree. This procedure results in a significantly longer tree—65 steps versus 46 for the shortest tree yet found. This can be accounted for by multiple independent origins of similar phenotypes. The program lists the number of times each trait must change, making it possible to identify the traits that account for the extra steps. The "guilty" steps are exactly those we previously recognized as likely to have arisen independently in different lineages.

Our operating hypothesis is that the regulatory phenotypes we have studied arise easily and can represent multiple, independent origins, and/or that superficially similar patterns with different underlying causes have been lumped together at the level of resolution so far employed. Indeed, our molecular analysis is consistent with independent origins of similar traits. For example, two species, *D. formella* and *D. hawaiiensis*, that have "lost" expression of ADH in larval midgut have apparent deletions at a similar position close to the structural gene (Rabinow and Dickinson, 1986). However, the deletions are of different size, consistent with two independent events. In addition, the two species differ with respect to other aspects of ADH expression.

If regulatory traits evolve too rapidly to be useful over the whole picture-winged phylogeny, they might resolve close relationships that are ambiguous in the chromosomal phylogeny. In this context, we have examined selected subgroups. As a test, we looked at the members of the adiastola group for which we have data. The chromosome phylogeny for this set of species is unambiguous. We usually obtain the "correct" relationships using the Wagner Parsimony Program, but two equally short "wrong" trees have been found. In the grimshawi subgroup, there is much more variability between runs and more deviations from the chromosome phylogeny. As noted above, *D. orthofascia* and *D. engyocracia* are usually placed in a monophyletic group with *D. formella* and *D. sylvarentis*, contrary to the chromosome phylogeny. Conversely, the expected monophyletic grouping of *D. bostrycha*, *D. disjuncta*, and *D. affinidisjuncta* is not evident.

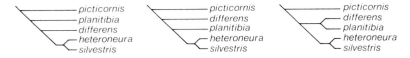

Figure 9.5. Possible phylogenetic relationships among species of the planitibia subgroup. All three arrangements are consistent with data on chromosome inversions. Data on gene regulation favor the middle arrangement.

In the planitibia subgroup, the chromosomal phylogeny is ambiguous and the three trees in Figure 9.5 are all possible. By rooting a Wagner tree at *D. picticornis*, we consistently obtain the middle pattern. We regard this as support, but certainly not definitive proof, for that resolution of the ambiguity. Addition of procedures for estimating the statistical significance of computer-generated phylogenies (Felsenstein, 1985) supports the significance of that resolution but confirms the impression that most phylogenies based on our regulatory data are not significant (Thorpe and Dickinson, 1988).

CONCLUSION

As indicated at the outset, this tour of ideas that have grown out of our initial work on Hawaiian *Drosophila* was highly speculative and provided more questions than answers. It is clear that some important questions will be difficult to answer. Nevertheless, I believe much more work on evolutionary comparisons of regulatory patterns will repay our efforts. The initial question concerning the relative importance of regulatory and structural changes in evolution remains unanswered. More comparative data, particularly comparisons of rapidly versus slowly evolving groups of organisms, could help answer that question. The same data could also have important implications for our understanding of developmental mechanisms and could be useful in phylogenetic analyses.

LITERATURE CITED

Abraham, I., and W. W. Doane. 1978. Genetic regulation of tissue-specific expression of amylase structural genes in *Drosophila melanogaster*. Proc. Nat. Acad. Sci. USA 75:4446–4450.

Allendorf, F. W., K. L. Knudsen, and R. F. Leary. 1983. Adaptive significance of differences in the tissue-specific expression of a phosphoglucomutase gene in rainbow trout. Proc. Natl. Acad. Sci. USA 80:1397–1400.

Bewley, G. C. 1981. Genetic control of the developmental program of 1-glycerol-3-phosphate dehydrogenase isozymes in *Drosophila melanogaster*: identification of a cis-acting temporal element affecting GPDH-3 expression. Dev. Genet. 2:113–129.

Carson, H. L., and K. Y. Kaneshiro. 1976. *Drosophila* of Hawaii: systematics and ecological genetics. Annu. Rev. Ecol. Systematics 7:311–345.

Carson, H. L., and J. S. Yoon. 1982. Genetics and evolution of Hawaiian *Drosophila*. In M. Ashburner, H. L. Carson, and J. N. Thompson, Jr. (eds.), The Genetics and Biology of *Drosophila*, Vol. 3b, pp. 297–344. Academic Press, New York.

Chovnick, A., W. Gelbart, M. McCarron, B. Osmund, E. P. M. Candido, and D. L. Baillie. 1976. Organization of the rosy locus in *Drosophila melanogaster:* evidence for a control element adjacent to the xanthine dehydrogenase structural element. Genetics 84:233–255.

Chovnick, A., W. Gelbart, and M. McCarron. 1977. Organization of the *rosy* locus in *Drosophila melanogaster.* Cell 11:1–10.

Clark, S. H., S. B. Daniels, C. A. Rushlow, A. H. Hilliker, and A. Chovnick. 1984. Tissue specific and pretranslational character of variants of the rosy locus control element in *Drosophila melanogaster.* Genetics 108:953–968.

Clarke, B., R. Camfield, Galvin, A., and C. Pitts. 1979. Environmental factors affecting the quantity of alcohol dehydrogenase in *Drosophila melanogaster.* Nature 280:517–518.

Dickinson, W. J. 1975. A genetic locus affecting the developmental expression of an enzyme in *Drosophila melanogaster.* Dev. Biol. 42:131–140.

Dickinson, W. J. 1978. Genetic control of enzyme expression in *Drosophila*: a locus influencing tissue specificity of aldehyde oxidase. J. Exp. Zool. 206:333–342.

Dickinson, W. J. 1980a. Evolution of patterns of gene expression in Hawaiian picture-winged *Drosophila*. J. Mol. Evol. 16:73–94.

Dickinson, W. J. 1980b. Tissue specificity of enzyme expression regulated by diffusible factors: evidence in *Drosophila* hybrids. Science 207:995–997.

Dickinson, W. J. 1980c. Complex cis-acting regulatory genes demonstrated in *Drosophila* hybrids. Dev. Genet. 1:229–240.

Dickinson, W. J. 1983. Tissue-specific allelic isozyme patterns and cis-acting developmental regulators. In M. C. Rattazzi, J. G. Scandalios, and G. S. Whitt (eds.). Isozymes, Vol. 9. pp. 107–122. Alan R. Liss, New York.

Dickinson, W. J. 1988. On the architecture of regulatory systems: evolutionary insights and implications. Bio Essays 8:204–208.

Dickinson, W. J., and H. L. Carson. 1979. Regulation of the tissue specificity of an enzyme by a cis-acting genetic element: evidence from *Drosophila* hybrids. Proc. Natl. Acad. Sci. USA 76:4559–4562.

Dickinson, W. J., and S. Gaughan. 1981. Aldehyde oxidases of *Drosophila*: contributions of several enzymes to observed activity patterns. Biochem. Genet. 19:567–583.

Dickinson, W. J., R. G. Rowan, and M. D. Brennan. 1984. Regulatory gene evolution: adaptive differences in expression of alcohol dehydrogenase in *Drosophila melanogaster* and *Drosophila simulans*. Heredity 52:215–225.

Doane, W. W. 1969. Amylase variants in *Drosophila melanogaster.* Linkage studies and characterization of enzyme extracts. J. Exp. Zool. 171:321–342.

Dobson, D. E., E. M. Prager, and A. C. Wilson. 1984. Stomach lysozymes of ruminants. J. Biol. Chem. 259:11607–11614.

Felsenstein, J. 1982. Numerical methods for inferring evolutionary trees. Q. Rev. Biol. 57:379–404.

Felsenstein, J. 1985. Confidence limits on phylogenies: an approach using the boot strap. Evolution 39:783–791.

Ferris, S. D., and G. S. Whitt. 1978. Phylogeny of tetraploid catostomid fishes based on the loss of duplicate gene expression. Syst. Zool. 17:189–206.

Ferris, S. D., and G. S. Whitt. 1979. Evolution of the differential regulation of duplicate genes after polyploidization. J. Mol. Evol. 12:267–317.

Fisher, S. E., and G. S. Whitt. 1978. Evolution of isozyme loci and their differential tissue expression. J. Mol. Evol. 12:25–55.

Garabedian, M. J., M.-C. Hung, and P. C. Wensink, 1985. Independent control elements that determine yolk protein gene expression in alternative *Drosophila* tissues. Proc. Natl. Acad. Sci. USA 82:1396–1400.

Gillespie, J. H., and C. H. Langley. 1974. A general model to account for enzyme variation in natural populations. Genetics 76:837–884.

Goldstein, D. J., C. Rogers, and H. Harris. 1982. Evolution of alkaline phosphatases in primates. Proc. Natl. Acad. Sci. USA 79:879–883.

Gould, S. J., and R. C. Lewontin. 1979. The spandrels of San Marco and the Panglossian paradigm: a critique of the adaptationist programme. Proc. R. Soc. Lond. B205:581–598.

Hafen, E., A. Kuroiwa, and W. J. Gehring. 1984. Spatial distribution of transcripts from the segmentation gene *fushi tarazu* during *Drosophila* embryonic development. Cell 37:833–841.

Hayashi, S., H. Kondoh, K. Yasuda, G. Soma, Y. Ikawa, and T. S. Okada. 1985. Tissue-specific regulation of a chicken δ-crystallin gene in mouse cells: involvement of the 5' end region. EMBO J. 9:2201–2207.

Heberlein, U., B. England, and R. Tjian. 1985. Characterization of *Drosophila* transcription factors that activate the tandem promoters of the alcohol dehydrogenase gene. Cell 41:965–977.

Hedrick, P. W., and J. F. McDonald. 1980. Regulatory gene adaptation: an evolutionary model. Heredity 45:83–97.

Hiromi, Y., A. Kuroiwa, and W. J. Gehring. 1985. Control elements of the *Drosophila* segmentation gene *fushi tarazu*. Cell 43:603–613.

Johnson, G. B. 1974. Enzyme polymorphism and metabolism. Science 184:28–37.

King, M. C., and A. C. Wilson. 1975. Evolution at two levels: molecular similarities and biological differences between humans and chimpanzees. Science 188:107–117.

Kojima, K., J. Gillespie, and Y. N. Tobari. 1970. A profile of *Drosophila* species' enzymes assayed by electrophoresis. I. Number of alleles heterozygosities and linkage disequilibrium in glucose-metabolizing systems and some other enzymes. Biochem. Genet. 4:627–637.

Kornberg, T., I. Siden, D. O'Farrell, and M. Simon. 1985. The engrailed locus of *Drosophila*: in situ localization of transcripts reveals compartment-specific expression. Cell 40:45–53.

Korochkin, L. I. 1980. Genetic regulation of isozyme patterns in *Drosophila* during development. Isozymes 4:159–202.

Laurie-Ahlberg, C. C. 1985. Genetic variation affecting the expression of enzyme-coding genes in *Drosophila:* an evolutionary perspective. Isozymes 12:33–88.

Laurie-Ahlberg, C. C., G. Maroni, G. C. Bewley, J. C. Lucchesi, and B. S. Weir. 1980. Quantitative genetic variation of enzyme activities in natural populations of *Drosophila melanogaster.* Proc. Natl. Acad. Sci. USA 77:1073–1077.

Laurie-Ahlberg, C. C., A. N. Wilton, J. W. Curtsinger, and T. H. Emigh. 1982. Naturally occurring enzyme activity variation in *Drosophila melanogaster*. I. Sources of variation for 23 enzymes. Genetics 102:191–206.

Levine, M., E. Hafen, R. L. Garber, and W. J. Gehring. 1983. Spatial distribution of *Antennapedia* transcripts during *Drosophila* development. EMBO J. 2:2037–2046.

Levis, R., T. Hazelrigg, and G. M. Rubin. 1985. Separable cis-acting control elements for expression of the white gene of *Drosophila*. EMBO J. 4:3489–3499.

Lewontin, R. C. 1974. The Genetic Basis of Evolutionary Change. Columbia University Press, New York.

MacIntyre, R. J. 1982. Regulatory genes and adaptation: past, present, and future. Evol. Biol. 15:247–285.

Markert, C. L., J. B. Shaklee, and G. S. Whitt. 1975. Evolution of a gene. Science 189:102–114.

Maroni, G., and Laurie-Ahlberg, C. 1983. Genetic control of *Adh* expression in *Drosophila melanogaster*. Genetics 105:921–933.

McDonald, J. F., and F. J. Ayala. 1978. Gene regulation in adaptive evolution. Can. J. Genet. Cytol. 20:159–175.

Paigen, K. 1979a. Genetic factors in developmental regulation. In J. G. Scandalios (ed.), Physiological Genetics. Academic Press, New York.

Paigen, K. 1979b. Temporal genes and other developmental regulators in mammals. In T. Leighton and W. F. Loomis (eds.), The Molecular Genetics of Development. Academic Press, New York.

Parker, H. R., D. P. Philipp, and G. S. Whitt. 1985a. Relative developmental success of interspecific *Lepomis* hybrids as an estimate of gene regulatory divergence between species. J. Exp. Zool. 233:451–466.

Parker, H. R., D. P. Philipp, and G. S. Whitt. 1985b. Gene regulatory divergence among species estimated by altered developmental patterns in interspecific hybrids. Mol. Biol. Evol. 2:217–250.

Powell, J. R. 1975. Protein variation in natural populations of animals. In T. Dobzhansky, M. K. Hecht, and W. C. Steere (eds.), Evolutionary Biology, Vol. 8, pp. 81–119.

Powell, J. R. 1979. Population genetics of *Drosophila* amylase. II. Geographic patterns in *D. pseudoobscura*. Genetics 92:613–622.

Powell, J. R. 1980. Population genetics of *Drosophila* amylase. III. Interspecific variation. Evolution 34:209–213.

Powell, J. R., and J. M. Lichtenfels. 1979. Population genetics of *Drosophila* amylase. I. Genetic control of tissue-specific expression in *D. pseudoobscura*. Genetics 92:603–612.

Powell, J. R., M. Rico, and M. Andjelkovic. 1980. Population genetics of *Drosophila* amylase III. Interspecific variation. Evolution 34:209–213.

Rabinow, L., and W. J. Dickinson. 1986. Complex cis-acting regulators and locus structure of *Drosophila* tissue-specific ADH variants. Genetics 112:523–537.

Rubin, G. M., and A. C. Spradling. 1982. Genetic transformation of *Drosophila* with transposable element vectors. Science 218:348–353.

Sampsell, B. M., and W. A. Held. 1985. Variation in the major urinary protein multigene family in wild-derived mice. Genetics 109:549–568.

Selander, R. K. 1976. Genic variation in natural populations. In F. J. Ayala (ed.), Molecular Evolution, pp. 21–45. Sinauer Press, Sunderland, MA.

Shaklee, J. B., and G. S. Whitt. 1981. Lactate dehydrogenase isozymes of gadiform fishes: divergent patterns of gene expression indicate a heterogeneous taxon. Copeia 1981(3):563–578.

Thorpe, P. A., and W. J. Dickinson. 1988. The use of regulatory patterns in constructing phylogenies. Syst. Zool. 37(2):97–105.

Valentine, J. W., and C. A. Campbell. 1975. Genetic regulation and the fossil record. Am. Sci. 63:673–680.

Waddington, C. H. 1942. Canalization of development and the inheritance of acquired characters. Nature 150:563–565.

Waddington, C. H. 1962. New Patterns in Genetics and Development. New York, Columbia University Press.

Wallace, B. 1963. Genetic diversity, genetic uniformity and heterosis. Can. J. Genet. Cytol. 5:239–253.

Whitt, G. S. 1983. Isozymes as probes and participants in developmental and evolutionary genetics. Isozymes 10:1–40.

Whitt, G. S., D. P. Philipp, and W. F. Childers. 1977. Aberrant gene expression during the development of hybrid sunfishes (Perciformes, Teleostei). Differentiation 9:97–109.

Wilson, A. C. 1976. Gene regulation in evolution. In F. J. Ayala (ed.), Molecular Evolution, pp. 225–234. Sinauer Press, Sunderland, MA.

Wilson, A. C., S. S. Carlson, and T. J. White. 1977. Biochemical evolution. Annu. Rev. Biochem. 46:573–649.

Wu, B. J., R. E. Kingston, and R. I. Morimoto. 1986. Human HSP70 promoter contains at least two distinct regulatory domains. Proc. Natl. Acad. Sci. USA 83:629–633.

Yamamoto, K. R. 1986. Hormone-dependent transcriptional enhancement and its implications for mechanisms of multifactor gene regulation. In L. Bogorad (ed.), Molecular Developmental Biology, pp. 131–148. Alan R. Liss, New York.

Zuckerkandl, E. 1963. Perspectives in molecular anthropology. In S. L. Washburn (ed.), Classification and Human Evolution, pp. 243–272. Aldine Publ., Chicago.

V
A CLASSICAL COUNTERPOINT

10

Modes of Speciation: The Nature and Role of Peripheral Isolates in the Origin of Species

EVIATAR NEVO

PERIPATRIC SPECIATION: PROBLEMS AND POSTULATES

How do new species arise? Are there multiple modes of speciation, or rather similar modes for different organisms, with different ecologies, demographies, and life histories? Species vary in kind, ecology, genetic system, demographic structure, and life history. It seems plausible therefore to assume that multiple modes may lead to the establishment of reproductive isolation—that is, to the origin of species. Nevertheless, the problem of origins of species is one of the oldest in evolutionary biology, and the relative importance of the multiple potential models is yet to be quantitatively estimated. The processes of speciation in animals have been reviewed recently by Mayr (1982a), who also devoted a lifetime to this problem, (Mayr, 1942, 1954, 1963, 1970, 1982a,b). Among other earlier reviews are those of Dobzhansky (1937) and Stebbins (1950). Recent ones include those of Grant (1971), Bush (1975), Endler (1977), White, (1978), Barigozzi (1982), Carson and Templeton (1984), and Barton and Charlesworth (1984).

In this chapter, I will specifically discuss the idea of "peripatric speciation" (Mayr, 1954, 1982a), that is, the notion that a new species may be formed allopatrically in peripherally semi-isolated or isolated populations. Proponents of this view (e.g. Mayr, 1954, 1982a; Carson and Templeton, 1984, and references therein) contend that under some circumstances, the founder event may set the stage for speciation by altering genetic conditions in the gene pool. By contrast, Charlesworth and Smith (1982, p. 235) conclude that "an important role of bottlenecks in population size as a cause of rapid speciation remains to be established." Mayr (1954), in his pioneering paper on the problem, assumed that a genetic

revolution takes place in the new population started by the founder. However, in a recent discussion of the problem Mayr (1982b, p. 1124) wrote:

> My systematic studies of literally thousands of peripherally isolated popu-
> lations during the preceding 25 years had shown that such a drastic change
> occurs only very occasionally. I did not claim in the least that every founder
> population experiences a genetic revolution. Neither did I claim that all or
> even most genes were genetically affected. All I claimed was that by chang-
> ing their genetic milieu the phenotypic expression and hence the selective
> values of many genes would be affected.

Based on our multidisciplinary studies conducted during the last 25
years on the active speciation and adaptive radiation of subterranean
mole rats in Israel (Nevo, 1979; reviewed in Nevo, 1982, 1985a, 1986a,b),
the discussion in this chapter will focus on the following propositions:

1. Peripheral isolates may provide cradles for speciation, besides other
   probable modes.
2. Genotypic and phenotypic variances occur at several organizational
   levels even in a small isolate and may provide the basis for the origin
   of new species.
3. Speciation does not depend on a genetic revolution.
4. The fixation of new adaptive genic and chromosomal mutations to
   the ecology of the isolate may provide ecophysiological superiority
   to the isolated population over its ancestors.
5. The origin and gradual establishment of reproductive isolation may
   be intertwined in evolution with adaptation to the new ecology of
   the isolate. The unique operation of evolutionary forces in a small
   isolate involving natural selection, genetic drift, segregation distortion, and inbreeding may enhance the initiation of reproductive isolation.
6. The success of a speciation event out of many extinct evolutionary
   experiments, particularly at the borders of the species range, depends on the availability of open ecological niches, which the ancestral parental type failed to colonize owing to its ecophysiological
   limitations.
7. Speciation, at least in the subterranean mole rats discussed here, is
   a gradual, not a punctuational process, and the establishment of
   complete reproductive isolation may take tens to hundreds of thousands of generations.

THE ACTIVE SPECIATION AND ADAPTATION
OF THE *SPALAX EHRENBERGI* COMPLEX IN ISRAEL

Subterranean mole rats of the *S. ehrenbergi* superspecies in Israel involve
four, morphologically indistinguishable, homozygous Robertsonian chromosome species each associated with and adapted to a specific climatic

regime. The four chromosomal species, $2n = 52, 54, 58$, and 60 (Wahrman et al., 1969, 1985; Nevo, 1985a), are distributed parapatrically and represent progressive late stages of chromosomal speciation. They reveal diverse patterns of incipient adaptive speciation, highlighting basic problems of speciation and adaptation. Their adaptive radiation in Israel occurred during Lower to Upper Pleistocene times, as documented by multifaceted evidence including fossils (Tchernov, 1968, 1987), immunological (Nevo and Sarich, 1974) and allozymic (Nevo and Shaw, 1972; Nevo and Cleve, 1978) distances, and major histocompatibility complex (MHC) differentiation (Nizetic et al., 1984, 1985). Evolutionary divergence time for the *S. ehrenbergi* complex, deduced from electrophoretic data, is within the last $250,000 \pm 20,000$ years and the $2n = 60$, the last derivative of speciation, originated presumably 75,000 years ago (Nevo and Cleve, 1978). These evolutionary divergence dates are in accordance with the hybrid zones (Nevo and Bar-El, 1976; Nevo, 1985a) and fossil evidences (Tchernov, 1968). The recent evidence derived from DNA–DNA hybridization (Catzeflis et al., 1988) suggests somewhat earlier divergence times, supporting an adaptive radiation starting in the Middle Pleistocene, as also indicated by the fossil record (Tchernov, 1987).

The chromosomal speciation of the *S. ehrenbergi* complex represents a remarkable case of ecological speciation in action. The climatic evolution of Israel displays, since Miocene times, a dessication trend causing desertification and savannization proceeding northward (Tchernov, 1968). The open country biota that followed the aridity trend provided the open habitats needed for the adaptive radiation and speciation of *Spalax* in Israel. The biogeographical distribution of the four chromosomal species in Israel is associated with increasing aridity and temperature stresses both southward ($2n = 52 \rightarrow 58 \rightarrow 60$) and eastwards ($2n = 52 \rightarrow 54$). Each chromosomal species is associated with a specific climatic regime characterized by a combination of humidity and temperature variables (Nevo, 1985): $2n = 52$ in the cool-humid Upper Galilee Mountains, $2n = 54$ in the cool-semidry Golan Plateau, $2n = 58$ in the warm-humid Lower Galilee Mountains and Central Israel, and $2n = 60$ in the warm-dry regions of Samaria, Judea, Northern Negev Mountains, Southern Jordan Valley, and Southern Coastal Plain (Fig. 10.1). The subterranean ecological niche provides a very effective buffering system, and yet, the subterranean microclimates of the four chromosomal species vary and they are correlated with the different macroclimates above ground (Nevo, 1985a). Thus each chromosomal species is adapted to its specific climatic microhabitat, by a multiple syndrome involving morphological, biochemical, genetical, physiological, ecological, and behavioral characteristics. The adaptive syndromes to three major ecological stresses involving adaptations to (1) the subterranean ecotope, (2) the climatic stresses, and (3) the unique burrow atmosphere have been extensively reviewed (Nevo, 1979, 1982, 1985a, 1986a,b).

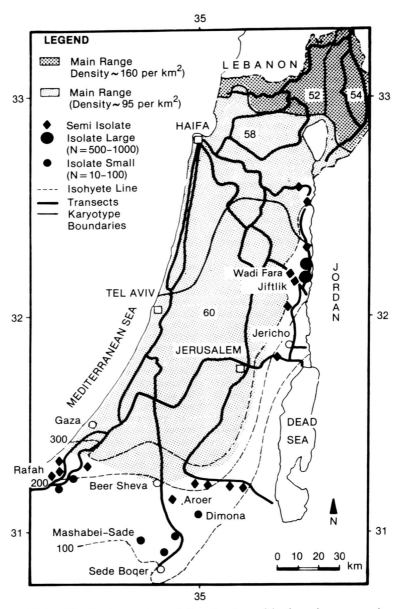

Figure 10.1. Distribution and population structure of the four chromosomal species of *Spalax ehrenbergi* complex in Israel: $2n = 52, 54, 58,$ and 60 and the peripheral semi-isolates and isolates of $2n = 60$.

THE ORIGIN AND EVOLUTION OF REPRODUCTIVE ISOLATION IN THE *SPALAX EHRENBERGI* SUPERSPECIES

The southward widening of the hybrid zones separating the four chromosomal species of *S. ehrenbergi*, from 320 m in the north to 2800 m in the south, highlights the terminalization of speciation in the complex. The four chromosomal species exhibit varying degrees of late final phases of speciation (Nevo and Bar-El, 1976; Nevo, 1985a). While hybrids are partly fertile (Nevo and Bar-El, 1976), postmating cytogenetic barriers to reproduction appear to exist in the hybrids of the *S. ehrenbergi* complex (Wahrman et al., 1985). Furthermore, premating positive-assortative mating operates primarily in the older species ($2n = 52$, 54, and 58), but not in $2n = 60$, the recent derivative of speciation, which is either 75,000 years old (Nevo and Cleve, 1978), or $200,000 \pm 40,000$ years old (Catzeflis et al., 1988). In the $2n = 60$, in contrast to the older species, mate selection based on vocal and chemical communications, is low (Heth and Nevo, 1981; Nevo et al., 1987). A deeper analysis indicates that females of $2n = 60$ are polymorphic for mate preference comprising negative, low positive, and high positive preferences for homospecific mates, in contrast to the monomorphic high positive preferences in the females of the older species (Beiles and Nevo, 1984). This pattern suggests that

1. The origin and evolution of premating isolating mechanisms start within populations as a transitory polymorphic state, evolving *gradually* into a monomorphic state.
2. The three-modal distribution of mate selection in females of $2n = 60$ is explicable even on the basis of one major gene, involving three genotypes with modifiers. Thus the establishment of reproductive isolation may be determined by a few changes in major genes without the necessity for any "genetic revolution" or a dependence on polygenic systems. Clearly, the future strengthening of reproductive isolation may incorporate additional, complementary systems including, among others, vocal, chemical, tactile, and aggression behavior (Nevo et al., 1986a; reviewed in Nevo, 1986b).
3. The accumulation and final establishment of an effective premating isolating mechanism seems to be, at least in mole rats, but possibly also in other taxa, a gradual and slow process. Extrapolation of the evidence derived from the hybrid zones and the evolution of ethological isolation suggests 300,000 years for a complete closure of a hybrid zone of mole rats (Nevo, 1985a). Hence the idea of punctuated equilibrium (Eldredge and Gould, 1972; Gould and Eldredge, 1977) is not supported by the active speciation of mole rats. Despite their relatively rapid speciation it is still a gradual process, not only in population genetics terms, but also geologically.

THE EVOLUTIONARY HISTORY OF ISOLATED POPULATIONS
OF S. EHRENBERGI

The semi-isolates and isolates at the periphery of $2n = 60$ (Fig. 10.1) represent relictual populations. The chromosomal species $2n = 60$ radiated adaptively from the southern coastal plain of Israel and northern Sinai westward into North Africa, reaching Libya, presumably during the moister Würm pluvial period (Lay and Nadler, 1972). The moist Würm, which is the pluvial Mediterranean equivalent of the last glacial period, began 70,000 years ago, around the estimated evolutionary divergence time of $2n = 60$ from its ancestor, $2n = 58$. The temperature drop of the last glacial period in the eastern Mediterranean region was significant. Horowitz (1979, and references therein) suggested a temperature drop of about 4–5°C, based on inferred composition of the vegetation from pollen spectra, which reflect a northern Mediterranean type. Similar conclusions were drawn from faunal profiles (Tchernov, 1968, 1982, and his references).

The temperature drop and consequent higher average rainfall during the Würm pluvial period in the desert regions enabled the Palaearctic flora and fauna to penetrate southward, deep into the southern Palaearctic desert belt (Tchernov, 1982). The postglacial desertification period left many of the Mediterranean and Irano-Turanian plant and animal species as glacial relicts disjunct from their main distributional ranges. Such relicts include plants (Zohary, 1973), insects, anurans, reptiles, birds, and mammals, few of which fully speciated in their relictual populations (Tchernov, 1982). The $2n = 60$ karyotype of the S. ehrenbergi superspecies persisted during the 70,000 years in a continuous range connecting the Coastal Plains of the Negev and Sinai deserts with northern Africa, but became disjunct in the post-Würm period (Lay and Nadler, 1972). The isolated populations of $2n = 60$ studied here may therefore be isolated more than 10,000 years. Although they did not speciate chromosomally into a new species, they display the population genetic model of semi-isolated and isolated populations potentially appropriate for peripatric speciation.

OBJECTIVES

The major objectives of this study were to answer the following questions: (1) What are the genetic structures of the semi-isolates and isolates as compared with the continuous central, marginal, and near-hybrid zone populations of their parental species, $2n = 60$, and with those of the older species, $2n = 52, 54,$ and 58? and (2) What are the phenotypic structures (morphological, physiological, and behavioral) of the semi-isolates and isolates as compared with their parental species and the older species?

DATA SETS

The present analysis includes 14 variables:

A. Genetical

1. Chromosome polymorphisms of p and qh of chromosome no. 1 (Nevo et al., 1988a)
2. Allozyme polymorphisms of 16 loci, 3 of which were polymorphic in $2n = 60$: esterase extracted from the liver *(Est-L)*, 6-phosphogluconate dehydrogenase *(6Pgd)*, and malate dehydrogenase *(Mdh)*
3. Mitochondrial DNA, tested by 18 restriction enzymes, 14 of which proved to be polymorphic (Yonekawa and Nevo, in preparation)
4. Ribosomal DNA polymorphisms of the 3' end of 28S rRNA genes (Suzuki et al., 1987)

B. Morphological (Nevo et al., 1986b, 1988b)

5. Body weight
6. Body length
7. Forelimb length
8. Hindlimb length

C. Physiological (reviewed in Nevo, 1986a)

9. Hematocrit concentration
10. Hemoglobin concentration
11. Carbon dioxide concentration in urine
12. Solids in urine

D. Behavioral

13. Food intake
14. Exploratory behavior (Heth et al., 1987)

Not all variables were recorded for each of the populations tested. Biological and practical limitations prohibited the analysis of large sample sizes in the semi-isolates and isolates.

THE EVIDENCE

The evidence for all populations and variables tested appear in Tables 10.1 to 10.3. Following are the major results pertaining to each of the above-mentioned variables, which will provide the comparative framework of continuous populations of the four chromosomal species (comprising three to four populations within the continuous range of each species: truly central, ecologically marginal, and near-hybrid zone populations) as compared to semi-isolates and isolates of $2n = 60$ (Table 10.1). The means and standard errors of representative genetic and phenotypic variables of the four chromosomal species, semi-isolates, and isolates appear in Table 10.2. The following major results are indicated:

Genetic Systems

Genetic Variation

Noteworthy, the semi-isolates and isolates harbor genetic polymorphism in several genetic systems. The semi-isolates harbor mitochondrial DNA (mtDNA), allozyme, and chromosome no. 1 polymorphisms (Nevo et al., 1988a); the isolates harbor polymorphisms in allozymes, several molecular systems at the DNA level, restriction fragment length polymorphisms (RFLPs), major histocompatibility complex (MHC) (Ben-Shlomo et al., 1988), aldolases (Nevo et al., submitted), and chromosomes (Table 10.2). The results for all three polymorphic systems will appear elsewhere, and for the trends in $2n = 52, 54, 58$, and 60, see Nevo and Shaw (1972) and Nevo and Cleve (1978). Here I present data of allozyme polymorphism (Table 10.2). It is noteworthy that average heterozygosity, based on 16 gene loci, is similar in the isolates, semi-isolates and in the three continuous populations of $2n = 60$. The isolated populations of Sede Boqer still retain genetic variation. Thus, despite their low effective population sizes, the isolates retain a considerable amount of genetic variation. Most importantly, we also found intraindividual variation in the chromosomal polymorphism, even in the isolated populations. Furthermore, in mtDNA, different semi-isolates and isolates habor different phenotypes.

Explanation of Genetic Variation by Climate

The genetic polymorphisms, at the chromosomal, allozymic, mtDNA, and nuclear DNA levels are correlated with climatic factors of water availability and temperature—that is, with the environment. This is true for the continuous, semi-isolated, and isolated populations. Significantly high levels of interpopulation variation were explained for a variety of genetic, chromosomal, and phenotypic (morphological, physiological, and behavioral) systems (Table 10.3). The average R^2 for two- to three-variable combinations, chosen out of 14 variables, was 0.703 ($SD = 0.154$) for the entire complex (i.e., $2n = 52, 54, 58$, and 60, semi-isolates and isolates). The multiple regression conducted only on the $2n = 60$ and its peripheral isolates and semi-isolates resulted in high and significant estimates of R^2 for many variables. Although the high R^2 may be inflated owing to the small number of populations involved, they appear to reflect

Table 10.1. The species and populations compared in this study

Chromosomal species (2n)	Sample size (n individuals)	Populations	
		Site	Type
Continuous populations			
52	3–42	Kerem Ben Zimra	Central
52	2–32	Qiryat Shemona	Marginal
52	2–44	Maalot	Near hybrid zone
54	4–29	Quneitra	Central
54	1–39	Hermon	Marginal
54	4–25	El Al	Near hybrid zone
58	3–57	Zippori	Central
58	1–25	Carmel	Central
58	2–25	Afiq	Marginal, near hybrid zone
58	2–41	Kabri	Near hybrid zone
60	2–59	Anza	Near hybrid zone
60	2–34	Jerusalem	Marginal
60	4–54	Lahav	Central
Semi-isolates			
	3–8	Wadi Fara	Eastern Samaria steppes
60	1–3	Jiftlik	Eastern Samaria steppes
	1–3	Ramat Hovav	Northern Negev steppes
	1–3	Aroer	Northern Negev steppes
	1–2	Rafah	Northern Negev steppes
Isolates			
60	2–19	Sede Boqer	Northern Negev Desert
	1	Dimona	Northern Negev Desert
	3–5	Mashabei Sade	Northern Negev Desert

lower—but nevertheless real—correlations with the environment. The climatic variables explaining both genotypic and phenotypic variances involve those of water availability, temperature, and their interaction.

Genetic Distances and Genic Differentiation Between Populations

The genetic distances and genetic differentiation, based on allozymic, chromosomal, and morphological distances, between the four chromosomal species and between the $2n = 60$ and the isolates are very small. The genetic distances (Nei, 1972), D, between the continuous, isolated, and semi-isolated populations are extremely low: mean $D = 0.007$ (range, 0.001–0.012). Likewise, genic differentiation (Nei, 1973), D_{st} proved remarkably low: mean $D_{st} = 0.003$ (range, 0.0–0.028). In other words, based on allozymic differentiation, the semi-isolates and isolates are very close

Table 10.2. Means and *SE* of genetic and phenetic variables of the four chromosomal species of *S. ehrenbergi* and the peripheral semi-isolates and isolates of $2n = 60$

	2n = 52	2n = 54	2n = 58	2n = 60	2n = 60	
	Continuous populations				Semi-isolates	Isolates
DNA polymorphism						
rDNA, no. of dif. frag.	6.00±0.0 (3)[a]	5.33±0.67 (3)	5.33±0.67 (3)	3.08±0.58 (3)	—	2.00 (1)
mtDNA KpnI. form B	—	—	—	0.00±0.00 (3)	0.25±0.25 (4)	1.00 (1)
mtDNA HpaII. form A	—	—	—	0.33±0.33 (3)	0.75±0.25 (4)	1.00±0.0 (2)
Allozyme variation						
Polymorphism, P-1%	0.083	0.250	0.125	0.146±0.021 (3)	0.047±0.030 (4)	0.063±0.0 (2)
Genetic diversity (He)	0.032	0.028	0.059	0.062±0.005 (3)	0.024±0.017 (4)	0.021±0.011 (2)
Av. heterozygosity (H)	0.008	0.014	0.015	0.035±0.030 (3)	0.023±0.014 (4)	0.037±0.038 (2)
Allele diversity (A)	1.125	1.292	1.167	1.250±0.0 (3)	1.078±0.059 (4)	1.063±0.0 (2)
6Pgd-b frequency	0.00	0.023	0.155	0.730±0.052 (3)	0.890±0.105 (4)	1.000±0.0 (2)
Chromosome polymorphism						
Heterogeneity-p	0.67±0.17 (3)	0.54±0.21 (2)	1.00±0.0 (4)	0.72±0.06 (3)	0.42±0.75 (4)	0.11±0.11 (2)
Heterogeneity-qh	0.67±0.17 (3)	0.42±0.08 (2)	1.00±0.0 (4)	0.48±0.24 (3)	0.50±0.22 (4)	0.67±0.33 (2)
Physiological variables						
Hematocrit (%)	52.50±1.67	50.40±2.27	51.60±1.47	49.97±2.75 (3)	48.00±2.10 (2)	46.40±1.82 (2)
Urine-CO_2 (g/l)	3.53±0.27 (3)	3.10±0.33 (3)	3.19±0.67 (3)	2.32±0.07 (2)	2.14±0.49 (4)	3.95±0.90 (3)
Morphological variables						
Female body weight (g)	121.3±1.45 (3)	136.3±5.36 (3)	116.7±3.93 (3)	105.7±3.76 (3)	111.1±14.87 (3)	98.5±12.72 (2)
Forelimb length (mm)	19.93±0.25	18.82±0.70	19.81±0.23	18.40±0.35	17.56±1.83 (4)	17.74±0.37 (3)
Behavioral variables						
Relative food intake[b]	0.50±0.04 (3)	0.49±0.07 (3)	0.54±0.02 (3)	0.69±0.07 (2)	0.74±0.06 (5)	0.75±0.16 (2)
Exploring new area[c]	5.55±0.77 (3)	4.64±0.19 (3)	5.72±0.72 (4)	4.85±0.41 (3)	1.26±0.38 (5)	2.25±1.39 (3)

Note: Data based on population means.　　[b]Food weight (g)/body weight (g).

[a]Number of populations in parentheses.　　[c]Total length of tunnel explored in 3 minutes (meters).

Table 10.3. Coefficients of multiple determination (R^2) with 14 dependent variables and 11 independent variables

	A. All populations							B. $2n = 60$ and peripheral isolates						
	N	X_1	X_2	X_3	R_1^2	R_2^2	R_3^2	N	X_1	X_2	X_3	R_1^2	R_2^2	R_3^2
I. DNA polymorphisms														
rDNA – No. of dif. frag.[a]	13	Rd	Tdd	Ev	0.642§	0.737§	0.779‡	4	Tj	Td	—	0.689ns	0.990§	—
mtDNA (HindII, form A)[b]	8							8	Tj	Rn	Ta	0.692‡	0.780‡	0.953‡
II. Allozyme polymorphisms														
Polymorphism P% (P-1%)	12	Rn	Td	Rd	0.472†	0.555‡	0.686†	9	Tdd	Td	Ta	0.334ns	0.534ns	0.856‡
Genetic diversity (He)	12	Tdd	Td	Ev	0.267*	0.431*	0.667‡	9	Tdd	—	Ta	0.566‡	—	—
Allelic diversity (A)	12	Rn	Td	Ev	0.484†	0.605†	0.710†	9	Rn	Td	Hu	0.583†	0.690†	0.807‡
6Pgd-a (frequency)	12	Rd	Hu	Ta	0.726§	0.800§	0.921§	9	Rn	Ev	Ta	0.301ns	0.450ns	0.648ns
III. Chromosomal polymorphisms														
Heterogeneity, p in chrom. 1	16	Rd	Tj	—	0.190*	0.666§	—	7	Rn	Alt	—	0.480*	0.859†	—
Heterogeneity, qh in chrom. 1	16	Rd	Tm	—	0.156ns	0.460†	—	7	—	—	—	—	—	—
IV. Physiological variables														
Hematocrit	9	Tdd	Alt	Tj	0.778‡	0.848‡	0.895‡	7	Tdd	Td	Ev	0.791†	0.931†	0.980†
Urine CO_2	18	Tj	Td	Alt	0.167*	0.305*	0.431†	9	Tj	—	—	0.392*	—	—
V. Morphological variables														
Average body weight	17	Td	Alt	Tm	0.254†	0.329*	0.615†	8	Td	Alt	Tm	0.814‡	0.869‡	0.920†
Forelimb length	11	Rn	Hu	Td	0.178ns	0.556†	0.757†	8	Ta	Tm	Tdd	0.431*	0.866‡	0.945‡
VI. Behavioral variables														
Relative food intake	18	Rd	Tj	Rn	0.408‡	0.469‡	0.696§	9	Tj	Ev	Tdd	0.614†	0.741†	0.817†
Exploratory behavior[c]	21	Tdd	Hu	Ev	0.434§	0.525§	0.601§	11	Tdd	Hu	Ev	0.330*	0.460†	0.686‡

Note: The 14 dependent variables include *genetic* (DNA, allozymic, and chromosomal polymorphisms) and *phenotypic* (morphological, physiological, and behavioral) variables. The 11 independent variables include climatic factors of water availability and temperature. The analysis includes 21 populations as follows: $2n = 52$ (3 of (A) all populations), $2n = 54$ (3), $2n = 58$ (4), $2n = 60$ (5), and isolates of $2n = 60$ (3). The table is comprised of two multiple regression analyses of (A) all species ($2n = 52, 54, 58, 60$ and its peripheral semi-isolates and isolates); and of (B) $2n = 60$, including continuous, semi-isolated, and isolated populations. *Symbols:* Independent variables include (1) *Geographical:* Alt, altitude (in meters). (2) *Temperature:* Tm, mean annual temperature; Ta, mean August temperature; Tj, mean January temperature; Td, mean seasonal temperature difference; Tdd, mean day-night temperature difference; Ev, mean annual evaporation. (3) *Water availability:* Rn, mean annual rainfall (in mm); Rd, mean number of rainy days; Hu, mean humidity at 14:00; Th, Thornthwaite's moisture index. [a]Number of different fragments [b]Phenotype A, by restriction enzyme HindII. [c]Total length of tunnel explored by experimental in 5 minutes (meters). *$p < .10$; †$p < .05$; ‡$p < .01$; §$p < .001$; ns$p > .10$; levels of significance. *Source:* Climatic values were taken from the Atlas of Israel (1970) and from multiyear records of the Meteorological Service of Israel.

to their ancestral continuous populations. Notably, very high genetic identity, I, between the four chromosomal species was recorded earlier (mean $I = 0.966$; range, 0.931–0.988) in Nevo and Cleve (1978).

Phenotypic Systems

For most morphological, physiological, and behavioral variables studied, the semi-isolates and isolates displayed considerable standard deviations (Table 10.2). Notably, hematocrit concentration decreases slightly, exploratory behavior and female body weight decrease distinctly, southward. This is a remarkable testimony that not only at the genetic, but most importantly also at the phenotypic levels, high interindividual variation exists in diverse quantitative characteristics.

Discriminant Analysis

Several studies of discriminant analyses (Statistical Package for the Social Sciences [SPSS-x], 1983) based on population means have been conducted, and one of them is represented graphically in Figure 10.2. The different discriminations are based on various combinations of variables including representatives of the genetic and phenotypic variables studied. Consequently, the analyses involve different population numbers since not all variables were recorded in each population. The following results are indicated:

1. In Figure 10.2, the several semi-isolates and isolates appear in the neighborhood of $2n = 60$, but are separated in space. Furthermore, the four chromosomal species appear in the diagram largely according to their spatial geographical relationships.
2. High level of correct population classifications are achieved.
3. Pairwise comparisons between all four species are highly significant, but the semi-isolates and isolates show only near-significant differences when compared with the continuous populations of $2n = 60$.

THEORETICAL CONSIDERATIONS

Classical Postulates of Peripatric Speciation

The biological species sensu Mayr (1954, 1963, 1982a) constitutes a coherent genetic unit that is unified by the stabilizing effects of gene flow, developmental homeostasis, and epistasis. The origin of new species therefore involves, according to Mayr, rapid and drastic genetic change, in a small founder population, leading from one well-integrated and stable condition to another, through a highly unstable period.

The idea of peripatric speciation was derived by Mayr (1982a) from the common observation that there is roughly an inverse relationship between

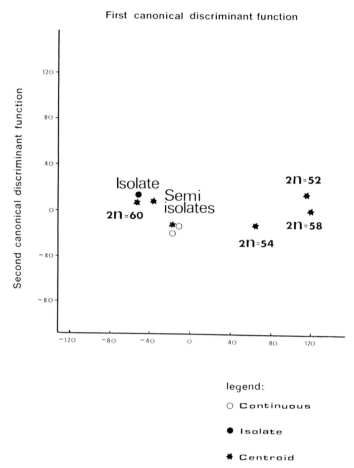

Figure 10.2.   Discriminant analysis of the four chromosomal species of the *S. ehrenbergi* superspecies and the peripheral semi-isolates and isolates of $2n = 60$. Variables used in the analysis: heterogeneity of "p" in chromosome 1; female weight at capture; exploration of new area, in 3 minutes; frequency of 6*Pgd*-a; H (= average heterozygosity; A (= mean number of alleles per locus).

the population size and rate of speciation. The idea, first stated by Mayr in 1954, suggests that the gene pool of either a founder or a relict peripheral population, can be rapidly reorganized, resulting in quick acquisitions of isolating mechanisms and an ecological shift. Thus, peripatric speciation involves populations that pass through a bottleneck in population size. Mayr envisaged peripatric speciation as a process involving the following aspects:

1. There is a considerable, if not drastic, loss of genetic variability in a founder population.
2. There is greatly increased homozygosity in the new population. This

will affect the selective value of many genes as well as the total internal balance of the genotype.

3. Conditions are sometimes provided for the occasional occurrence of a veritable genetic revolution.

4. Such founder populations can pass quickly through a condition of heterozygosity in those cases in which the heterozygotes are of lowered fitness. This is often the case with chromosomal rearrangements.

5. Such unbalanced populations may be ideally situated to shift into new niches.

6. The genetic revolution is a population phenomenon and not that of individuals.

A similar theory for founder-induced speciation was suggested by Carson (1968, 1975, 1982). It is based on a major reorganization of polygenic balances, only after a drastic reduction in effective population size. Carson has emphasized that genome reorganization is not instantaneous in any sense and may take several hundreds or several thousands of generations. His theory draws much from Wright's shifting balance theory (Wright, 1931, 1970, 1982). He has stressed the relaxation of selection pressures during population flushes and crashes, and has underlined the importance of sexual selection during speciation.

Templeton's theory (1980, 1982) emphasizes "transilience" as the important passage of a population, from one equilibrium to another through an unstable intermediate state. Templeton suggests that reproductive isolation is more likely to evolve through stochastic changes at a few major loci, followed by coadaptations at modifier loci, rather than in a single "genetic revolution." The stochastic alterations at the major loci result in a drastic fitness reweighting of the pleiotropic effects associated with the major genes. The strong selective forces that emerge after the founding event depend on the altered genetic environment determined by one or a few major loci. Speciation is regarded as a consequence of the peak-shift resulting from the above process. Notably, whereas the founder–flush model emphasizes the ecological context of the founder population, the transilience model emphasizes the importance of the genetic environment as defined by epistasis and pleiotropy.

Kaneshiro's theory (1976, 1980, 1983) suggests that ethological isolation evolves in small founder populations through the loss of components of male courtship behavior because of drift. Ancestral females will discriminate *against* males of the derived populations, whereas derived females discriminate *for* them, thus exhibiting asymmetrical sexual isolation.

To what extent does the evidence presented here for the *s. ehrenbergi* superspecies support the above-mentioned theory of peripatric speciation as made explicit by the theories of Mayr, Carson, Templeton, and Kaneshiro? I will discuss each of Mayr's points specifically, and add comments related to the other theories.

Genetic Variability

Although there is, at least in some systems, a loss of genetic variability in the semi-isolates and isolates (Table 10.2), it is by no means a drastic and an overall major genome reorganization. The average genetic distances based on 16 gene loci between the isolates and the continuous populations are very small ($D = 0.012 \pm 0.001$), and genic differentiation is minimal ($D_{st} = 0.003$). All the genetic systems studied in the semi-isolates and sometimes in the isolates display polymorphisms: chromosome no. 1 (p and qh), allozymes *(Est-L* and *6Pgd)*, and mtDNA. Furthermore, the phenotypic characteristics display considerable levels of standard deviations in morphological, physiological, and behavioral characters. Likewise, the populations of the semi-isolates are not far in their means from those of the continuous populations (Table 10.2). Thus, the idea of a drastic loss of genetic variation or of a major reorganization of polygenic balances does not characterize the isolates and semi-isolates of *S. ehrenbergi.*

It is also noteworthy that loss of heterozygosity at enzyme loci is not observed in the rapidly speciating Hawaiian *Drosophila* (Sene and Carson, 1977; Craddock and Johnson, 1979). I suspect that this represents a general pattern. It will be desirable to test in depth as many isolates as possible to find the extent to which the pattern of *S. ehrenbergi* is a general one. Theory predicts that the effects of a founder event on genetic variability, either on heterozygosity at neutral loci (Lewontin, 1965; Nei et al., 1975; Chakraborty and Nei, 1977; Templeton, 1980) or on the additive genetic variance of neutral quantitative characters (Bulmer, 1980; Lande, 1980), can be quite small. Barton and Charlesworth (1984, p. 141), suggest that "the bottleneck effect, if it exists at all, is not very significant in evolution." Notably, Carson's and Templeton's models of founder effect require high levels of genetic variation in the founder population and predict little chance of speciation if a significant drop in levels of genetic variation occurs. Their models are incompatible with genetic revolution in the founder population (Carson and Templeton, 1984).

Increased Homozygosity

This aspect was in fact partly dealt with above. Again, although a higher level of homozygosity may characterize some genetic systems, it is noteworthy that it does not affect others. Moreover, some genotypes and phenotypes are unique to the isolates, whereas others represent a natural continuation of the gene frequencies or means of quantitative characters of the continuous populations of $2n = 60$ or even that of the older species (e.g., p and qh in chromosome no. 1, *6Pgd, Est-L)*. In other words, the same polymorphisms prevalent in the continuous populations appear, at least partly, in the semi-isolates and isolates, either in lower (p of chromosome 1) or higher (qh of chromosome 1, and *6Pgd)* frequency. For example, *6Pgd* displays fixation of allele *b* in the isolates, whereas allele *a* is nearly fixed in the northern species ($2n = 52$ and 54). Thus, homozygosity is not the rule in the isolates, and it does not appear to affect the selective value of "many genes as well as the total internal balance of the genotype," as suggested by Mayr (1982a).

Apparently extreme homozygosity is unlikely, even in very small populations, and if it does occur, these few individuals are likely to become extinct rather than provide the basis for new species. Speciation depends on a substantive amount of available genetic variation in the founder population to respond to selection, as correctly argued by Barton and Charlesworth (1984) and Carson and Templeton (1984). The organismal characteristics in mole rats involving morphological, physiological, and behavioral characteristics seem to exhibit the same pattern. These characteristics show either an increase or a decrease with respect to the population means exhibited by the continuous populations of their ancestral species (i.e., $2n = 60$) or even the means characterizing the older species (i.e., $2n = 52$, 54, and 58). This pattern does not support the idea of a major reorganization of polygenic balances (Carson, 1982) and is in accordance with quantitative genetic theory (Bulmer, 1980; Lande, 1980).

Genetic Revolution

No signs of "genetic revolution" are observed in the semi-isolates or isolates of $2n = 60$ as compared with their parental species in any of the genetic or phenetic characteristics studied here (Table 10.2). Nor, for that matter, is there any hint of genetic revolution in the entire *S. ehrenbergi* complex ($2n = 52$, 54, 58, and 60) in chromosomal, allozymic, and DNA polymorphism levels (mtDNA, rDNA, MHC) or in the primary structure of amino acids of hemoglobin (Nevo, 1986a, and references therein). It is very obvious from the diverse levels studied in *S. ehrenbergi* that speciation in this, and probably in many other cases, does not involve major genic or genomic changes. This may be true not only at the structural level, but also in genetic changes at the regulatory level and in polygenic systems accompanying speciation. The principal role of genetic revolution—that is, to break the old coadapted gene complex—is apparently not realized either in the four chromosomal species or in the isolates of mole rats.

Carson's idea (1975) of a temporary relaxation of selection during phases of population crashes or flushes so that rare gamete types increase in frequency is not realized in the isolates of *S. ehrenbergi*. Conversely, it appears that strong selection for "arid adaptations" operates in the isolates. This deduction derives from two facts: (1) Allele frequencies and means of quantitative characters reveal patterns similar to the continuous populations of $2n = 60$. (2) The chromosomal, allozymic, and mtDNA polymorphisms in the isolates are correlated with and predictable by a combination of humidity and temperature variables. This pattern appears to be in accordance with theoretical expectations (Barton and Charlesworth, 1984, pp. 142–143) which challenge the population flushes and crashes and the changes in background variability as triggers of genomic reorganization and adaptive peak-shift. It is appropriate, however, to stress that a variety of experimental systems involving laboratory and natural populations seem to imitate some of the predictions of the founder-induced speciation models involving founder–flush and transilience theories (for experimental studies and their evaluation, see Carson and Templeton, 1984, pp. 122–126). It would be desirable to test these hypotheses also in other taxa since they are at present largely based on the speciation of Hawaiian *Drosophila*.

The Heterozygosity Bottleneck

The heterozygotes found in the semi-isolated and isolated populations of *S. ehrenbergi* in the Samaria and Northern Negev steppes and deserts seem to be adaptive at both the chromosomal and genic levels. This deduction derives from their correlation with climatic factors of water availability and temperature. Such extensive correlations, if demonstrated as parallel phenomena in many unrelated species, may be considered not only as correlations but also as adaptive causations (Nevo, 1983). However, as will be discussed shortly, the peripheral *small* isolates are indeed—as Mayr, Carson, and Templeton suggested—ideal sites to pass quickly through a condition of heterozygosity with lowered fitness, characterizing some chromosomal rearrangements. The chromosomal heterozygotes may *initiate* postmating reproductive isolation by means of the alternative new homokaryotype they may generate (but see below, on a case of heterosis of Robertsonian chromosomal heterozygotes).

Shift into New Niches

I fully agree that such peripheral isolates may be optimally situated to shift into new niches. This is so, not because they seemingly underwent a genetic revolution, but because of their multiple adaptations to the frontier, open environment that was earlier and is currently close to their ancestors. In all systems studied here, they do not display a drastic qualitative change but rather a quantitative tendency of either increase or decrease in gene frequencies, intimately associated with their local environment, as revealed by the multiple regression analysis (Table 10.3). Because they essentially now inhabit a desert environment, they are potentially the best candidates for colonizing increasingly more arid environments. Clearly, some qualitative changes in several genetic systems (e.g., mtDNA in *S. ehrenbergi*) may occur in addition to the quantitative changes mentioned earlier. These, however, do not constitute any genetic revolution neither of individuals nor of the population, but contribute to the ecophysiological superiority of the isolate over its ancestors, in the more arid environment.

The peripatric mode of speciation was probably the mechanism for the entire *S. ehrenbergi* superspecies. This conclusion is based on the multiple quantitative and qualitative comparisons of the four chromosomal species and on their distinct parapatric distribution (Nevo, 1985a, 1986a,b). I suggest that adaptation and speciation are intimately interwined in the *S. ehrenbergi* superspecies, as possibly in numerous other cases. New species may arise peripatrically since the isolated populations can accumulate *old* and *new* genetic changes that permit them to outcompete their ancestors in the new open niches at the geographical frontier they inhabit.

The Origin and Evolution of Postmating Reproductive Isolation in S. ehrenbergi

Our evidence derived from mole rats (Nevo and Bar-El, 1976; Wahrman et al., 1985) and theoretical considerations (Charlesworth and Smith, 1982; Barton and Charlesworth, 1984; pp. 144–147) suggest that genetic revolutions producing strong reproductive isolation by chromosomal or genic barriers in a single founder event are most unlikely. By contrast, reproductive isolation appears to evolve slowly and gradually by a sequential buildup on both empirical and theoretical grounds. The initiation and establishment of *partial* chromosomal isolation may indeed occur faster in the isolate, but the gradual accumulation of premating reproductive isolation which improves mate recognition, making it strong and effective, may take tens to hundreds of thousands of generations (Heth and Nevo, 1981).

The Origin and Evolution of Premating Ethological Isolation in S. ehrenbergi

It is important to realize that, at least in the *S. ehrenbergi* case, the chromosomal species $2n = 60$ has not yet achieved distinct ethological isolation against its ancestors. In fact, we have shown that the origin and evolution of assortative mating are slow and gradual (Heth and Nevo, 1981) and begin as a female polymorphic preference distribution (Beiles et al., 1984). If this pattern is indeed general, then reproductive isolation may start to build up even on the basis of one major gene. The slow and gradual evolution of ethological isolation in mole rats proceeds presumably from a polymorphic state involving females of three phenotypes—one preferring males of the parental population, another preferring males of the derivative species, and a middle undecisive phenotype. This evolution is presumably based on a combination of natural and sexual selection and is apparently not triggered by genetic drift as assumed in Kaneshiro's theory. (For corroborating but also contradictory theoretical aspects of the evolution of post- and premating reproductive isolation, see Nei [1976]. For other theoretical discussions of speciation, see also Maynard-Smith [1966], Templeton [1980], and Felsenstein [1981].)

The Evolutionary Forces Operating in Isolated Populations

What are the major evolutionary forces contributing to the dynamic chromosomal and/or genic evolutionary change leading to speciation in the peripheral isolates? The main evolutionary forces operating, singly or combinatorially, where chromosomal mutations lead to homozygous fixation involve the following: (1) natural selection, (2) genetic drift, (3) segregation distortion, (4) recombination modification, (5) inbreeding (Bengtsson and Bodmer, 1976; Lande, 1979; Hedrick, 1981), and (6) kinfounding (Hedrick and Levin, 1984). Lande (1979) based his model on the evidence that in most diploid organisms, inversion and translocation heterozygosity, as well as homozygosity, rarely produce phenotypic effects, other than reducing heterozygote fertility. On this basis he proposed a model in a subdivided population in which occur spontaneous chromosomal rearrangements having a heterozygote disadvantage. He concluded that the initial fixation of such rearrangements in a deme can happen only by random genetic drift "because they are selected against when in the minority, but favored when in the majority." Because interdeme migration of a few individuals per generation can prevent the local establishment of a new rearrangement, they generally arise in isolated populations. Spontaneous inversions, reciprocal translocations, and Robertsonian changes occur at rates between 1 : 1,000 to 1 : 10,000 per gamete per generation in a variety of animals. The common fixation of inversions and Robertsonian changes, as in the *S. ehrenbergi* superspecies, entailing possibly several percentage points of heterozygote disadvantage, implies that during speciation of many animal taxa effective deme sizes have been in the range of a few tens to a few hundreds of individuals. This conclusion seems to be in accordance with the speciation pattern of *S. ehrenbergi*. Nevertheless, the question arises whether genetic drift is the only framework for these evolutionary changes.

Apparent chromosomal heterosis has been proposed for the subterranean mammal *Geomys bursarius major* (Patton et al., 1980). They provide evidence indicating that in the case of *G. bursarius*, Robertsonian chromosomal polymorphism is maintained by differential viabilities of the three chromosomal morphs, with the heterozygote being favored. This conclusion appears particularly significant in view of the genic homoselection prominent in subterranean mammals (Nevo, 1979). The *Geomys* evidence suggests that at least in some cases, the fitness in a heterozygote may outweigh meiotic problems. If so, the heterozygote meiotic bottleneck may not invariably be an impediment to chromosomal evolution as prior discussions, including Lande's (1979) model, suggested. In fact, we have recently demonstrated extensive chromosomal polymorphisms in chromosome no. 1 of all four chromosomal species of *S. ehrenbergi*, including the semi-isolates and isolates. These polymorphisms appear to be adaptive, and maintained, at least partly, by natural selection (Nevo et al., 1988). A canalization model of chromosomal evolution that emphasizes the importance of the adaptive nature of the karyotype has been recently described (Bickham and Baker, 1979). Chromosomal polymorphisms such as those described above in *Geomys* and *Spalax,* whose advantages outweigh the effects of small population size and high potential of inbreeding, may be important for speciation.

It would be desirable to assess the relative roles and evolutionary rates played by the different mechanisms leading to homozygous chromosomal fixation—that is, random drift, segregation distortion, viability advantage, and recombination modification. In contrast to Lande's (1979) conclusion, Bengtsson and Bodmer (1976) took the view that "fixation of chromosome mutations by drift only occurs under special, and presumably very rare circumstances." From his mathematical analysis, Hedrick (1981) proposed that theoretically there are four potentially important mechanisms: (1) meiotic drive alone, (2) meiotic drive plus genetic drift, (3) inbreeding plus a selective advantage of the new homokaryotype, and (4) inbreeding plus genetic drift. He believes that other factors, such as the selective advantage of the new homokaryotype alone, appear to be of relatively less potential importance. Walsh (1982) stated that "in populations with small or moderate effective population size, in the absence of strong homozygote advantage or drive, chromosomally induced isolation can occur only by the fixation of many weakly underdominant rearrangements." Notably, in his more recent analysis of quantitative genetic models of phenotypic evolution during allopatric speciation in small isolated populations, Lande (1980) concluded that "data on the strength of natural selection and the spontaneous mutability of quantitative characters, in conjunction with the models, provide a feasible microevolutionary mechanism for such substantial and geologically rapid phenotypic evolution in small isolated populations."

The Genetic and Phenotypic Structures of Mole Rats in the Isolates: Nature and Role in Speciation

Our data on the peripheral semi-isolates and isolates of the $2n = 60$ chromosomal species indicate that demes with effective sizes of the order of a hundred individuals or less, can maintain enough genetic variability. The latter involves genotypic variation at the protein and DNA levels, and phenotypic variation at the morphological, physiological, and behavioral levels. This variation enables substantial future evolution in the intertwined processes of both adaptation and speciation. Those isolates that do not become extinct, as apparently most of them do, and enter a new adaptive zone, can either develop new adaptations to the new zone, and/ or speciate. Neither adaptation nor speciation demands a genetic revolution (i.e., a substantive and extensive genic and/or genomic change). In fact, by its initial marginal origin, the peripheral isolate already involved a multiple syndrome of adapatations at all levels, enabling it to survive in the periphery. Owing to the nature of the small isolate, additional new adaptations may arise and become rapidly established there. Furthermore, if initial pre- or postmating isolating mechanisms arise, based on relatively few major chromosomal and/or genic mutations, reproductive isolation may emerge, and may evolve gradually again *without major genomic reorganization*. We have earlier argued that if an assortative mating locus combines with sexual selection of the frequent male adapted optimally to the local environment of the isolate, then speciation and adaptation will be tightly linked in the evolution of mole rats (Beiles et al., 1984).

It is noteworthy that the scenario described above is based neither on any genetic revolution as described by Mayr (1954, 1963; see also his restricting qualification of a genetic revolution, 1982b, p. 1124) nor on major reorganization of polygenic balances (Carson, 1982). Nor does it necessarily depend on transilience (Templeton, 1980), which is based on drift-induced shifts in allele frequencies at one or more major loci that have many pleiotropic effects. Speciation in mole rats does not seem to reflect a "quantum evolution" as described by Simpson (1944, 1953), or the "punctuated equilibrium" model of Gould and Eldredge (1977).

The important features of the scenario in mole rats are:

1. The isolate harbors enough genetic variability for future evolution.
2. The genetic variability of the isolate involves a considerable amount of *genetic variation adaptive to its local environment*, representing a quantitative extension of, as well as a qualitative novelty to, its stressful ecology.
3. No signs of drastic genic and/or genomic changes qualifiable as a genetic revolution are discernible in the isolated population at either the genotypic or the phenotypic level, tested hitherto. By contrast, most of its genetic structures and genome organization are very similar to those of its ancestral species, as well as to those of the older ancestral species ($2n = 52, 54$).
4. By the nature of the population (i.e., isolation and small effective population size) and by its ecological locale (i.e., extreme ecological marginality), it provides an optimal site for novel evolutionary experiments, a point emphasized by Mayr (1954).
5. The novel experimentation may involve chromosomal, genic, and DNA polymorphisms, which are at least partly adapted to the new ecology.
6. Few genetic changes—either chromosomal changes, such as new homozygous Robertsonian karyotypes, or a polymorphic assortative mating locus—in mole rats could initiate the evolution of reproductive isolation.
7. The evolution of a new species does not necessarily involve a "quantum evolution" or "punctuated equilibrium." On the contrary, the new isolate does evolve novel adaptations to its unique environment, but the evolution of both pre- and postmating reproductive

isolation appear to be gradual and slow, in population genetics terms as well as geologically.

8. The above scenario does not imply major genomic organization and uniquely rapid evolution. The operating mechanisms at the microevolutionary levels leading gradually to macroevolutionary change involve primarily natural selection coupled with genetic drift. The orienting mechanism, however, even in the small population is natural selection. However, as rightly stressed by Wright (1931), genetic drift may prove to be a constructive as well as a detrimental agent in driving evolutionary change, by exposing new genetic combinations to the scrutiny of natural selection.

How does this scenario relate to earlier and recent (Barton and Charlesworth, 1984; Carson and Templeton, 1984) discussions of speciation? It is in accordance with the idea that peripherally isolated populations, or founder events, may provide potential cradles of peripatric speciation, as argued by Mayr (1954, 1982a) and Carson and Templeton (1984), but challenged by Barton and Charlesworth (1984) on theoretical and some empirical grounds. Yet, I wish to emphasize that we are still totally ignorant as to the relative importance of this peripatric mode versus all other possible mechanisms, including allopatric, stasipatric, or sympatric modes of speciation (see Bush and Howard, 1986). The peripatric mode of speciation suggested here for mole rats, and apparently also applicable to many other organisms, does not depends on genetic revolution (Mayr, 1954), on genetic transilience (Templeton, 1980, 1981, 1982), or on a drastic reorganization of coadapated gene combination as argued in the founder–flush model of speciation (Carson, 1968, 1975, 1982). Whereas genetic drift may be an enhancer or promoter of speciation, the major orienting force in evolutionary divergence is natural selection, based on the available genetic diversity within and between populations.

The novel genetic changes needed for the adaptations to the new environment of the isolate, and then for radiating into the open ecological niche, are rather limited. They are based primarily on *previously existing* genetic polymorphisms and on polygenically determined quantitative traits that undergo both quantitative and qualitative evolutionary change. Reproductive isolation may evolve either at the early evolutionary history of the isolate, or later, after the multiple adaptive syndrome emerges. In any event, it does not evolve in a burst, but rather is built up slowly and gradually as a by-product of the divergence of gene pools (Heth and Nevo, 1981). Evolutionary divergence usually proceeds by an extension of microevolution into macroevolution without any need for major genomic changes or reorganization, as rightly argued by Charlesworth et al. (1982). In other words, speciation may not be substantially distinct from phyletic evolution, except in the gradual acquisition of reproductive isolation. Adaptation and speciation appear to be intertwined and explicable by the combination of the well-known Neo-Darwinian evolutionary forces: natural selection, mutation (in its broadest sense), genetic drift, migration, and reproductive isolation.

CONCLUSIONS AND PROSPECTS

Rates of Speciation

Despite their relatively rapid chromosomal evolution which certainly provided momentum to their speciation, mole rats do not reflect the phenomenon of punctuated equilibrium (Gould and Eldredge, 1977). The evidence derived from natural hybridization and the rate of accumulation of ethological assortative mating is discordant with punctuationism. Or, should we regard 300,000 generations as a burst? Similar theoretical conclusions for the gradual build up of reproductive isolation are also argued by Barton and Charlesworth (1984). Furthermore, while mole rat morphology appears to be conservative (Nevo et al., 1986b, 1988b), genetics, physiology, and behavior reflect distinct diversity, far from reflecting any stasis (reviewed in Nevo, 1986a,b). Mole rat speciation may have been rapid, but not punctuational. It certainly fits into the framework of Neo-Darwinian evolution. The latter conceives of a widespread spectrum of speciation and evolutionary rates from the gradual to the punctuational extremes (see Simpson, 1944; and a recent critical review in Bengtsson, 1980, and in Charlesworth et al., 1982; see also Newman et al., 1985).

Genetic Differentiation During Speciation

The idea that speciation depends on genetic revolution, which presumably liberates the species from its past epistatic constraints (Mayr, 1954), has been reviewed critically by Barton and Charlesworth (1984) and by Carson and Templeton (1984). Both reviews conclude that genetic revolutions are unlikely to occur during speciation. Earlier, I argued on allozymic evidence that genetic revolution does not accompany the active speciation of mole rats (Nevo and Shaw, 1972; Nevo and Cleve, 1978), pocket gophers (Nevo et al., 1974), or other rodents (Nevo, 1985b). Similar conclusions have been drawn for other organisms by Ayala (1975) and Avise (1976). (For a comparative summary of genetic distances in the vertebrates see Avise and Aquadro, 1982). We now have additional evidence derived from DNA polymorphisms including rDNA, mtDNA, and MHC, as well as from the primary structure of hemoglobins (reviewed in Nevo, 1986a), supporting the conclusion that on the basis of hitherto analyzed genotypic and phenotypic systems no genetic revolution was involved in mole rat speciation.

Linkage-group conservation appears to be the prevailing rule in vertebrate evolution (Ohno, 1984). Conservation of mammalian X-linkage groups in toto originally proposed by Ohno et al. (1964) has now been extended to large blocks of autosomal linkage groups as well (Ohno, 1973, 1984). Chromosomal homologies consisting of large blocks of autosomal linkage groups have been found between very remote species such as the human and the domestic cat (Nash and O'Brien, 1982), and the human and the mouse (Nadeau and Taylor, 1984); some linkage groups extend also from fish to human (Ohno, 1984). If substantiated in the future, such conservation of large blocks of autosomal linkage groups from fish to mammals may provide indisputable genetic evidence against the idea of genetic revolution. Speciation may indeed be accompanied by relatively few genomic changes. Even future evolutionary divergence, after the establishment of reproductive isolation, which involves the accumulation of genetic changes, apparently does not need dramatic changes at the DNA level. The molecular evidence of *Spalax* to date, involving variation of both protein and DNA, supports this conclusion. A recent discussion on molecular biological mechanisms of speciation appears in Rose and Doolittle (1983). The possible impacts of molecular drive (Dover, 1982) on DNA family turnover, and the evolution of chromosomes are discussed by Dover et al. (1984). The three basic mechanisms of interchromosomal transfer are unequal exchange, gene conversion, and transposition. The extent of their genetic potential for speciation through concerted evolution is yet to be determined by studying diverse taxa undergoing active speciation.

Micro- and Macroevolutionary Processes

The evidence derived from actively speciating taxa such as Hawaiian drosophilids (Carson and Templeton, 1984), African cichlids (Greenwood, 1979, 1984), and mole rats (Nevo, 1982, 1985a, 1986a,b, and the present discussion) does not lend support to a distinct dichotomy of micro- from macroevolution. There appears to be no empirical evidence or theoretical reason to believe that macroevolution is a distinct evolutionary process from microevolution. No extraordinary genetic or phenotypic processes appear to be involved in the origin of species. Although speciation in Hawaiian drosophilids is often associated with radical evolutionary transformation which renders species uniqueness in morphology, behavior, and ecology, Carson's and Templeton's models of founder-induced speciation do not dissociate micro- from macroevolution. Conversely, the relatively rare founder-event mode of speciation can sometimes have great macroevolutionary significance (Carson and Templeton, 1984; Templeton, 1986).

Neo-Darwinian theory is supported by the available evidence and can also accommodate the phenomena at the molecular level (Fitch, 1982). Even molecular-drive mechanisms (Dover, 1982; Dover et al., 1984), which can be both random and directional in activity, leading to concerted evolution, are not violating Darwin's fundamental concepts. Like genetic drift, molecular drive cannot alone orient evolutionary processes of adaptation and speciation. Molecular mechanisms of unequal exchange, gene conversion, and transposition, as well as other mechanisms, may be considered mutations in a broader sense, thus enlarging the reservoir of genetic variation upon which the Darwinian selection acts when the environment changes. Darwinism underscores the gradual transformation of a species over space and/or time by natural selection operating on the genetic variability in the population. Neither the neutral theory of molecular evolution (Kimura, 1983), nor the new molecular turnover mechanisms that are expanding our knowledge of genes, gene families, and mutations, imply non-Darwinian evolution. The expansion of our knowledge at the molecular and organismal levels can be accommodated under the Darwinian worldview, in contrast to Gould's assertions (Gould, 1982). The real challenge to Darwinism is not in the need for a major theoretical change, but in the critical multidisciplinary analyses of diverse taxa at all organizational levels, both within and between species. For speciation theory, such critical analyses should be conducted on *many* organisms currently undergoing active speciation, in an attempt to assess the *relative importance* of the different modes of speciation in nature.

Multiple Modes of Speciation

Doubtlessly, other modes of speciation exist beside peripatric speciation (see table 1 in Barton and Charlesworth, 1984, for additional speciation modes, and see also Bush and Howard, 1986). It is indeed inconceivable that the major innovation of organic nature—that is, *diversity*—is limited to adaptation and not exploited also in speciation. Our ignorance here is collosal: We do not know how many mechanisms of speciation exist in nature, and what their relative importance is. Although peripheral isolates appear potential cradles of speciation, other plausible mechanisms certainly exist in nature (Bush, 1975; Endler, 1977; Lande, 1982; Bush and Howard, 1986). In order to approach a more predictable and falsifiable speciation theory, it is imperative to explore multidisciplinarily, at all levels from the molecular to the organismal, a *variety* of ecological, demographic, life history, and taxonomic cases. Such an intensive and extensive experimental approach, *particularly in many actively speciating groups*, could provide us with the indispensable genotypic and phenotypic data needed to select from the potential alternatives the probable mechanism of speciation for each specific case. Only then will the relative importance of the diverse potential mechanisms of speciation approach an educated estimation, and can some limited predictability be attached to speciation theory.

SUMMARY

Subterranean mole rats of the *S. ehrenbergi* superspecies in Israel represent a dynamic case of ecological speciation in action (Nevo, 1985a). The complex consists of four chromosomal species ($2n$ = 52, 54, 58, and 60), which display progressive stages of late chromosomal speciation. Their adaptive radiation in Israel from the Lower Pleistocene to recent times is closely associated with fossoriality and increasing aridity. Each of the four chromosomal species is adapted genetically and phenotypically to its own climatic regime characterized by a unique combination of humidity and temperature. Populations are largely continuously distributed in their main ranges, but become semi-isolated particularly in the peripheral steppes and deserts surrounding the chromosomal species $2n$ = 60.

The present study compares and contrasts 14 genetic and phenotypic variables of the continuous populations of $2n$ = 52, 54, and 58, but particularly of $2n$ = 60 with its peripheral post-Würm relictual semi-isolates and isolates. Separated from the main range by several kilometers to tens of kilometers of inhospitable steppic and desert environments, they consist of low effective population sizes (N = 10 to several hundred individuals, about 100 in the main isolate studied). The results indicate that:

1. Semi-isolates and isolates harbor genetic polymorphisms and phenotypic variances that are similar in nature to those of the continuous populations, but lower in their levels of polymorphism.
2. Chromosomal, allozymic, and mtDNA polymorphisms are correlated with climatic factors of water availability and temperature.
3. Based on currently available molecular and organismal evidence, no signs of genetic revolution or major genome reorganizations are discernible in the isolates.

These results suggest that peripatric speciation in the isolates may occur primarily because of their multiple genetic and phenotypic adaptations to their new ecological open frontier, and to the unique combination of evolutionary forces operating in small isolated populations. These forces, involving natural selection, genetic drift, segregation distortion, and inbreeding, may facilitate, singly or in combination, the *initiation* of reproductive isolation. The latter evolves slowly and gradually until complete reproductive isolation is established, thereby intimately combining adaptation and speciation and merging micro- into macroevolution without any sign of drastic punctuationism. The relative role of peripatric speciation among actual and potential modes of speciation is still unknown; its determination will depend on future multidisciplinary evaluation of diverse taxonomic, ecological, demographic, and life history cases, particularly of *actively speciating taxa*.

ACKNOWLEDGMENTS

This chapter is dedicated to the late M. Avrahami, a dear friend, and collaborator for many years in the attempts to understand the evolution of mole rats. My deep gratitude is extended to S. Simson, G. Heth, and A. Beiles, longtime collaborators in studying *Spalax*. Special thanks are due to Avigdor Beiles for his continuous assistance in statistical analysis and discussions pertinent to this and other studies, and to Diane Kaplan and Rachel Ben-Shlomo for commenting on the manuscript. I am indebted to the Israel Discount Bank Chair of Evolutionary Biology, and to the Ancell-Teicher Foundation for Genetics and Molecular Evolution, established by Florence and Theodore Baumritter of New York, for financial support.

LITERATURE CITED

Avise, J. C. 1976. Genetic differentiation during speciation. In F. J. Ayala (ed.), Molecular Evolution, pp. 106–122. Sinauer Assoc., Sunderland MA.

Avise, J. C., and C. F. Aquadro. 1982. A comparative summary of genetic distances in the vertebrates. Evol. Biol. 15:151–185.

Ayala, F. J. 1975. Genetic differentiation during the speciation process. Evol. Biol. 8:1–78.

Barigozzi, C. (ed.). 1982. Mechanisms of speciation. Alan R. Liss, New York.

Barton, N. H., and B. Charlesworth. 1984. Genetic revolutions, founder effects, and speciation. Annu. Rev. Ecol. Systematics 15:133–164.

Beiles, A., G. Heth, and E. Nevo. 1984. Origin and evolution of assortative mating in actively speciating mole rats. Theor. Popul. Biol. 26:265–270.

Bengtsson, B. O. 1980. Role of karyotype evolution in placental mammals. Heredity 92:37–47.

Bengtsson, B. O., and W. F. Bodmer. 1976. On the increase of chromosomal mutations under random mating. Theor. Popul. Biol. 9:260–281.

Ben-Shlomo, R., F. Figueroa, J. Klein, and E. Nevo. 1988. MHC class II DNA polymorphisms within and between chromosomal species of the *Spalax ehrenbergi* superspecies in Israel. Genetics 119:141–149.

Bickham, J. W., and R. J. Baker. 1979. Canalization model of chromosome evolution. Bull. Carnegie Mus. Nat. Hist. 13:70–83.

Bulmer, M. G. 1980. The Mathematical Theory of Quantitative Genetics. Oxford University Press, Oxford.

Bush, G. 1975. Modes of animal speciation. Annu. Rev. Ecol. Systematics 6:339–364.

Bush, G. L., and D. J. Howard. 1986. Allopatric and non-allopatric speciation: assumption and evidence. In S. Karlin and E. Nevo (eds.), Evolutionary Processes and Theory, pp. 411–438. Academic Press, New York.

Carson, H. L. 1968. The population flush and its genetic consequences. In R. C. Lewontin (ed.), Population Biology and Evolution, pp. 123–137. Syracuse University Press, Syracuse, NY.

Carson, H. L. 1975. The genetics of speciation at the diploid level. Am. Natur. 109:73–92.

Carson, H. L. 1982. Speciation as a major reorganization of polygenic balances.

In C. Barigozzi (ed.), Mechanisms of Speciation, pp. 411–433. Alan R. Liss, New York.

Carson, H. L., and A. R. Templeton. 1984. Genetic revolutions in relation to speciation: the founding of new populations. Annu. Rev. Ecol. Systematics 15:97–131.

Catzeflis, F. M., E. Nevo, J. E. Ahlquist and C. G. Sibley. 1989. Relationships of the chromosomal species in the Eurasian mole rats of the *Spalax ehrenbergi* group as determined by DNA-DNA hybridization, and an estimate of the spalicid-murid divergence time. Jour. Molec. Evol. (in press).

Chakraborty, R., and M. Nei. 1977. Bottleneck effects on average heterozygosity and genetic distance with the stepwise mutation model. Evolution 31:347–356.

Charlesworth, B., and D. B. Smith. 1982. A computer model of founder effect speciation. Genet. Res. Cambridge 39:227–236.

Charlesworth, B., R. Lande, and M. Slatkin. 1982. A Neo-Darwinian commentary on macroevolution. Evolution 36:474–498.

Craddock, E. M., and N. E. Johnson. 1979. Genetic variation in Hawaiian *Drosophila*. V. Chromosomal variation and allozymic diversity in *Drosophila silvestris* and its homosequential species. Evolution 33:137–155.

Dobzhansky, T. 1937. Genetics and the Origin of Species. Columbia University Press, New York.

Dover, G. A. 1982. Molecular drive: a cohesive mode of species evolution. Nature 299:111–117.

Dover, G. A., M. Trick, T. Strachan, E. S. Coen, and S. D. M. Brown. 1984. DNA family turnover and the coevolution of chromosomes. Chromosomes Today 8:229–240.

Eldredge, N., and S. J. Gould. 1972. Punctuated equilibria: an alternative to phyletic gradualism. In T. J. M. Schopf (ed.), Models in Paleobiology, pp. 82–115. Freeman, Cooper, San Francisco.

Endler, J. A. 1977. Geographic Variation, Speciation, and Clines. Princeton University Press, Princeton, NJ.

Felsenstein, J. 1981. Skepticism towards Santa Rosalia, or why are there so few kinds of animals? Evolution 35:124–138.

Fitch, W. 1982. The challenges to Darwinism since the last centennial and the impact of molecular studies. Evolution 36:1133–1143.

Gould, S. J. 1982. Darwinism and the expansion of evolutionary theory. Science 216:380–387.

Gould, S. J., and N. Eldredge. 1977. Punctuated equilibria: the tempo and mode of evolution reconsidered. Paleobiology 3:115–151.

Grant, V. 1971. Plant Speciation. Columbia University Press, New York.

Greenwood, P. H. 1979. Macroevolution—myth or reality? Biol. J. Linn. Soc. 12:293–304.

Greenwood, P. H. 1984. African cichlids and evolutioanry theories. In A. A. Echelle and J. Kornfield (eds.), Evolution of Fish Species Flocks, pp. 141–154. University of Maine at Orono Press, Orono, ME.

Hedrick, P. W. 1981. The establishment of chromosomal variants. Evolution 35:322–332.

Hedrick, P. W., and D. A. Levin. 1984. Kin-founding and the fixation of chromosomal variants. Am. Natur. 124:789–797.

Heth, G., and E. Nevo. 1981. Origin and evolution of ethological isolation in subterranean mole rats. Evolution 35:254–274.

Heth, G., E. Nevo, and A. Beiles. 1987. Adaptive exploratory behaviour: differential patterns in species and sexes of subterranean mole rats. Mammalia 51:27–37.

Horowitz, A. 1979. The quaternary of Israel. Academic Press, New York.

Kaneshiro, K. Y. 1976. Ethological isolation and phylogeny in the planitibia subgroup of Hawaiian *Drosophila*. Evolution 30:740–745.

Kaneshiro, K. Y. 1980. Sexual isolation, speciation and direction of evolution. Evolution 34:437–444.

Kaneshiro, K. Y. 1983. Sexual selection and direction of evolution in the biosystematics of Hawaiian *Drosophila*. Annu. Rev. Entomol. 28:161–178.

Kimura, M. 1983. The Neutral Theory of Molecular Evolution. Cambridge University Press, Cambridge.

Lande, R. 1979. Effective deme sizes during long-term evolution estimated from rates of chromosomal rearrangement. Evolution 33:234–251.

Lande, R. 1980. Genetic variation and phenotypic evolution during the allopatric speciation. Am. Natur. 116:463–479.

Lande, R. 1982. Rapid origin of sexual isolation and character divergence in a cline. Evolution 36:213–223.

Lay, D. M., and C. F. Nadler. 1972. Cytogenetics and origin of North African *Spalax* (Rodentia: Spalacidae). Cytogenetics 11:279–285.

Lewontin, R. C. 1965. Comments to Baker's paper. In H. G. Baker and G. L. Stebbins (eds.), The Genetics of Colonizing Species, pp. 481–484. Academic Press, New York.

Maynard Smith, J. 1966. Sympatric speciation. Am. Natur. 100:637–650.

Mayr, E. 1942. Systematics and the Origin of Species. Columbia University Press, New York.

Mayr, E. 1954. Change of genetic environment and evolution. In J. Huxley, A. C. Hardy, and E. B. Ford (eds.), Evolution as a Process, pp. 157–180. Allen & Unwin, London.

Mayr, E. 1963. Animal Species and Evolution. Harvard University Press, Cambridge, MA.

Mayr, E. 1970. Populations, Species, and Evolution. Harvard University Press, Cambridge, MA.

Mayr, E. 1982a. Processes of speciation in animals. In C. Barigozzi (ed.), Mechanisms of Speciation, pp. 1–19. Alan R. Liss, New York.

Mayr, E. 1982b. Speciation and macroevolution. Evolution 36:1119–1132.

Nadeau, J. H., and B. T. Taylor. 1984. Lengths of chromosomal segments conserved since divergence of man and mouse. Proc. Natl. Acad. Sci. USA 81:814–818.

Nash, W. G., and S. J. O'Brien. 1982. Conserved regions of homologous G-banded chromosomes between orders in mammalian evolution: carnivores and primates. Proc. Natl. Acad. Sci. USA 79:6631–6635.

Nei, M. 1972. Genetic distance between populations. Am. Natur. 106:283–292.

Nei, M. 1973. Analysis of gene diversity in subdivided populations. Proc. Natl. Acad. Sci. USA 70:3321–3323.

Nei, M. 1976. Mathematical models of speciation and genetic distance. In S. Karlin and E. Nevo (eds.), Population Genetics and Ecology, pp. 723–765. Academic Press, New York.

Nei, M., T. Maruyama, and R. Chakraborty. 1975. The bottleneck effect and genetic variability in populations. Evolution 29:1–10.

Nevo, E. 1979. Adaptive convergence and divergence of subterranean mammals. Annu. Rev. Ecol. Systematics 10:269–308.

Nevo, E. 1982. Speciation in subterranean mammals. In C. Barigozzi (ed.), Mechanisms of Speciation, pp. 191–218. Alan R. Liss, New York.

Nevo, E. 1983. Population genetics and ecology: The interface. In D. S. Bendall (ed.), Evolution from Molecules to Men, pp. 287–321. Cambridge University Press, Cambridge.

Nevo, E. 1985a. Speciation in action and adaptation in subterranean mole rats: patterns and theory. In V. Sbordoni (ed.), Animal Speciation and Contact Zones. Special volume Bull. Zool. 52:33–63.

Nevo, E. 1985b. Genetic differentiation and speciation in spiny mice, Acomys. Acta Zool. Fenn. 170:131–6.

Nevo, E. 1986a. Mechanisms of adaptive speciation at the molecular and organismal levels. In S. Karlin and E. Nevo (eds.), Evolutionary Processes and Theory, pp. 439–474. Academic Press, New York.

Nevo, E. 1986b. Evolutionary behavior genetics in active speciation and adaptation of fossorial mole rats. Proc. Intern. Meet. Variability and Behavioral Evolution Rome 23–26 Nov., 1983. Acad. Nazionale Lincei 259:39–109.

Nevo, E., and H. Bar-El. 1976. Hybridization and speciation in fossorial mole rats. Evolution 30:831–840.

Nevo, E., and H. Cleve. 1978. Genetic differentiation during speciation. Nature 275:125–126.

Nevo, E., and C. Shaw. 1972. Genetic variation in a subterranean mammal. Biochem. Genet. 7:235–241.

Nevo E., and V. Sarich. 1974. Immunology and evolution in the mole rat, Spalax. Israel J. Zool. 23:210–211.

Nevo, E., Y. J. Kim, C. Shaw, and S. C. Thaeler, Jr. 1974. Genetic variation, selection and speciation in Thomomys talpoides pocket gophers. Evolution 28:1–23.

Nevo, E., G. Heth, and A. Beiles. 1986a. Aggression patterns in adaptation and speciation of subterranean mole rats. J. Genet. 65:65–78.

Nevo, E., G. Heth, A. Beiles, and E. Frankenberg. 1987. Geographic dialects in blind subterranean mammals: the role of vocal communication in active speciation. Proc. Natl. Acad. Sci. USA 84:3312–3315.

Nevo, E., A. Beiles, G. Heth, and S. Simson. 1986b. Adaptive differentiation of body size in speciating mole rats. Oecologia, 69:327–333.

Nevo, E., E. Tchernov, and A. Beiles. 1988b. Morphometrics of speciating mole rats: adaptive differentiation in ecological speciation. Z. zool. Syst. Evolut.-forsch. 26(4):286–314.

Nevo, E., M. Corti, G. Heth, A. Beiles, and S. Simson. 1988a. Chromosomal polymorphisms in subterranean mole rats: origins and evolutionary significance. Biol. J. Linn. Soc. 33:309–22.

Nevo, E., K. Joh, K. Hori, and A. Beiles 1988b. Aldolase DNA polymorphism in subterranean mole rats: genetic differentiation and environmental correlates. (Submitted)

Newman, C. M., J. E. Cohen, and C. Kipnis. 1985. Neo-Darwinian evolution implies punctuated equilibria. Nature 315:400–401.

Nizetic, D., F. Figueroa, H. J. Miller, B. Arden, E. Nevo, and J. Klein. 1984. Major histocompatibility complex of the mole rat. I. Serological and biochemical analysis. Immunogenetics 20:443–451.

Nizetic, D., F. Figueroa, E. Nevo, and J. Klein. 1985. Major histocompatibility complex of the mole rat. II. Restriction fragment polymorphism. Immunogenetics 22:55–62.

Ohno, S. 1973. Ancient linkage groups conserved in human chromosomes and the concept of frozen accidents. Nature 244:259–262.

Ohno, S. 1984. Linkage group conservation and the notion of 24 primordial vertebrate groups. Chromosomes Today 8:268–278.

Ohno, S., W. Becak, and M. L. Becak. 1964. X-autosome ratio and the behaviour pattern of individual X-chromosomes in placental mammals. Chromosoma 15:14–30.

Patton, J. C., R. J. Baker, and H. H. Genoways. 1980. Apparent chromosomal heterosis in a fossorial mammal. Am. Natur. 116:143–146.

Rose, M. R., and W. Doolittle. 1983. Molecular biological mechanisms of speciation. Science 220:157–167.

Sene, F. M., and H. L. Carson. 1977. Genetic variation in Hawaiian *Drosophila*. IV. Allozymic similarity between *D. silvestris* and *D. heteroneura* from the island of Hawaii. Genetics 86:187–198.

Simpson, G. G. 1944. Tempo and Mode in Evolution. Columbia University Press, New York.

Simpson, G. G. 1953. The Major Features of Evolution. Columbia University Press, New York.

Statistical Package for the Social Sciences [SPSS-x]. 1983. User's Guide. McGraw-Hill, New York.

Stebbins, G. L. 1950. Variation and Evolution in plants. Columbia University Press, New York.

Suzuki, H., K. Moriwaki and E. Nevo. 1987. Ribosomal DNA (rDNA) spacer polymorphism in mole rats. Mol. Biol. Evol. 4:602–610.

Tchernov, E. 1968. Succession of Rodent Faunas During the Upper Pleistocene in Israel. Verlag P. Parey, Hamburg and Berlin.

Tchernov, E. 1982. Faunal responses to environmental changes in the Eastern Mediterranean during the last 20,000 years. In J. L. Bintiff and W. Van Zeist (eds.), Palaeoclimates, Palaeoenvironments and Human Communities in the Eastern Mediterranean Region in Later Prehistory, pp. 105–187. Oxford: BAR Int. Series 133.

Tchernov, E. 1987. The age of the Ubeidiya formation, and Early Pleistocene hominid site in the Jordan Valley, Israel. Isr. J. Earth Sci. 36:3–30.

Templeton, A. R. 1980. The theory of speciation via the founder principle. Genetics 94:1011–1038.

Templeton, A. R. 1981. Mechanisms of speciation—a population genetic approach. Annu. Rev. Ecol. Systematics 12:23–48.

Templeton, A. R. 1982. Genetic architectures of speciation. In C. Barigozzi (ed.), Mechanisms of Speciation, pp. 105–112. Alan R. Liss, New York.

Templeton, A. R. 1986. The relation between speciation mechanisms and macroevolutionary patterns. In S. Karlin and E. Nevo (eds.), Evolutionary Processes and Theory, pp. 365–390. Academic Press, New York.

Wahrman, J., R. Goitein, and E. Nevo. 1969. Mole rat *Spalax:* evolutionary significance of chromosome variation. Science 164:82–84.

Wahrman, J., C. Richler, R. Gamperl, and E. Nevo. 1985. Revisiting *Spalax:* mitotic and meiotic chromosome variability. Israel. J. Zool. 33:15–28.

Walsh, J. B. 1982. Rate of accumulation of reproductive isolation by chromosome rearrangements. Am. Natur. 120:510–532.

White, M. J. D. 1978. Modes of Speciation. W. H. Freeman, San Francisco.

Wright, S. 1931. Evolution in Mendelian populations. Genetics 16:97–190.

Wright, S. 1970. Random drift and the shifting balance theory of evolution. In K. J. Kojima (ed.), Mathematical Topics in Population Genetics, pp. 1–31. Springer Verlag, New York.

Wright, S. 1982. The shifting balance theory and macroevolution. Annu. Rev. Genet. 16:1–19.

Zohary, M. 1973. Geobotanical Foundations of the Middle East. Fischer Verlag, Stuttgart.

VI
FOUNDER EFFECTS
AND SEXUAL SELECTION

11

The Effects of Founder–Flush Cycles on Ethological Isolation in Laboratory Populations of *Drosophila*

JEFFREY R. POWELL

In several brilliantly argued contributions, Carson advances an un-
orthodox view . . .

TH. DOBZHANSKY (1972)

Can an aberration-free, sexually reproducing and cross-fertilizing
diploid species produce a single fertilized propagule which is competent to produce a new species? . . . if single founders are involved
in some way in the formation of species it should be possible through
laboratory manipulations, to bring the process under close observa-
tional and experimental scrutiny.

H. L. CARSON (1971)

As the above quotation indicates, Professor Carson clearly recognized
that one aspect of speciation via founder events is that it likely occurs
very rapidly relative to more "classical" models of speciation. Indeed,
the need for a rapid speciation process to explain the large Hawaiian *Drosophila* fauna at least partially stimulated the development of the theory.
Another crucial observation was that patterns of chromosomal variation
(inversions) were most easily reconcilable with the hypothesis that species of Hawaiian *Drosophila* were derived from one another via small
populations, perhaps even single gravid females (Carson, 1970).

Thirteen years ago, I began a laboratory experiment with "an
aberration-free, sexually reproducing and cross-fertilizing diploid species" *(Drosophila pseudoobscura)* in an attempt to determine if speciation
via founder events can be studied in controlled situations. The results
from these studies have been published elsewhere (Powell, 1978; Dodd
and Powell, 1985) and will not be presented in detail here; these papers
can be consulted for experimental details as well as for the "raw" data
on which the following discussion is based. Rather, I wish to discuss here
five aspects of the results that have not been stressed previously. These
are (1) the evolution of premating isolation in the absence of postmating

isolation; (2) the possible importance of free recombination; (3) the importance of the population flush; (4) the marked asymmetry in the induced isolation; and (5) comparison of the level of stochastically induced isolation to that obtained by selection.

EXPERIMENTAL POPULATIONS

Figure 11.1 summarizes the experiments. A few details not indicated in this figure are important. Founder events involve single pair matings—that is, a sample of four genomes. Flushes (rapid population increases) were in very large population cages and reached peaks of about 20,000 adults. The maintenance population cages support populations of 2 to 5,000 adults. When the populations were switched to bottles, they were maintained in at least two half-pint bottles, and at least 50 individuals were transferred when changing bottles; no problems were encountered in maintenance, and thus no obvious bottlenecks have occurred since March 1975. Two types of controls were used. The original hybrid population was maintained as three replicates (Origin, A, and B) which were handled the same way as the experimental founder populations were except that there were no founder-flush cycles. The second set of controls consisted of inbred lines started from the Origin but not allowed to undergo flushes; the results from these latter controls are reported in Powell (1978), and further relevant data are in Powell and Morton (1979).

RESULTS

The results of mating preference tests among these populations are summarized in Table 11.1. Mating data were analyzed by chi-square 2×2 contingency tests to determine if there were significant deviations from random mating. A second statistic (I) was used to determine the direction of deviation. This statistic is the number of homogamic (same strain) matings minus the number of heterogamic matings over the total number of matings (Merrell, 1950). A value of zero indicates random mating; positive values indicate positive assortative mating; and negative values indicate negative assortative mating. H. Levene (in Malogolowkin et al., 1965) derived the standard error for I. None of the controls (Origin, A, and B inter se, and the inbred lines inter se) gave any indication of reproductive isolation. Some combinations involving the derived founder populations did show significant isolation, more than expected by chance. Overall, there is indication of isolation, as the majority (78%) of the combinations involving founders yielded positive I values (bottom Table 11.1).

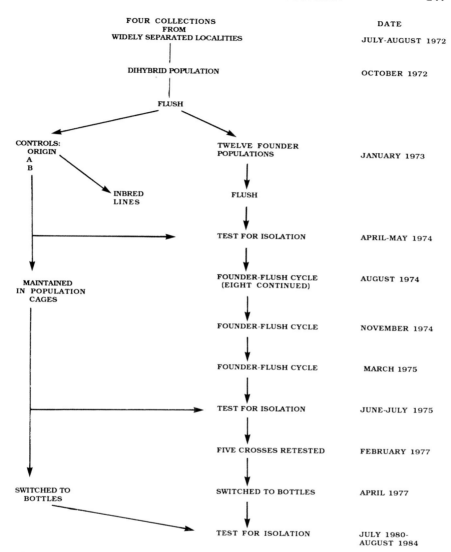

Figure 11.1. Schematic representation of laboratory experiments on the founder–flush speciation theory. Details are described in text and in Powell (1978) and Dodd and Powell (1985).

Premating Isolation in the Apparent Absence of Postmating Isolation

One of the classic models of speciation, most clearly explicated by Dobzhansky (e.g., 1941), envisions premating (ethological) isolation to evolve as a reinforcement of previously evolved postmating isolation. In the pres-

Table 11.1.   Results of mating tests in three different years

Year		No. tests	No. with significant chi-square tests		
			$.05 > p > .01$	$p < .01$	No. involving population 1
1974	control:control[a]	3	0	0	
	control:derived	12	0	0	
	derived:derived	45	0	1	1
1975	control:control	3	0	0	
	control:derived	8	1	0	1
	derived:derived	28	4	6	4
1980s	control:control	3	0	0	
	control:derived	18	2	2	3
	derived:derived	20	3	1	3

Combined[b]	I values		
	Positive	Zero	Negative
1975 and 1980s	58	4	12

Note: All significant tests were in the direction of positive assortative mating. I's cannot be calculated from the 1974 data.

[a]The controls referred to are Origin, A, and B.

[b]The lower part of the table sums all the tests involving *derived* populations; that is, controls with controls are *not* included.

ent studies, this is clearly not the case. We have tested for postmating isolation in these laboratory populations and have been unable to detect any; we looked for F_1 and F_2 sterility as well as F_2 and backcross "breakdown" (morphological abnormalities). On the other hand, several pairwise combinations have exhibited significant premating isolation (Table 11.1; also Table 11.5, below). Thus it is clear that premating isolation can evolve without detectable postmating isolation. This has also been observed in Hawaiian *Drosophila* (Ahearn et al., 1974).

What could have "caused" the premating isolation to evolve? Clearly reinforcement is not involved, because the populations were never in contact before the tests for isolation were done. Nor is it reasonable to suppose that adaptation to different environments induced behavioral changes; as far as could be controlled, all populations experienced the same environment. The most reasonable hypothesis is that the isolation is the result of stochastic processes. There are at least two major stochastic processes involved: (1) the sampling of four genomes from the previous population, and (2) the subsequent recombination among these four genomes. Thus, different founder populations will have different arrays of genotypes upon which selection can act. Any trait for which there is genetic variance in the original population may evolve differently in the founder populations. It is well established that genetic variance exists for a number of aspects of *Drosophila* mating behavior (reviewed by Petit and Ehrman, 1969; Spiess, 1970; and Spieth and Ringo, 1983). Thus it is not inconsistent with what is known about the genetics of mating behavior in *Drosophila* that, by chance, some founder populations evolved detectably different mating propensities. (As discussed later, it is not my view that the behavioral isolation evolved solely as a product of chance events; sexual selection is a crucial component.)

Possible Importance of Free Recombination

Drosophila pseudoobscura is highly polymorphic for paracentric inversions in the third chromosome; these inversions suppress recombination in heterozygotes. Only one of the founder populations, that designated 1, was homokaryotypic from the start; thus this population was the only one to experience free recombination throughout the experiment. It is also the most isolated population in the sense that it was involved in more cases of statistically significant isolation than was any other population. This is especially striking in the 1980s data (Table 11.1). Was it chance that the single population experiencing the most recombination evolved the greatest degree of isolation? To be sure, from this single experiment it is impossible to draw any firm conclusions. Nevertheless, the result is consistent with the notion that recombination is important in generating variation that might lead to the first stages of speciation. Sudden homokaryotypy (via a founder event) in a normally heterokaryotypic species, may promote the breakup of "closed genetic systems" (sensu Carson, 1975).

The Importance of Flushes

The main difference between the derived founder populations in the present study and inbred lines is that the founder populations were allowed to undergo rapid expansion in a very short period: from two individuals to 20,000 in two or three generations. This occurred in each of four founder–flush cycles. It has generally been thought (e.g., Ford and Ford, 1930; Carson, 1975) that during periods of rapid population expansion, selection is relaxed. During this time, genotypes may survive and reproduce that would never have done so under more stressful conditions. Offspring produced from such normally maladapted genotypes would be rare or absent in the ancestral population and may represent unique opportunities for selection. Phrased in terms of Wright (1982), the period of the flush is a time when adaptive landscapes become flatter and a population may be able to traverse a valley or saddle and reach a previously unattainable peak.

Asymmetrical Isolation

Kaneshiro (1976, 1980) proposed the following hypothesis, based on observations of mating propensities of Hawaiian *Drosophila:* If a population (or species) derived via a founder event exhibits ethological isolation from the ancestral population, the isolation will be asymmetrical. Females of the ancestral population will discriminate against males of the derived population, whereas females of the derived population will accept both types of males equally. In most natural situations, it is difficult to test this hypothesis as it is impossible to know unambiguously which populations are derived and which are ancestral. In the experiments being discussed here, there is no such ambiguity. Therefore it is instructive to ask whether asymmetry exists between the ancestral controls (Origin, A, and B) and the derived founder populations. Kaneshiro (1980) analyzed the data obtained in 1975 (presented in Powell, 1978) and found that indeed there was evidence for asymmetry in the predicted direction. A summary of the more recently collected data (in Dodd and Powell, 1985) is presented in Table 11.2. For all three controls, the asymmetry is in the direction predicted by Kaneshiro. Derived males are almost exactly half as successful with ancestral females as are ancestral males. Derived females accept both types of males about equally. This last point is especially important; it indicates that the lack of success of derived males with ancestral females is not due to some type of general inbreeding depression.

There is another, more natural, situation with the obscura group of *Drosophila* which lends itself to a test of the Kaneshiro hypothesis. *D. subobscura,* natively an Old World species, was recently introduced into South America (first detected in 1978) and has subsequently been found in North America. Brncic and Budnik (1984) have performed mating tests among strains from Europe and South America. Table 11.3 summarizes their results. Again the asymmetry is in the direction predicted and of nearly the same degree as observed in the laboratory studies. South American males are about one half as successful with European females as are European males. South American females accept both types of males about equally.

The issue of the interpretation of asymmetrical isolation is a controversial one, and therefore, before leaving this subject, I wish to make a few further points. First, it should be emphasized that Kaneshiro's development of his interpretation is empiracally based. Because of the geological history of the Hawaiian archipelago, very strong inferences can be drawn about the direction of evolution of Hawaiian *Drosophila*—that is, *which* species are derived from *which* ancestors. When looked for, the direction of asymmetry has been consistent with Kaneshiro's hypothesis (Kaneshiro, 1976, 1980; Ahearn, 1980). Another important observation on Hawaiian *Drosophila* concerns the demonstration that there exists genetic variance for the induction of asymmetrical isolation. Kaneshiro and Carson (1982) were able to select high discrimination and low discrimination

Table 11.2. Number of matings between control populations (Origin, A, and B) and all founder populations combined (the latter designated OT for others)

	Female	
Male	O	OT
O	139	130
OT	68	100
	A	OT
A	104	78
OT	51	82
	B	OT
B	105	67
OT	48	64

Source: Dodd and Powell (1985), 1980s tests, by permission.

Table 11.3. Results of mating tests between European (ancestral) and South American (derived) strains of *D. subobscura*

	Female	
Male	European	S. American
European	326	211
S. American	175	241

Source: Brncic and Budnik (1984), by permission.

lines from a single isofemale line of *D. silvestris* in only two generations! This is a truly remarkable indication of the level of genetic variance for mating behavior that may exist in some species.

Although all of the results cited here support the Kaneshiro hypothesis, arguments have been made and data presented in favor of exactly the opposite interpretation of asymmetrical isolation and the direction of evolution. Watanabe and Kawanishi (1979) hypothesized that derived species need to be more discriminatory to "protect" themselves from being swamped by the presumed numerically larger ancestral species. As pointed out to me by Bruce Wallace (personal communication), the Watanabe interpretation holds only if the ancestral and derived species remain at least partly in contact during the crucial stages of speciation. The data cited by Watanabe and Kawanishi are on species groups that are primarily or exclusively continental and are today largely sympatric. For such groups, there is at least the potential for partial contact during early stages of speciation. For the cases cited above in support of the Kaneshiro hypothesis, the populations or species are or were entirely allopatric during the early stages of divergence. This is certainly true for the experimental work and is almost certainly true of the natural populations as well. It is interesting to speculate that asymmetrical behavioral isolation may evolve quite differently, depending upon the biogeographical context (cf. Giddings and Templeton, 1983).

Comparisons with Selection Experiments

As argued above, the ethological isolation among populations in the founder–flush experiments was likely induced by stochastic processes. It is interesting to compare the degree of isolation so induced with what can be obtained by selection. Probably the most extensive attempt to select for ethological isolation between strains of *Drosophila* is that of Dobzhansky with *D. paulistorum* (Dobzhansky and Pavlovsky, 1971; Dobzhansky et al., 1976). Using mutants of two strains called New Llanos and Georgetown, they selected against hybrids every generation for 132 generations. A summary of their results is in Table 11.4. Complete isolation was never attained; after an initial gain up to an I of $+0.84$, there was a leveling off to about $+0.60$.

As pointed out earlier, in the laboratory founder–flush experiments, founder population 1 was the most isolated from other populations. Table 11.5 shows the isolation coefficients for this population. The level of isolation approaches, in some cases, that obtained by Dobzhansky after 132 generations of selection. The combination of populations 1 and 6 has consistently yielded high isolation; this was the one combination that gave significant isolation after only one founder–flush cycle in 1974 (Fig. 11.1 and Table 11.1). Thus over a period of nearly 10 years, this combination has displayed isolation indices of $+0.47$, $+0.53$, and $+0.40$.

Ehrman (1983 and references therein) has carried out experiments with *D. melanogaster* wherein hybridization between two strains was selected against "naturally" because of chromosomal rearrangements. The results are remarkably similar to those of Dobzhansky. Behavioral isolation of a moderate degree ($I = 0.4$ to 0.6) can evolve fairly rapidly—in the case of Ehrman's work, 14 generations. Thereafter the degree of isolation fluctuates around these values for several years.

At first blush it seems remarkable that stochastically induced isolation is nearly as great as that obtained by many generations of selection. However, considering the protocols of the experiments, perhaps it is not so surprising. The importance of bottlenecks, flushes, and recombination were emphasized above. In Dobzhansky's selection experiments, none of these factors would have occurred to any great extent. Rather large populations were used, and no purposeful bottlenecks or flushes occurred. Further, *D. paulistorum* is highly polymorphic for inversions in all arms of all chromosomes. Apparently, no attempt was made to control for lev-

Table 11.4.   Results of selection for ethological isolation
between two strains of D. *paulistorum*

Generation	$I \pm SE$
0	+0.12 ± 0.10
12	+0.52 ± 0.08
27	+0.59 ± 0.07
50	+0.84 ± 0.05
64	+0.82 ± 0.05
90	+0.74 ± 0.06
100	+0.65 ± 0.05
118	+0.63 ± 0.07
132	+0.63 ± 0.07

Source: Dobzhansky et al. (1976), by permission.

Table 11.5.   Results of mating test of founder
population 1 with controls (O, A, and B) and other numbered
founder populations

Population 1 with:	$I \pm SE$	
	1977	1980s
O	+0.28 ± 0.011	+0.28 ± 0.13
A	—[a]	+0.21 ± 0.11
B	—[a]	+0.41 ± 0.12
2	+0.52 ± 0.10	+0.27 ± 0.13
3	+0.29 ± 0.12	+0.29 ± 0.12
4	+0.31 ± 0.11	+0.13 ± 0.11
5	+0.16 ± 0.12	0.00 ± 0.12
6	+0.53 ± 0.11	+0.40 ± 0.10
7	+0.13 ± 0.12	+0.25 ± 0.12
8	+0.22 ± 0.14	+0.33 ± 0.12

[a]Tests A and B were not performed in 1977.

els of inversion polymorphism in the experiments. However, it is likely
that recombination was suppressed at least partially and perhaps rather
greatly; likewise with Ehrman's strains that had chromosomal rearrangements. It would be of considerable interest to repeat the selection experiments with the addition of steps designed to perturb genetic balances—
for example, with small bottlenecks, flushes, free recombination, etc.

DISCUSSION AND CONCLUSIONS

In the above sections, I have emphasized the stochastic aspect of the
induction of sexual isolation via founder–flush cycles. Here I would like
to point out explicitly that sexual selection must also play a crucial role.
Carson (1986) has pointed out that evolutionary biologists have perhaps

not appreciated the large role that sexual selection may play in evolution especially in regard to speciation. My thinking on this subject is greatly influenced by this paper.

I use the term "sexual selection" in a broad sense. It is the selective pressure for the sexes to locate and recognize each other and successfully copulate. Thus sexual selection puts pressure on a species to do at least three things: First, some mechanism must evolve to bring the sexes into physical proximity or contact. Second, the sexes (or at least one of the sexes) must be able to determine that they belong to the same species; that is, if they go through the trouble of copulating, there must be some assurance that viable and fertile offspring will result. Finally, sexual selection may act to refine mate recognition to the point that the sexes not only produce offspring, but also recognize individuals with whom they will produce the most offspring of the highest fitness. (For experimental verification that such refinements occur, see Partridge, 1980.) The strength of the selection and the opportunity to evolve these three aspects of mate recognition systems will, in turn, depend upon at least three factors:

1. The *density of conspecific individuals.* If conspecifics are very rare—say, to the point where any individual may encounter a conspecific of the opposite sex only once in its life—then selection to refine mate recognition would be weak. On the other hand, in dense populations, individuals may "sample" many members of the opposite sex and thus potentially evolve more discriminatory behavior.
2. The *number and density of closely related species.* If there are no or few nonconspecific individuals in the area, then selection pressure will not act to evolve elaborate species recognition systems.
3. The *degree of perceptable genetic variance in fitness.* To evolve intraspecific discrimination, the mate (or mates) must somehow "know" that copulating with some genotypes is better than with others.

In this context, it is instructive to think about founder–flush events. First, in a newly colonized territory, at least initially the density of the population will be low and thus the probability of encountering a conspecific will be low. Second, there may be no (or relatively few) closely related species present. Finally, because of the recent bottleneck, the genetic variance in fitness may be lessened (although just how much genetic variance is lost in a founder event depends upon several factors and may not be great if pleiotropy is common; see Carson and Templeton, 1984). Thus the three factors outlined above which can affect the intensity of sexual selection (population density, number of other species, and perceptable genetic variance in fitness) are such that sexual selection will act to simplify mate recognition systems. Under such conditions a complex system not only is not necessary but also may be outright detrimental. For example, if a female requires courtship by several males before mating, she may never mate, owing to a low density of males.

This scenario suggests that the asymmetry in isolation comes about by a new context of sexual selection. This contrasts with the initial suggestion by Kaneshiro (1976) that the asymmetry may be due to a loss of courtship "elements," due to the bottleneck. More recently Kaneshiro (1983, and Chapter 13, this volume) has expressed views similar to those reached above.

Two other points are worthy of mention. First, this effect may well be short-lived. As the density of the newly founded population increases, there is an opportunity for increased selection for more complex mate-recognition systems. Closely related species may move into the territory. There will be a gradual accumulation of intraspecific genetic variance in fitness. All these factors lead to increased intensity of sexual selection and increased complexity of the mate recognition system. Second, the simplification of mate recognition systems in the derived population or species would not be likely to occur if the derived population maintained contact with the ancestral species or was sympatric with a number of closely related species. In fact, as argued by Watanabe and Kawanishi (1979), just the opposite may occur in such circumstances.

Finally, I return to the issue of the stochastic element in the evolution of premating isolation via founder–flush cycles. Above, I used the term "stochastically induced isolation." I chose the word "induced" to indicate that stochastic events were the initiating factors, not the sole causative factors. An old established population or species will have been subjected to strong selection to evolve a stable, genetically buffered (or "homeostatic" sensu Lerner, 1954; or "closed" sensu Carson, 1975) mate recognition system. To break out of the established system requires a major perturbation of the genetic system. This may be a founder–flush cycle accompanied by free recombination, sudden complete homozygosity (Carson et al., 1977), or hybridization (Wallace et al. 1983). After such perturbations, sexual selection will be strong to establish a new mate-recognition system based on the genetic variants available. What genotypes are available from which to mold a new system will be largely the result of chance. Although one can predict that some new system will evolve, one cannot predict what it will be. Returning to Wright's (1982) landscape metaphor, populations are selected to sit on a peak that represents an established, buffered mate-recognition system. When a drastic event occurs, the population may be knocked off the peak and begin to climb another. The nature of the new peak will be determined largely by the stochastic events (i.e., genetic drift and recombination) that generate the genotypes available for sexual selection.

ACKNOWLEDGMENTS

From the inception of these experiments, Professor Hampton L. Carson has been most generous with ideas, advice, and encouragement. Several colleagues read versions of this paper and made useful comments; these were Adalgisa Caccone, Hampton Carson, Lee Ehrman, Kenneth Kaneshiro, Eliot Spiess, Alan Templeton, Bruce Wallace, and Marvin Wasserman. To all I offer thanks. Diane

M. B. Dodd collected most of the data from the 1980s. Financial support was provided by the National Science Foundation and a Genetics Training Grant from the National Institutes of Health.

LITERATURE CITED

Ahearn, J. M. 1980. Evolution of behavioral reproductive isolation in a laboratory stock of *Drosophila silvestris*. Experientia 36:63–64.

Ahearn, J. M., H. L. Carson, Th. Dobzhansky, and K. Y. Kaneshiro. 1974. Ethological isolation among three species of the planitibia subgroup of the Hawaiian *Drosophila*. Proc. Natl. Acad. Sci. USA 71:901–903.

Brncic, D., and M. Budnik. 1984. Experiments on sexual isolation between Chilean and European strains of *Drosophila subobscura*. Experientia 40:1014–1016.

Carson, H. L. 1970. Chromosome tracers of the origin of species. Science 168:1414–1418.

Carson, H. L. 1971. Speciation and the founder principle. Stadler Symp. 3:51–70.

Carson, H. L. 1975. Genetics of speciation. Am. Natur. 100:83–92.

Carson, H. L. 1986. Sexual selection and speciation. In S. Karlin and E. Nevo (eds.), Evolutionary Processes and Theory, pp. 391–409. Academic Press, New York.

Carson, H. L., and A. R. Templeton. 1984. Genetic revolutions in relation to speciation phenomena. Annu. Rev. Ecol. Systematics 15:97–132.

Carson, H. L., L. T. Teramoto, and A. R. Templeton. 1977. Behavioral differences among isogenic strains of *Drosophila mercatorum*. Behav. Genet. 7:189–197.

Dobzhansky, Th. 1941. Genetics and the Origin of Species, 2nd ed. Columbia University Press, New York.

Dobzhansky, Th. 1972. Species of *Drosophila*. Science 177:664–669.

Dobzhansky, Th., and O. Pavlovsky. 1971. Experimentally created incipient species of *Drosophila*. Nature 230:289–292.

Dobzhansky, Th., O. Pavlovsky, and J. R. Powell. 1976. Partially successful attempt to enhance reproductive isolation between semispecies of *Drosophila paulistorum*. Evolution 30:201–212.

Dodd, D. M. B., and J. R. Powell. 1985. Founder–flush speciation: an update of experimental results with *Drosophila*. Evolution 39:1388–1392.

Ehrman, L. 1983. Fourth report on natural selection for the origin of reproductive isolation. Am. Natur. 121:290–293.

Ford, H. D., and E. B. Ford. 1930. Fluctuation in numbers and its influence on variation in *Melitaea aurinia*. Trans. R. Entomol. Soc. Lond. 78:345–351.

Giddings, L. V., and A. R. Templeton. 1983. Behavioral phylogenies and the direction of evolution. Science 220:372–378.

Kaneshiro, K. Y. 1976. Ethological isolation and the phylogeny in the planitibia subgroup of Hawaiian *Drosophila*. Evolution 30:740–745.

Kaneshiro, K. Y. 1980. Sexual isolation, speciation, and the direction of evolution. Evolution 34:437–444.

Kaneshiro, K. Y. 1983. Sexual selection and the direction of evolution in the biosystematics of Hawaiian Drosophilidae. Annu. Rev. Entomol. 28:161–178.

Kaneshiro, K. Y., and H. L. Carson. 1982. Selection experiments on mating behavior in *Drosophila silvestris*. Genetics 100:s34.

Lerner, I. M. 1954. Genetic Homeostasis. Oliver and Boyd, Edinburgh.

Malogolowkin-Cohen, Ch., A. Solima-Simmons, and H. Levene. 1965. A study of sexual isolation between certain strains of *Drosophila paulistorum*. Evolution 19:95–103.

Merrell, D. J. 1950. Measurement of sexual isolation and selective mating. Evolution 4:326–331.

Partridge, L. 1980. Mate choice increases a component of fitness in fruit flies. Nature 283:290–291.

Petit, C., and L. Ehrman. 1969. Sexual selection in *Drosophila*. Evol. Biol. 3:177–224.

Powell, J. R. 1978. The founder–flush speciation theory: an experimental approach. Evolution 32:465–474.

Powell, J. R., and L. Morton. 1979. Inbreeding and mating patterns in *Drosophila pseudoobscura*. Behav. Genet. 9:425–429.

Spiess, E. B. 1970. Mating propensity and its genetic basis in *Drosophila*. In M. Hecht and W. Steere (eds.), Essays in Honor of Theodosius Dobzhansky, pp. 315–380. Appleton-Century-Crofts, New York.

Spieth, H. T., and J. M. Ringo. 1983. Mating behavior and sexual isolation in *Drosophila*. In M. Ashburner, H. L. Carson, and J. Thompson (eds.), Genetics and Biology of *Drosophila*, Vol. 3c, pp. 224–284. Academic Press, New York.

Wallace, B., M. W. Timm, and M. P. Strambi. 1983. The establishment of novel mate recognition systems in introgressive hybrid *Drosophila* populations. Evol. Biol. 16:467–488.

Watanabe, T. K., and M. Kawanishi. 1979. Mating preference and the direction of evolution in *Drosophila*. Science 205:906–907.

Wright, S. 1982. Character change, speciation, and the higher taxa. Evolution 36:427–443.

12

Origin of *Drosophila* of the Sonoran Desert Revisited: In Search of a Founder Event and the Description of a New Species in the Eremophila Complex

WILLIAM B. HEED

Island archipelagos historically have been the chief areas of interest for the study of speciation, and many of our concepts stem from these studies; however, it is not clear whether the concepts also apply to processes occurring in continental lowlands. One of the more promising approaches to gaining insights into the origin of new balanced genetic systems has been the comparison of speciation events on islands versus continents with the same kinds of organisms. In recognition of the significant contribution of Hampton L. Carson to our understanding of the complex mechanisms of speciation, the *Drosophila* of the Sonoran Desert, and their close relatives, are analyzed here, using the principles of speciation thus far garnered from the *Drosophila* of the Hawaiian Islands.

The single most important consideration to emerge from the Hawaiian *Drosophila* program is the concept of the ethospecies and its origin by a founder event (Carson, 1978). Briefly stated, an "ethospecies" is a neospecies that differs from its relatives chiefly by its sexual behavior and by diagnostic characters related to sexual selection. The discovery of lek behavior in the Hawaiian *Drosophila* (Spieth, 1974) presaged the mechanism and importance of sexual selection in this group of flies.

As an adjunct to the concept of ethospecies, Kaneshiro (1976) discovered, in mate-choice experiments, that ancestral females from geologically older islands resist mating with derived males from younger islands but that the reciprocal mating is not inhibited. In other words, females recognize older male courtships from which they originally evolved, but not new male courtships with which they have not coevolved. According to Kaneshiro (1983), the causal factors promoting this asymmetrical mating behavior are genetic drift and relaxation of the intensity of sexual selection in the small founding population, leading to simplified courtship requirements of the female. This, in turn, permits the male to simplify his courtship (but see Templeton, 1979). The emphasis here is on genetic drift, which of course is only part of the explanation for the founder event (Carson and Templeton, 1984); therefore, the identification of asymmetrical courtship patterns in organisms outside the Hawaiian archipelago may suggest recent genetic drift in the "derived" species or population, but not especially true founder events.

More recent efforts have been made to examine the extent to which the Kaneshiro pattern may determine direction in phylogeny (Giddings and Templeton, 1983; Kaneshiro, 1983). Their findings are generally in accord with the pattern, but there are exceptions, one of which will be discussed later. The present report is yet another attempt to test the Kaneshiro hypothesis on a group of *Drosophila* in which reasonable evidence exists concerning their relative phylogenetic positions.

One of the more enlightening applications of the founder event to processes occurring outside the Hawaiian archipelago applies to the Paleozoic, Mesozoic, and recent benthic assemblages of marine organisms as described by Jablonski and his co-workers (Jablonski and Bottjer, 1983, Jablonski et al., 1983; Valentine and Jablonski, 1983) and reviewed in Carson and Templeton (1984). In brief, there is paleontological evidence suggestive of rare founder-induced speciation events emanating from widespread species with high dispersal rates and a low degree of population structuring, but not from parent species in which planktonic larvae are absent and in which there is evidence of subdivision. These patterns are in accord with the restrictions established on theoretical grounds by Templeton (1980a,b).

The utilization of paleontological data and interpretation as models for neontological findings is a precarious but necessary exercise for those who believe that micro- and macroevolutionary events are fundamentally a difference in geological time. The fossil data and its subsequent interpretation (references above) tell us that the modern onshore species with extensive dispersal, and presumed gene flow, are slow to speciate and have "extinction resistant" properties giving rise to taxa with relatively low diversity, whereas the archaic offshore species with limited dispersal have higher speciation-extinction rates and are more speciose. This pattern is also evident on a microscale in the *Drosophila* of the Sonoran Desert and their relatives from southern Mexico and the West Indies.

The probable origins of the four unrelated endemic desert *Drosophila* were discussed and compared by Heed (1982). A division was made between the two host plant specialists, *D. pachea* and *D. nigrospiracula*, and the two host plant generalists, *D. mettleri* and *D. mojavensis*, in regard to the distribution pattern of their respectively related species. In the former instance, there are disjunct distributions between the specialists in the desert and their relatives in southern Mexico, whereas the distributions of the species related to the generalists overlap them in the southern and eastern parts of the desert and also extend into southern Mexico.

Long-distance dispersal was not emphasized for any of the four endemics as a means for invading the area where the desert now exists. That possibility is explored here.

Most recently a new species related to *D. mettleri* has been collected in the West Indies on the islands of Hispanola and Jamaica. Extensive hybridization studies were conducted with the two species and *D. eremophila*, another related species from Mexico. In addition, hybridization tests have been completed with *D. nigrospiracula* and other members of the anceps complex. Both sets of hybridization tests are reported with special attention to asymmetrical mating patterns and their possible significance.

CHARACTERISTICS OF SONORAN DESERT *DROSOPHILA*

Applying the conditions outlined in the preceding discussion to the endemic *Drosophila* of the Sonoran Desert, one may conveniently divide the species into those for which founder events are possible and those for which they are improbable (Fig. 12.1). *D. mojavensis* belongs in the latter category because of (1) its low dispersal rate among cacti, relative to other desert flies (Johnston, 1974, 1978); (2) its positive reproductive isolation indices across the Gulf of California (Wasserman and Koepfer, 1977; Zouros and D'Entremont, 1980); (3) large differences in allozyme frequencies across the Gulf and in the Mojave Desert (Zouros, 1973; Heed, 1978); and (4) the qualitative and quantitative differences of inversion polymorphisms among and within populations (Johnson, 1980). The shift in host plant use between populations is also indicative of considerable isolation within the species (Heed and Mangan, 1986). Furthermore, *D. mojavensis* is a member of the large mulleri complex, which contains 26 species (Was-

	MONO/OLIGOPHAGOUS ON CACTI	POLYPHAGOUS ON CACTI
* FOUNDER EVENT POSSIBLE	D. PACHEA D. NIGROSPIRACULA	D. METTLERI
FOUNDER EVENT IMPROBABLE		D. MOJAVENSIS

*1. ISOLATION INDICES BETWEEN POPULATIONS NEUTRAL OR NEGATIVE.
*2. NO EVIDENCE OF SUBDIVISION WITH MEDIUM TO HIGH DISPERSAL RATE.
*3. MEMBER OF NON-SPECIOSE GROUP OR COMPLEX.

Figure 12.1. Population structure and host plant specificity in the *Drosophila* of the Sonoran Desert, showing probability of initial establishment as a founder event.

serman, 1982). These attributes provide an opportunity for speciation events that occur and reoccur but are thought to produce major new genetic reorganizations only rarely (Carson and Templeton, 1984).

The other three species characteristic of the Sonoran Desert are more likely to have been the product of a founder event, or to be capable of initiating one, because they exhibit genetic and ecological characteristics more or less opposite to those of *D. mojavensis* (Fig. 12.1). All three species are members of species groups or species complexes with low diversity (but see below); they show little chromosomal (Cooper, 1964; Ward et al., 1975) or allozyme variability (Rockwood-Sluss et al., 1973; Sluss, 1975) that would indicate significant population subdivision; the species that have been measured have high dispersal rates among cacti (Johnston and Heed, 1976; Johnston, 1978) and the reproductive isolation across the gulf is either neutral or negative (Markow et al., 1983). These characteristics point to very large populations with a substantial amount of gene flow over their distributions, and it is under these conditions that a founder event is more likely to ensue (Carson and Templeton, 1984), and by inference, to have given rise to the present three desert species.

In reality, it is difficult to determine whether or not *D. nigrospiracula* and *D. mettleri* actually belong to a speciose group. Both forms are members of the large repleta group, but the group is divided into subgroups and complexes on the basis of morphology, reproductive isolation, and inversion differences. *D. nigrospiracula* is a member of the anceps complex, which includes only two other species, *D. anceps* and *D. leonis* (Wasserman, 1982). *D. mettleri* is a member of the eremophila complex (Wasserman, 1982) which also includes only two other species, *D. eremophila* and *D. micromettleri*, a new species described in the Appendix to this chapter. Thus on the basis of only immediately related forms, *D. nigrospiracula* and *D. mettleri* may be considered members of nonspeciose clusters. However, in the context of the repleta species group as a whole, that would not be true.

The Nannoptera Species Group

D. pachea is a member of the cactophilic nannoptera species group in which there are only three other species known to date. The biological characteristics of the species in the group have been reviewed as an aid to understanding their limited diversity as compared to the more than 80 species in the repleta species group (Heed and Mangan, 1986). The conclusion was made that the species could be relatively old and represent the relicts of a once widespread and successful lineage that was possibly displaced by the cactophilic repleta group. Another interpretation of course would be that most of the species arose by an infrequent founder event. However, the cytological characteristics of the four species (Ward and Heed, 1970) are not the ones recommended by Carson and Templeton (1984) to permit founder events to occur most readily. The highly derived metaphase chromosomes, with several fixed pericentric inversions, an X-autosome fusion, and large additions of heterochromatin, suggest a genetic system that would be unresponsive to the opportunity for the release of new genetic variation.

Even so, the few hybridization tests available (Russell et al., 1977) are asymmetrical and in the direction predicted by the Kaneshiro test if *D. pachea*, living in the Sonoran Desert, is considered a derived species (Table 12.1). The crosses were made in pair matings and large mass matings of variable number. Spermathecae and ventral receptacles were examined for sperm after 10–14 days in crosses 1 and 2, and after 6–10 days in crosses 3, 4, and 5. Unfortunately, females of the reciprocal mating to cross 5 were not examined. Species W is *D. pachea's* closest relative in southern Mexico (Heed, 1982), and crosses 3 and 4 show that the "derived" females *(D. pachea)* accept "ancestral" males (i.e., species W) but not the reverse. Also a few males of *D. nannoptera* were accepted by *D. pachea* females. *D. nannoptera* is considered to be the ancestral species for the nannoptera group on anatomical grounds (L. H. Throckmorton, personal communication). These observations are congruent with the distributional and chromosomal evidence discussed by Heed (1982), in which *D. pachea* is considered a derived species, and a pre-*D. pachea* population is considered to have entered the desert region by migration from southern Mexico. That it could have entered as a long-distance founder was not considered as a viable alternative at the time. The species has a disjunct distribution of more than 550 kilometers with its relatives in the south.

The Anceps Complex

Table 12.2 illustrates the results of two sets of small mass matings between *D. nigrospiracula* from Tucson and its relatives. The strain of *D. anceps* was collected near Zumpango, Guerrero, whereas *D. leonis* was collected near Guadalajara, Jalisco. All flies were aged for one to two weeks before mating. The first series was run for 19 days, whereas the second series was run for 25 days. The majority of females remaining

Table 12.1. Hybridization tests in the nannoptera species group

Females × males	Inseminated	Virgin
1. Species W × *D. nannoptera*	0	100
2. *D. nannoptera* × species W	0	100
3. Species W × *D. pachea*	0	24
4. *D. pachea* × species W	14[a]	7
5. *D. pachea* × *D. nannoptera*	3[a]	12

[a]All sperm were in the spermathecae and were nonmotile.
Source: Russell et al. (1977), by permission.

Table 12.2. Hybridization tests in the anceps complex
(15–25 pairs per vial in two replications)

Females × males	Courtships	Matings observed	Eggs	F_1 larvae and adults	Parentals inseminated	Parentals virgin
1. *D. nigrospiracula* × *D. nigrospiracula*	(a)	28	many	yes	16	1
2. *D. leonis* × *D. leonis*	(a)	7	many	yes	18	8
3. *D. anceps* × *D. anceps*	(a)	14	many	yes	12	5
4. *D. nigrospiracula* × *D. leonis*	(d)	0	many	0	0	9
5. *D. nigrospiracula* × *D. anceps*	(e)	5	many	0	0	18
6. *D. leonis* × *D. nigrospiracula*	(c)	0	many	0	0	9
7. *D. leonis* × *D. anceps*	(b)	0	many	0	0	7
8. *D. anceps* × *D. nigrospiracula*	(e)	0	few	0	0	3
9. *D. anceps* × *D. leonis*	(d)	0	many	0	0	3

Symbols: (a) Vigorous and persistent; females usually accept. (b) Vigorous and persistent; females usually refuse. (c) Vigorous with attempted mounts but not persistent; females do not accept. (d) Sporadic licking and wing flicking; females walk away. (e) Brief; males break off early.

alive after these times were dissected to detect the presence of sperm in the ventral receptacle. In the case of *D. nigrospiracula* homogamic matings, 12 of the 16 dissections with sperm also showed the presence of a "sperm sac" in the uterus or extruded from the uterus. The "sperm sac" is a thin-walled, opaque, open-ended bag that usually can be extruded from the uterus by pressure on the cover slip. It is almost always densely packed with active sperm which can readily be seen spilling out of the bag when it is removed from the uterus. The relation of this structure to the insemination reaction found in many species of *Drosophila* (Patterson and Stone, 1952) is not clear. The possibility also exists that the sac may be a spermatophore, in which case it would be produced by the male. In the positive cases of *D. anceps* and *D. leonis* homogamic matings, sperm was present in the ventral receptacle but no "sperm sacs" were seen, nor was there any positive identification of an insemination reaction.

No systematic study was made of copulation times, but it is apparent that *D. leonis* mates in the least time, the mean and standard deviation for 15 observations being 85.0 ± 25.21 seconds for this species. Two counts on *D. anceps* gave 420 and 430 seconds of mating time, whereas one count on *D. nigrospiracula* gave 380 seconds. Three other mating times for the latter species were between 5 and 6 minutes each.

Even though no progeny were produced and very few matings were observed in the interspecific crosses, the data in Table 12.2 may be grouped by male behavior into three categories: Exemplifying the first and second categories, *D. nigrospiracula* and *D. anceps* exhibit vigorous courtship with *D. leonis* females and very brief courtship with each other's females. In the latter case, the courtship is cut short as though the females had the ability to suppress the advances of the males. The suppression is not completely effective, however, because five copulations were subsequently noted when one of the crosses of *D. nigrospiracula* females with *D. anceps* males was changed to fresh food. These matings must have been ineffectual because no sperm was found in the 18 females dissected. The third category of male behavior is exhibited by males of *D. leonis* with the other two species, and it may be most accurately described as sporadic with much grooming and with male-to-male chasing in the intervening time.

These observations, coupled with short copulation times, indicate that *D. leonis* differs behaviorally from the other two species. The only point relating to the Kaneshiro hypothesis is the pseudocopulation of *D. anceps* males with *D. nigrospiracula* females. If derived females have lowered discrimination, then *D. nigrospiracula* could be considered derived. The meagerness of the data, however, do not impart confidence in the statement. The strong sexual isolation in the anceps complex as a whole suggests it to be a relatively old lineage whose females rarely recognize the courtship patterns of males of related species. This conclusion is in agreement with the evidence from metaphase comparisons and distributional patterns among the three species (Heed, 1982).

The Eremophila Complex

Hybridization Tests

The results of the hybridization tests in the eremophila complex are most interesting not only because of the asymmetry in the frequency of matings (Table 12.3) but also because of the postmating effects on fecundity and fertility (Tables 12.3 and 12.4) and the large variance in time of copulation between the species and their hybrids (Table 12.5). The tests in Table 12.3 were conducted in 1983 with *D. micromettleri* from Fond Parisien, Haiti, and Hatillo, Dominican Republic. The crosses were repeated in 1985 with the strain from Hatillo and also one from Port Henderson, Jamaica. The Jamaican strain hybridized better with the other two species. *D. eremophila* was obtained from the Drosophila Center in Bowling Green and was originally collected near Guayalejo, Tamaulipas. *D. mettleri* was collected near Tucson. The hybrid matings were made in small mass cultures by flies of various ages and were continued for periods ranging from one week to one month, depending on the condition of the cross. Some of the crosses were repeated many times (crosses 7 and 8). The majority of the females remaining alive after the tests were dissected to detect sperm in the ventral receptacle. "Sperm sacs" (spermatophores?) were found extruded from the uterus in many of the dissections and also were discovered on the surface of the medium in many of the crosses where they were deposited by the females. "Sperm sacs" (spermatophores?) were also found in all homogamic matings. These structures appear to be identical to those discovered in *D. nigrospiracula*.

D. mettleri has a much larger body size than the other two species, whereas *D. eremophila* is only slightly larger than the species from the West Indies. Large *D. mettleri* females have a 20% longer thorax and a 28% longer total body length than large *D. eremophila* females. Large *D. mettleri* males have a 12% longer thorax and a 10% longer body than *D. eremophila* males. The courtship of *D. mettleri* males can be very aggressive and persistent with females of the other species under certain conditions. *D. eremophila* males have motile sperm two hours after emergence and have been seen to court vigorously 3½ hours after emergence. This observation is significant since males of many species in the repleta group delay sexual maturity for seven days or more. *D. eremophila* and *D. micromettleri* females were isolated as pupae in order to ensure virginity.

What is of interest in Table 12.2 is the almost complete absence of insemination when males of *D. eremophila* are crossed to females of the other two species (crosses 6 and 8). In both crosses, the males usually courted vigorously but the females were unreceptive. The females in cross 8 originated from either Hatillo, Dominican Republic, or Port Henderson, Jamaica, in approximately equal numbers, and they both resisted *D. eremophila* males. The homogamic matings of *D. eremophila* compared favorably with those of the other two species in the number of mat-

Table 12.3. Hybridization tests in the eremophila complex

Females × males	No. of trials	Courtship	Matings observed	Eggs	F_1	Parentals inseminated	Parentals virgin
1. 27 D. eremophila × 22 D. eremophila	3	(a)	15	many	yes	8	0
2. 18 D. mettleri × 15 D. mettleri	2	(a)	17	many	yes	7	0
3. 29 D. micromettleri × 39 D. micromettleri	4	(a)	21	many	yes	4	0
4. 70 D. eremophila × 72 D. mettleri	5	(a)	18	360	15L,1P	12	1
5. 53 D. eremophilus × 84 D. micromettleri	6	(b)	8	1,680	6L,2F	11	22
6. 43 D. mettleri × 51 D. eremophila	4	(c)	1	740	0	0	33
7. 217 D. mettleri × 268 D. micromettleri	12	(a)	64	11,830	189L,88F,25M	42	54
8. 186 D. micromettleri × 132 D. eremophila	14	(b)	6	750	1L,1P	1	66
9. 82 D. micromettleri × 75 D. mettleri	6	(d)	5	110	0	29	14

Symbols: (a) Vigorous with attempted mounts; females usually receptive. (b) Vigorous to none; females usually unreceptive. (c) Vigorous with attempted mounts; females unreceptive. (d) Brief, and not especially vigorous. L, larvae; P, pupae; F, adult female; M, adult male.

Table 12.4. Hybridization tests in the eremophila complex

Female × male[a]	Trials	Eggs	Progeny		Generation of the progeny[b]
			♀♀	♂♂	
1. 25 ME/MI × 40 MI	2	3178	196	130	F_2BC
2. 99 (ME/MI × MI) × 83 MI[c]	6	48	0	0	F_3BC
3. 43 (ME/MI × MI) × 30 ME[d]	5	300		99[e]	F_3BC
4. 15 ME/MI × 10 ME	1	185	15[e]	10[e]	F_2BC
5. 15 (ME/MI × ME) × 10 (ME/MI × ME)	1	653	7[e]	6[e]	F_3IN
6. 7 (ME/MI × ME) × 6 (ME/MI × ME)	1	650		157[e]	F_4IN

[a]ME, *D. mettleri*; MI, *D. micromettleri*.

[b]BC, backcross; IN, inbred.

[c]Female reluctant to accept males, much decamping; males persistent.

[d]Female reluctant to accept; males court very vigorously.

[e]Fertile inter se.

ings observed, the 100% insemination rate, and the production of progeny. The conclusion is justified that *D. eremophila* males are
unacceptable to the females of the other two species using the strains
available. The most successful males are *D. mettleri* males, which have
an average insemination rate of 73% (crosses 4 and 9) compared to 41%
for *D. micromettleri* males (crosses 5 and 7). Thus the order of male success is *D. mettleri* > *D. micromettleri* >> *D. eremophila*. On the other
hand, the most receptive females are those of *D. eremophila*, with an
average of 50% insemination (crosses 4 and 5), whereas the other two
species average 30% between them (crosses 6–9). The order of female
receptivity is *D. eremophila* > *D. mettleri* = *D. micromettleri*. This difference is caused by the incompatible crosses with *D. eremophila* males.
Otherwise average female receptivity of *D. mettleri* and *D. micromettleri*
is identical to *D. eremophila* (50%). Under the assumption that the Kaneshiro hypothesis is useful in instances other than in the Hawaiian archipelago, *D. eremophila* should be considered the derived species.

Table 12.3 shows that the only hybrid cross to produce a significant
number of progeny is *D. mettleri* females with *D. micromettleri* males
(Jamaica strain). The hybrid males were sterile. The egg-to-adult viability
of 1% is almost certainly exaggerated on the low side, but there are no
controls for comparison. Viability increased to 10.3% when the hybrid
females were backcrossed to *D. micromettleri* males (cross 1) and to
13.5% (cross 4) when backcrossed to *D. mettleri* males (Table 12.4).
These values are similar even though egg production (mean eggs per female) with *D. micromettleri* males was 10 times greater than with *D. mettleri* males (cross 1 vs. 4). The difference in the backcross resides in the
fertility of the progeny (F_2BC). The progeny from cross 1 are effectively
sterile with *D. micromettleri* males (cross 2) but fertile with *D. mettleri*
males (cross 3). The progeny from cross 4 were fertile interse. Thus fertility resides in *D. mettleri* males in both cases. The sterility associated
with *D. micromettleri* is difficult to explain. Notice the extreme bottleneck in the F_3 inbred progeny (cross 5).

The low fecundity, in crosses 4, 6, 8, and 9 of Table 12.3 and in crosses
3 and 4 of Table 12.4, can be explained by the apparent absence of mating
with *D. eremophila* males and the overly aggressive behavior of *D. mettleri* males. In the latter case, inseminated females often have swollen uteri
and ovipositors remindful of the effects of an insemination reaction, and
they become partly immobile. No eggs were deposited under these conditions. Furthermore, *D. mettleri* males continue to court, and the females eventually become stuck on their backs on the surface of the food.
The relation of this "insemination reaction" to the "sperm sac" is unclear. They are probably not the same thing.

Mating Times

Table 12.5 lists the mean copulation (i.e., mating) times that were recorded for the various crosses when they were first initiated in small mass
matings. Even though there was no control for age of the males, which

Table 12.5. Timed copulations in the eremophila complex

Females × males		N	Mean seconds ± SD
1. (micromettleri × eremophila)	× D. micromettleri	2	122.5 ± 53.03
2. D. mettleri	× D. micromettleri	13	137.69 ± 60.16
3. (mettleri/micro. × micro.)	× D. micromettleri	16	157.69 ± 29.86
4. (mettleri × micromettleri)	× D. micromettleri	20	194.75 ± 40.80
5. (mettleri/micro. × micro.)	× (mettleri/micro. × micro.)	14	206.0 ± 36.96
6. D. micromettleri	× D. micromettleri	9	220.0 ± 29.47
7. (mettleri × micromettleri)	× (mettleri × micromettleri)	4	230.0 ± 21.60
8. D. eremophila	× D. eremophila	5	328.44 ± 44.10
9. (mettleri × micromettleri)	× D. mettleri	12	352.92 ± 51.37
10. D. micromettleri	× D. eremophila	6	363.33 ± 76.53
11. (mettleri/micro. × micro.)	× D. mettleri	9	372.22 ± 116.81
12. D. mettleri	× D. mettleri	13	440.92 ± 113.92
13. D. micromettleri	× D. mettleri	5	481.0 ± 71.19

[a,b,c,d] Nonsignificant differences among crosses within each letter, determined by using the Tukey Kramer multiple comparison ($\alpha = 0.05$). Cross 1 is not included in the analysis. The ANOVA is shown below:

	F	a	b	c	d
	32.819	2–7	7–8	8–11	10–13

Source	Sum of squares	d.f.	Mean square	F
Between groups	1475044.743	11	134094.977	32.819
Within groups	465794.685	114	4085.918	

may account for the high variance in some of the matings with *D. mettleri*, these data are interesting because of the large differences among the species. Variation in mating times listed in Table 12.5 permits the following observations:

1. Mating time is mostly male-influenced.
2. *D. mettleri* and *D. eremophila* males have the longest mean times (crosses 8 to 13), whereas *D. micromettleri* males mate in the shortest times (crosses 1, 2, 3, 4, and 6).
3. Mean homogamic mating times in *D. mettleri* and *D. micromettleri* differ by a factor of 2.0 (cross 6 vs. 12). The mean homogamic mating time for *D. eremophila* (cross 8) is almost exactly intermediate to those of the other two species. Thus there is little evidence for an effect of body size because the two small species, *D. eremophila* and *D. micromettleri*, have essentially the same body size.
4. Reciprocal mating times between *D. mettleri* and *D. micromettleri* differ by a factor of 3.5 and are thus very asymmetrical (cross 2 vs. 13). This is also a male effect since cross 2 does not differ significantly from *D. micromettleri* homogamics (cross 6), whereas cross 13 does not differ significantly from *D. mettleri* homogamics (cross 12).
5. Mating time is not correlated with the production of progeny. Compare Tables 12.3 through 12.5

In summary, mating time is species specific and male dominated. That copulation time is regulated mostly by the male has also been observed by Kaul and Parsons (1965) for different inversions in *D. psuedoobscura*.

Distribution and Density

Figure 12.2 shows the distribution of the three species in the eremophila complex. One of the changes from the previous map (Heed, 1982) is the addition of *D. micromettleri* in Jamaica and Hispanola. Also a new chromosome phylogeny has been established. The metaphase chromosomes of the new species, *D. micromettleri*, are very similar to *D. mettleri*, except that there is a satellite on the X chromosome in the former species. In addition, two new inversions have been discovered in the fixed condition in the third and fifth chromosomes of three strains of *D. mettleri* (Alfredo Ruiz, personal communication). Thus the phylogeny reads as four inversions fixed from primitive I which the three species have in common, and this must be the ancestral population. *D. eremophila* and *D. mettleri* are fixed for two unique inversions each. This makes *D. micromettleri* unchanged in gene sequence since the time when it speciated from the ancestral population (or it is the ancestor). There is little cytological evidence that *D. eremophila* is more derived than the other two species except for the addition of slightly more heterochromatin on the dot chromosome. The chief point here, however, is the similarity in metaphase chromosomes of the two geographically marginal species. *D. mett-*

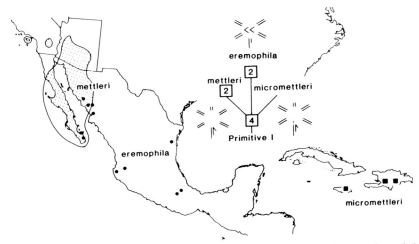

Figure 12.2. Distribution of the three species in the eremophila complex of the repleta species group of *Drosophila*. The numbers in phylogeny refer to the number of fixed inversions within and between the species.

leri in the Sonoran Desert is separated from *D. micromettleri* in Jamaica by over 3000 kilometers. The evidence from hybrids also indicates they are more similar to each other than either is to *D. eremophila*.

In order to learn more about this interesting distribution, the available banana-baited collecting records for the three species in the eremophila complex were examined (Tables 12.6 and 12.7). The use of a common attractant helps make the data comparable over years and among localities. The information in Table 12.6 shows that *D. eremophila* is a very rare species in the lowlands of western Mexico. It averages 1% of the total flies in 10 collections. It is more abundant in the Tehuacan region of southeastern Mexico (13% in two collections). By contrast, *D. mettleri* and *D. micromettleri* are well represented in collections in their respective geographical regions (Table 12.7). The former species averages 25% in four collections, whereas *D. micromettleri* averages 30% in three collections. Thus it is plausible to consider the geographically marginal species as living in continuously dense populations, whereas the central species is more often subjected to a series of bottlenecks. The soil-breeding habit may foster this pattern if conditions are often suboptimal in most of central Mexico.

The idea that all three species in the eremophila complex actually breed in the soil is an inference from the data on *D. mettleri*. This species has been reared from soil underneath several different species of columnar cacti in nature on a number of occasions, as well as under experimental conditions in the laboratory and in the field (Heed, 1977; Fogleman et al., 1981a, 1982; Fogleman, 1984). The information on the breeding sites of the other two species has been restricted to laboratory experiments (Fo-

Table 12.6. Banana-baited collections of *D. eremophila* in Mexico

Arizona no.	Year	Locality	Date	No. of species[a]	Total flies[a]	Percent *D. eremophila*
	1963–4	Acatlan, Puebla[b]		19	522	23.0
	1963–4	Tehuacan, Puebla[b]		21	631	4.4
A233	1969	Navojoa, Sonora	Jan. 25–26	9	1621	1.3
A236	1969	Los Mochis, Sinaloa	Jan. 27	10	1314	0.5
A239	1969	Empalme, Sonora	Jan. 28	9	544	0.2
A272 and A273	1970	Alamos and Navojoa, Sonora	Mar. 27–29	10	638	0.9
A657	1976	Navojoa, Sonora	Oct. 19	15	1687	1.0
A793	1981	San Bartolo, Baja California Sur[c]	Jan. 9	10	101	1.0
A789	1981	Puerto Balandra, Baja California Sur	Jan. 10–11	10	345	0.3
A806	1981	Tomatlan, Jalisco	July 25–26	12	246	1.2
A876	1984	Las Bocas, Sonora	Nov. 1–2	7	1131	0.3

[a]*D. melanogaster* and *D. simulans* are not included.

[b]From Wasserman (1967).

[c]Aspirated from fruits of Pitaya Agria (*Stenocereus gummosus*).

Table 12.7. Banana-baited collections of *D. mettleri* (above) and *D. micromettleri* (below)

Arizona no.	Year	Locality	Date	No. of species[a]	Total flies[a]	Percent
A226	1969	Tucson, Arizona[b]	Jan. 15–19	6	1956	25.6
A265	1970	Kino Bay, Sonora	Feb. 27–Mar. 1	6	309	7.4
A388	1972	Desemboque, Sonora	May 14–16	6	696	19.5
A826	1981	Santa Catalina Island, California	Oct. 31–Nov. 1	6	711	37.3
ORV 1	1982	Fond Parisien, Haiti	May 6–7	13	923	2.3
ORV 4	1982	Hatillo, Dominican Republic	May 11–12	11	1656	55.4
ORV 24	1983	Pt. Henderson, Jamaica	Nov. 23	7	793	7.2

[a]*D. melanogaster* and *D. simulaus* not included.
[b]From Johnston and Heed (1971).

gleman, 1984). However, because the morphology of the larval mouth hooks are very similar and distinctive for the three species, the hypothesis that *D. eremophila* and *D. micromettleri* are also soil breeders is realistic. Furthermore, there are no records of the two species having been reared from any cactus tissue.

DISCUSSION

The chief emphasis in the Hawaiian *Drosophila* program has been on cytological analysis, sexual behavior, and the process of speciation (Carson et al., 1970; Carson and Kaneshiro, 1976; Carson and Yoon, 1982). By comparison, studies in the Sonoran Desert have emphasized the coadapation between the trophic resources (physical and chemical composition of the cacti and their microorganisms) and the two feeding stages of the flies (larvae and adults) as a means to study host plant selection and other forms of natural selection (Kircher and Heed, 1970; Heed, 1978; Fogleman et al., 1981b; Starmer, 1982; Starmer et al., 1982; Heed and Mangan, 1986). Speciation studies have been limited in the desert region because of its low diversity, the difficulty of locating and recognizing related species, and the lack of hybridization tests. The one exception to this fact has been the ongoing emphasis on the sibling species, *D. arizonensis* and *D. mojavensis* (references in Heed, 1982).

However, the realization that there are only four endemic species in the desert, and that each one occupied the area independently, demonstrates that a winnowing process probably occurred in the past, and must have been intimately connected with the act of speciation for the survivors. This concept has been termed "adaptive infiltration" by Heed and Mangan (1986), and it stands in high contrast to the concept of adaptive radiation as applied to the Hawaiian flies. Adaptive infiltration signifies that the populations in the area had to speciate in order to survive increasingly stressful conditions. All species presently existing on the edge of the desert have very large distributions in Mexico and/or the United States. They have not permanently adapted to the climatically rigorous areas of the desert probably because of gene flow from nondesert regions. It is difficult to apply Carson's concept of the founder event to desert conditions since it starts with a nonadaptive phase—a stage of disorganization—presumably in a benign environment (Carson, 1975, 1982). However, the concept of genetic transilience (Templeton 1980a,b) appears to be less restrictive in this respect, but Templeton's model applies primarily to the genetic environment rather than the physical one. Alternatively, a founder event can be followed by adaptive divergence (Templeton 1980a,b) rather than by a genetic transilience in which the primary concern is with the physical environment and geographical isolation.

In reality, it may not be possible to distinguish, by hybridization tests, between a founder event leading to speciation through a genetic transilience and one leading to speciation through an adaptive divergence, especially if one is dealing with older species rather than neospecies. *D. mojavensis* is the only desert species that qualifies as a neospecies (ethospecies?) because of its low degree of postmating isolation with *D. arizonensis* (Zouros, 1981). As described earlier, however, *D. mojavensis* is not expected to have undergone a genetic transilience because of its population structure. Even so, it could have experienced a series of bottlenecks during its origination, and the asymmetry in mating behavior described by Wasserman and Koepfer (1977; 1980) and Markow (1981) is actually in the direction predicted by the Kaneshiro test if the assumption is made that *D. mojavensis* evolved in peninsular Baja California from *D. arizonensis* (or a pre-*arizonensis*) population on the mainland. Wasserman and Koepfer (1977) recorded a sexual isolation index of $+0.308 \pm .040$ *(SE)*, when peninsular *D. mojavensis* females were tested with *D. arizonensis* males from mainland Mexico. The isolation index for the reciprocal cross was $+0.779 \pm .031$ *(SE)*.

The concept of the origination of *D. mojavensis* on peninsular Baja California (where *D. arizonensis* is found only sparingly in the Cape Region) from an early *D. arizonensis* population in mainland Mexico has been inhibited until recently by the presence of inversion 3a which is fixed homozygous in *D. mojavensis* and all closely related species except *D. arizonensis* (Wasserman, 1982). The recent discovery of inversion 3a "hidden" by a newer inversion that reversed this sequence almost precisely in *D. arizonensis* (A. Ruiz, personal communication) clears the way for the interpretation presented earlier. *D. arizonensis* is widespread in Mexico and it also extends into Guatemala (Heed, 1982). *D. mojavensis* occupies the Sonoran Desert and immediate adjacent areas. The two species presently coexist on the mainland of Sonora, Mexico, and Southern Arizona, United States, presumably owing to a secondary invasion of *D. mojavensis* from Baja California. If the two present-day species eventually were to be found equally derived from a common ancestor then the Kaneshiro test would not be definitive.

The Kaneshiro hypothesis however certainly does not explain the asymmetry in mating behavior between *D. mojavensis* in peninsular Baja California and Sonora, Mexico, as described by Wasserman and Koepfer (1977) and Zouros and D'Entremont (1980), under the assumption that the species subsequently invaded Sonora across the midriff islands of the gulf, as it certainly appears to have done. Females of the derived Sonora populations discriminate to a significant degree against the ancestral peninsular males, but there is no discrimination in the reciprocal cross. An alternative explanation for this phenomenon is character displacement for sexual isolation, resulting from the presence of *D. arizonensis* (Wasserman and Koepfer, 1977; Zouros and D'Entremont, 1980; Markow et al., 1983). A pleiotropic effect initiated by the shift to a different host plant accompanying the invasion of the mainland by *D. mojavensis* also could have significant consequences for behavior (Markow et al., 1983). Another force, such as "social selection" as described by West-Eberhard (1983), also may be at work.

The only other desert species approaching the status of a neospecies is *D. mettleri* because of its semifertile relationship with *D. micromettleri*. Postmating barriers are much more prevalent in this example, however, in contrast to the *D. mojavensis* case. It is significant that the two "neo-species" in the desert and nearby areas are also the polyphagic ones in regard to host plant selection (Heed and Mangan, 1986). The Kaneshiro test suggests that *D. mettleri* is an ancestral species, as previously discussed with much less evidence (Heed, 1982). The rarity of *D. eremophila* in coastal western Mexico suggests that bottlenecking could be a continuous process for this species which could have induced the changes in mating behavior. Recent long-distance dispersal between the Sonoran Desert and the Greater Antilles, or vice versa, might explain the genetic and phenotypic similarity between *D. mettleri* and *D. micromettleri*. The difference in receptivity of the females (Table 12.3) would suggest the Antillean species to be the derived one.

An alternative explanation to long-distance dispersal is that both geographically marginal species became ancestral independently; that is, each has changed more slowly in certain characters from a single widespread species than has *D. eremophila* in central Mexico. The precedence for the observation of conservative characters existing in geographically marginal populations, as well as in marginal species, has been established on cytological grounds in the cardini species group by Heed and Russell (1971) and therefore, this synopsis is not unreasonable.

Interestingly, a high degree of asymmetry in terms of copulation is apparent in the reciprocal matings between the two marginal species: *D. mettleri* males are much slower (by a factor of 3.5) than are males of *D. micromettleri*. The difference may be attributed to the interaction of *D. mettleri* with *D. nigrospiracula*, which are both associated with the same cacti in the Sonoran Desert. The striking phenotypic resemblance in size and coloration between the two species is matched here by similar copulation times. The similarities (i.e., convergence) may have been prompted by social interactions similar to the kinds described by West-Eberhard (1983). Thus, even though *D. mettleri* is considered geneologically ancestral to *D. eremophila*, the species has unique apomorphic characters such as the two fixed paracentric inversions, large body size, and long copulation time. The latter two characters do not appear to be directly correlated, as mentioned previously.

D. pachea, although cytologically complex, has a favorable population structure for a founding event. Too little is known about the population structure of other members of the nannoptera species group to make a comment. Even so, the asymmetry detected in several of the matings of this species with its relatives in southern Mexico is in agreement with the Kaneshiro test, if one considers *D. pachea* as a derived species. This is very likely to be the case from all other information summarized by Heed

(1982). As with *D. mojavensis,* bottlenecking could have instigated the asymmetry. The very close biochemical adaptation of this species to its host plant *(Lophocereus schottii)* in the Sonoran Desert (Kircher and Heed, 1970) does not make long-distance dispersal, with a subsequent founder event, an attractive alternative.

D. *nigrospiracula* is certainly not a neospecies, nor is there sufficient data on hybridization to make a judgment in the direction of phylogeny. This species has the greatest disjunct distribution of any of the desert *Drosophila* with its relatives in southern Mexico (Heed, 1982). *D. nigrospiracula* is known to mate repeatedly with high frequency (Markow, 1982). The detection of sperm sacs (spermatophores?) in this species, as well as in the three species of the eremophila complex, may have significance in this respect. If the female can actively dispose of the sac, she may be able to manipulate her supply of sperm to a certain degree and thus select for high-priority males.

CONCLUSIONS

Where the data are available, and with certain assumptions, the endemic *Drosophila* of the Sonoran Desert, with one clear exception, are in agreement with the Kaneshiro hypothesis regarding direction in phylogeny as originally postulated from other evidence. Two species are probably derived *(D. pachea* and *D. mojavensis);* one species is probably ancestral *(D. mettleri)* and one species, *D. nigrospiracula,* is not indicated from the lack of hybridization data. Furthermore, all speciation events appear to occur by adaptive divergence. The concept of adaptive infiltration has been discussed.

The one clear exception to the Kaneshiro criterion concerns the asymmetry in mating behavior between *D. mojavensis* from Baja California (ancestral populations) and that from the mainland (derived populations), which is the reverse from expected, owing to other factors that ostensibly override the original effect. The other factors could include (1) character displacement, (2) social selection, and/or (3) pleiotropy. There also exists the possibility that the Kaneshiro effect never was in operation because of a low but continuous amount of gene flow between the two regions at the beginning of their separation and possibly into present time.

The reproductive behaviors of two of the desert species, *D. mojavensis* and *D. mettleri,* are believed to have been affected by interaction with other species. In the one case, the species are genetically closely related *(D. mojavensis* and *D. arizonensis).* In the other case, the species are ecologically closely related *(D. mettleri* and *D. nigrospiracula).* Detailed comparisons between the two sets of species should be very rewarding for the field of evolutionary behavior.

LITERATURE CITED

Carson, H. L. 1975. The genetics of speciation at the diploid level. Am. Natur. 109:83–92.

Carson, H. L. 1978. Speciation and sexual selection in Hawaiian *Drosophila*. In P. F. Brussard (ed.), Ecological Genetics: the Interface, pp. 93–107. Springer-Verlag, New York.

Carson, H. L. 1982. Speciation as a major reorganization of polygenic balances. In C. Barigozzi (ed.), Mechanisms of Speciation, pp. 411–433. Alan R. Liss, New York.

Carson, H. L., and K. Y. Kaneshiro. 1976. *Drosophila* of Hawaii: systematics and ecological genetics. Annu. Rev. Ecol. Systematics 7:311–345.

Carson, H. L., and A. R. Templeton. 1984. Genetic revolutions in relation to speciation phenomena: the founding of new populations. Annu. Rev. Ecol. Systematics 15:97–131.

Carson, H. L., and J. S. Yoon, 1982. Genetics and evolution of Hawaiian *Drosophila*. In M. Ashburner, H. L. Carson, and J. N. Thompson, Jr. (eds.), The Genetics and Biology of *Drosophila*, Vol. 3b, pp. 298–344. Academic Press, London.

Carson, H. L., D. E. Hardy, H. T. Spieth, and W. S. Stone. 1970. The evolutionary biology of the Hawaiian *Drosophilidae*. In M. L. Hecht and W. C. Steere (eds.), Essays in Evolution and Genetics in Honor of Theodosius Dobzhansky, pp. 437–543. Appleton-Century-Crofts, New York.

Cooper, J. W. 1964. Genetic and cytological studies of *Drosophila nigrospiracula* in the Sonoran Desert. Master's thesis, University of Arizona, Tucson.

Fogleman, J. 1984. The ability of cactophilic *Drosophila* to utilize soaked soil as larval substrates. Dros. Info. Serv. 60:106–107.

Fogleman, J. C., K. R. Hackbarth, and W. B. Heed. 1981a. Behavioral differentiation between two species of cactophilic *Drosophila*. III. Oviposition site preference. Am. Natur. 118:541–548.

Fogleman, J. C., W. T. Starmer, and W. B. Heed. 1981b. Larval selectivity for yeast species by *Drosophila mojavensis* in natural substrates. Proc. Natl. Acad. Sci. USA. 78:4435–4439.

Fogleman, J. C., W. B. Heed, and H. W. Kircher. 1982. *Drosophila mettleri* and senita cactus alkaloids: fitness measurements and their ecological significance. Comp. Biochem. Physiol. 71:413–417.

Giddings, L. V., and A. R. Templeton. 1983. Behavioral phylogenies and the direction of evolution. Science 220:372–378.

Heed, W. B. 1977. A new cactus-feeding but soil-breeding species of *Drosophila* (Diptera; Drosophilidae). Proc. Entomol. Soc. Wash. 79:649–654.

Heed, W. B. 1978. Ecology and genetics of Sonoran Desert *Drosophila*. In P. F. Brussard. (ed.), Ecological Genetics: The Interface, pp. 109–126. Springer-Verlag, New York.

Heed, W. B. 1982. The origin of *Drosophila* in the Sonoran Desert. In J. S. F. Barker and W. T. Starmer (eds.), Ecological Genetics and Evolution: The Cactus–Yeast–*Drosophila* Model Systems, pp. 65–80. Academic Press, Sydney.

Heed, W. B., and J. S. Russell. 1971. Phylogeny and population structure in island and continental species of the cardini group of *Drosophila* studied by inversion analysis. In M. R. Wheeler (ed.), Studies in Genetics VI. Univ. Texas Publ. 7103:91–130.

Heed, W. B., and R. L. Mangan, 1986. Community ecology of the Sonoran Desert *Drosophila*. In M. Ashburner, H. L. Carson, and J. N. Thompson, Jr. (eds.), The Genetics and Biology of *Drosophila*, Vol. 3e, pp. 311–345. Academic Press, London.

Jablonski, D., and D. J. Bottjer. 1983. Soft bottom epifaunal suspension-feeding assemblages in the late Cretaceous: implications for the evolution of benthic paleocommunities. In M. J. S. Tevesz and P. L. McCall. (eds.), Biotic Interactions in Recent and Fossil Benthic Communities, pp. 747–812. Plenum, New York.

Jablonski, D., J. J. Sepkoski, Jr., D. J. Bottjer, and P. M. Sheehan. 1983. Onshore–offshore patterns in the evolution of Phanerozoic shelf communities. Science 222:1123–1125.

Johnson, W. R. 1980. Chromosomal polymorphism in the desert-adapted species, *Drosophila mojavenisis*. Ph.D. dissertation, University of Arizona, Tucson.

Johnston, J. S. 1974. Dispersal in natural populations of the cactophilic *Drosophila pachea* and *D. mojavensis*. Genetics 77(Suppl.):32–33.

Johnston, J. S. 1978. Dispersal behavior and biological control of insects. In R. H. Richardson (ed.), The Screwworm Problem: Evolution of Resistance to Biological Control, pp. 113–127. University of Texas Press, Austin.

Johnston, J. S., and W. B. Heed. 1971. A comparison of banana and rotted cactus as a bait for desert *Drosophila*. Dros. Info. Serv. 46:96.

Johnston, J. S., and W. B. Heed. 1976. Dispersal of desert-adapted *Drosophila*: The saguaro-breeding *D. nigrospiracula*. Am. Natur. 110:629–651.

Kaneshiro, K. Y. 1976. Ethological isolation and phylogeny in the planitibia subgroup of Hawaiian *Drosophila*. Evolution 30:740–745.

Kaneshiro, K. Y. 1983. Sexual selection and direction of evolution in the biosystematics of Hawaiian Drosophilidae. Annu. Rev. Entomol. 28:161–178.

Kaul, D., and P. A. Parsons, 1965. The genotypic control of mating speed and duration of copulation in *Drosophila pseudoobscura*. Heredity 20:381–392.

Kircher, H. W., and W. B. Heed. 1970. Phytochemistry and host plant specificity in *Drosophila*. In C. Steelink and V. C. Runeckles (eds.), Recent Advances in Phytochemistry, Vol. 3, pp. 191–209. Appleton-Century-Crofts, New York.

Markow, T. A. 1981. Courtship behavior and control of reproductive isolation between *Drosophila mojavensis* and *Drosophila arizonensis*. Evolution 35:1022–1026.

Markow, T. A. 1982. Mating systems of cactophilic *Drosophila*. In J. S. F. Barker and W. T. Starmer (eds.), Ecological Genetics and Evolution: The Cactus–Yeast–*Drosophila* Model System, pp. 273–287. Academic Press, Sydney.

Markow, T. A., J. C. Fogleman, and W. B. Heed. 1983. Reproductive isolation in Sonoran Desert *Drosophila*. Evolution 37:649–652.

Patterson, J. T., and W. S. Stone. 1952. Evolution in the Genus *Drosophila*, Macmillan, New York.

Rockwood-Sluss, E. S., J. S. Johnston, and W. B. Heed. 1973. Allozyme genotype–environment interaction relationships. I. Variation in natural populations of *Drosophila pachea*. Genetics 73:135–146.

Russell, J. S., B. L. Ward, and W. B. Heed. 1977. Sperm storage and hybridization in *D. nannoptera* and related species. Dros. Info. Serv. 52:70.

Sluss, E. S. 1975. Enzyme variability in natural populations of two species of cactophilic *Drosophila*. Ph.D. dissertation, University of Arizona, Tucson.

Spieth, H. T. 1974. Mating behavior and evolution of the Hawaiian *Drosophila*. In M. J. D. White (ed.), Genetic Mechanisms of Speciation in Insects, pp. 94–101. Australia and New Zealand Book Co., Sydney.

Starmer, W. T. 1982. Associations and interactions among yeasts, *Drosophila* and their habitats. In J. S. F. Barker and W. T. Starmer (eds.), Ecological Genetics and Evolution: The Cactus–Yeast–*Drosophila* Model System, pp. 159–174. Academic Press, Sydney.

Starmer, W. T., H. J. Phaff, M. Miranda, M. W. Miller, and W. B. Heed. 1982. The yeast flora associated with the decaying stems of the columnar cacti and *Drosophila* in North America. Evol. Biol. 14:269–295.

Templeton, A. R. 1979. Once again, why 300 species of Hawaiian *Drosophila?* Evolution 33:513–517.

Templeton, A. R. 1980a. The theory of speciation via the founder principle. Genetics 94:1011–1038.

Templeton, A. R. 1980b. Modes of speciation and inferences based on genetic distances. Evolution 34:719–729.

Valentine, J. W., and D. Jablonski. 1983. Speciation in the shallow sea: general patterns and biogeographic controls. In R. W. Sims, J. H. Price, and P. E. S. Whalley (eds.), Evolution, Time and Space: The Emergence of the Biosphere, pp. 201–226. Academic Press, New York.

Ward, B. L., and W. B. Heed. 1970. Chromosome phylogeny of *Drosophila pachea* and related species. J. Hered. 61:248–258.

Ward, B. L., W. T. Starmer, J. S. Russell, and W. B. Heed, 1975. Correlation of climate and host plant morphology with a geographic gradient in an inversion polymorphism of *Drosophila pachea*. Evolution 28:565–575.

Wasserman, M. 1967. Collections of *Drosophila* from central Mexico. Dros. Info. Serv. 42:67–68.

Wasserman, M. 1982. Evolution in the repleta group. In M. Ashburner, H. L. Carson, and J. N. Thompson, Jr. (eds.), The Genetics and Biology of *Drosophila*, Vol. 3c., pp. 61–139. Academic Press, London.

Wasserman, M., and H. R. Koepfer. 1977. Character displacement for sexual isolation between *Drosophila mojavensis* and *Drosophila arizonensis*. Evolution 31:812–823.

Wasserman, M., and H. R. Koepfer. 1980. Does asymmetrical mating preference show the direction of evolution? Evolution 34:1116–1124.

West-Eberhard, M. J. 1983. Sexual selection, social competition, and speciation. Q. Rev. Biol. 58:153–183.

Zouros, E. 1973. Genic differentiation associated with the early stages of speciation in the mulleri subgroup of *Drosophila*. Evolution 27:601–621.

Zouros, E. 1981. The chromosomal basis of viability in interspecific hybrids between *Drosophila arizonensis* and *Drosophila mojavensis*. Can. J. Genet. Cytol. 23:65–72.

Zouros, E., and C. J. D'Entremont. 1980. Sexual isolation among populations of *Drosophila mojavensis:* response to pressure from a related species. Evolution 34:421–430.

ACKNOWLEDGMENTS

I wish to thank Jack Fell and Tom Starmer for initiating and administering the Caribbean trips on the ORV Cape Florida; Captain Bob Morgan and his crew for their enthusiastic cooperation and interest in the collections; Jim Gibbons, Rosentiel School of Marine and Atmospheric Science, University of Miami, for his concern and approval; Alfredo Ruiz very kindly analyzed the chromosomes of several of the species; Bill Etges, Tom Starmer, and Andy Beckenbach for their valuable comments on the manuscript; and Pupulio Sseskimpi for assistance on the statistics. The research was supported by The National Science Foundation.

APPENDIX:

Drosophila micromettleri Heed, New Species

External Characters of Imagines

In ♂ ♀, the arista has 3 dorsal and 2 ventral rays in addition to the terminal fork. The face and second antennal segment are brown, the third antennal segment is tinged with black. The carina is wider below and slightly sulcate. The frons is velvety chestnut brown, is divided medially by a narrow gray pollinose stripe, and has about 12 short irregularly placed bristles centrally clustered and 8 short bristles along the anterior margin. The base of the verticals and orbitals and the margin of the ocellar triangle are pollinose gray, except for an interruption by a brown spot at the base of the verticals and the posterior reclinates. The anterior proclinate bristle is about the same length as the posterior reclinate. The midorbital length is $\frac{1}{3}$ to $\frac{3}{7}$ the length of the posterior. The cheek is wide–about $\frac{1}{3}$ the greatest diameter of the eye—and is brown. The eye is dark red (maroon). There is one strong oral with 2–3 suborals.

The mesonotum ground color is a pollinose lichen-gray with a dark-brown spot at the base of the acrostichal bristles. Many spots are fused, forming an irregular pattern. The acrostichal bristles are in 8 irregular rows. The anterior scutellars are convergent. The pleural region is dark gray. The anterior sternopleural is about $\frac{4}{5}$ the length of the posterior; the midsternopleural is minute, or up to $\frac{1}{4}$ the length of the posterior. The legs of the male are tan and darkened more on the fore femur and apically on the other femora and at the base of the basitarsi. The females are similar, but all femora are darkened. The wing is clear, the costal index is about 2.3; the 4th vein index is about 1.8; the 4th costal index is about 1.1; the 5th vein index is about 1.5. The third costal section has heavy bristles on the basal $\frac{1}{3}$. The abdomen has wide blackish bands, slightly interrupted on the middorsal line and extending to the lateral margins. The body length (etherized) is 2.1–2.5 mm (♂) or 2.6–2.9 (♀); the wing length is 2.0–2.1 (♂) or 2.1–2.4 (♀).

Internal Characters of Imagines and Genitalia

The posterior tips of the Malpighian tubules are fused. The testis has 3–4 inner and 2–3 outer coils; it is yellow, darkening to orange-yellow with age. The claspers are rounded, with about 9 primary teeth and no secondary teeth (Fig. 12.3). The aedeagus is rounded, with a short gonapophysis having two small bristles in tandem (Fig. 12.3). The stalks of the spermathecae are long; the capsule is small, weakly sclerotized, and somewhat crumpled. The ventral receptacle is of medium length, with about 4 basal coils and about 10 irregular folds distally. The ovipositor is black and slightly blunted, with 6 teeth on the apical margin, 9 teeth on the ventral margin, and 2–3 teeth on the side of the plate, in addition to one dorsal and one ventrally directed vibrissa.

Egg

There are two pair of filaments.

Larva

The third instar mouth hooks are narrow with a broad arch and no teeth.

Puparium

The puparium is tan; the horns, including the yellow spiracles, are short—about ⅑ the length of the pupa. The spiracles number 8–10 and are slightly shorter than the stalk.

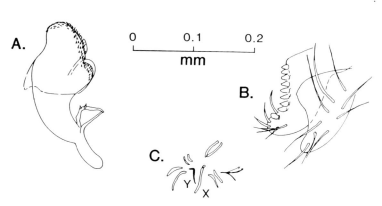

Figure 12.3. *Drosophila micromettleri,* Hatillo, Dominican Republic. (**A**) Lateral view of aedeagus with attached gonopod. (**B**) Clasper and lower section of genital arch. (**C**) Larval brain cell metaphase (not to scale).

Chromosomes

The female larval brain metaphase shows 6 pairs of rods. The longest rod is the X with satellites. The shortest rod is the dot chromosome with heterochromatin added; it is $\frac{1}{3}$–$\frac{1}{2}$ the length of the X. Male metaphase (Fig. 12.3) shows a J-shaped Y, approximately $\frac{2}{3}$ the length of the X. The salivary chromosomes show 5 arms and a dot.

Distribution and Ecology

The species is known only from the dry areas of Hispaniola and Jamaica, Greater Antilles. It was collected near Fond Parisien, Pont Beudet, and Gonaives, Haiti, and near Hatillo (east of Azua), Dominican Republic, on May 7–12, 1982. It was collected near Port Henderson, Jamaica, November 23, 1983.

D. micromettleri readily comes to banana bait in thorn forests and cactus outcroppings on both islands with a variety of cacti in the vicinity. The more abundant cacti are Stenocereus hystrix and Opuntia moniliformis. The species is believed to breed in the wet soil beneath these cacti in the same manner as D. mettleri in the Sonoran Desert.

Relationship

This is the third member of the eremophila complex of the mulleri subgroup of the repleta species of the subgenus Drosophila and is most closely related to D. mettleri Heed. It may be distinguished by its smaller size and by the slightly different shape of the aedeagus.

Types

The holotype male and the paratypes from Hatillo, Dominican Republic, have been deposited in the American Museum of Natural History, New York. Other paratypes have been deposited in the National Drosophila Species Resource Center, Bowling Green, Ohio.

13

The Dynamics of Sexual Selection and Founder Effects in Species Formation

KENNETH Y. KANESHIRO

In a recent paper, Mayr (1982) made the following statement: "Speciation . . . now appears as the key problem of evolution. It is remarkable how many problems of evolution cannot be fully understood until speciation is understood. . . ." There has been renewed interest in the process of speciation, partly due to the recent arguments advanced by proponents of the punctuated equilibrium model of macroevolution. Numerous theoretical papers have been presented to address the controversy among proponents of gradualism and those of saltationism, but it is not the intent of the author of this chapter to belabor the issues involved in such a debate. Nevertheless, as put forth by West-Eberhard in a recent paper (1983), "A scientific understanding of organic diversity means understanding the multiplication of forms during the long history of life on earth. And this requires understanding the process by which new species originate—the process of speciation."

THE ORIGIN OF REPRODUCTIVE ISOLATION

It is generally held that the accumulation of genetic differences that result in reproductive isolation between daughter populations is the most important feature of the speciation process. There are two classical theories concerning the acquisition of reproductive isolation. According to one theory, that of Muller (1942) and Mayr (1963), isolation barriers arise as incidental by-products of natural selection during spatial isolation and not as a direct result of selection for reproductive isolation. That is, genetic divergence as a result of gradual genetic changes to accommodate different environmental conditions plays an important role in the origin of both premating and postmating barriers to interspecific hybridization. Although reinforcement of these intrinsic barriers may occur following breakdown of spatial isolation between the daughter populations, this would merely involve the strengthening of existing barriers. According to the theory proposed by Fisher (1930) and elaborated by Dobzhansky (1940), genetic barriers are formed only incompletely during allopatry, and are perfected through natural selection only following breakdown of extrinsic barriers and the secondary encounter of the daughter species. Here, it is proposed that genetic divergence during allopatry results in some form of intrinsic barrier such as hybrid inferiority and that selection acts against those parental genotypes that hybridize. Thus, hybridization actually strengthens interspecific isolation barriers, and premating barriers such as behavioral and ecological differences evolve as a direct response to natural selection against hybridization to prevent gametic wastage. According to this theory then, a significant part of the speciation process occurs as direct selection for reproductive isolation following sympatry, whereas in the Muller–Mayr theory, most of the changes that may be involved in preventing interspecific hybridization arise in allopatry as incidental by-products of natural selection.

Although there have been numerous empirical studies (see reviews in Mayr, 1970; Dobzhansky, 1970) that appear to lend support to the reinforcement model proposed by Fisher and Dobzhansky, the incidental model of reproductive isolation proposed by Muller and Mayr has gained considerable support in recent times. Paterson (1982) has pointed out that convincing evidence exists for the allopatric mode of speciation, a view generally accepted by most evolutionists (Futuyma and Mayer, 1980; Kaneshiro, 1980). It follows then that there would be few opportunities for natural selection to play a direct role in the development of reproductive barriers between incipient populations since such populations are spatially separated during the crucial stages of species formation. Thus, isolation barriers that may exist between biological species may have arisen simply as a pleiotropic effect of natural selection on some other phenotypic trait and not as the result of selection for isolation per se.

SPECIATION VIA THE FOUNDER PRINCIPLE

"Can an entire[ly] new species arise as a population descended from a single individual of an ancestral species?" (Carson, 1971). Although it was Mayr (1942) who first proposed the concept of the founder principle and defined it as "the establishment of a new population by a few original founders (in an extreme case, by a single fertilized female) which carry only a small fraction of the total genetic variation of the parental population. . ." (Mayr, 1963), the idea of the founder principle actually originated with Darwin's work on the Galapagos Islands. As Darwin contemplated the biota of these islands, he recognized that the fauna was derived from a few stray colonists from the adjacent continent a few hundred miles away. The fact that the fauna and flora of the Galapagos are truly endemic, and not just slight variations of continental forms, indicates that there was not a continuous influx of migrants from the mainland. The flow of genetic material between the islands and the mainland was restricted, and new populations were more likely than not to be founded by a small number of individuals or perhaps by a solitary fertilized female. These notions certainly were a major influence on Darwin's thinking and in his development of evolutionary theory (Darwin, 1859).

Mayr (1963) suggested that severe drift conditions resulted in a reduction in levels of genetic variability which then led to rearrangements of the coadapted gene complexes. Mayr referred to this phenomenon as the "genetic revolution" that accompanies founder events. He argued that these conditions led to increased homozygosity, triggering a genetic revolution that broke up the coadapted gene complexes present in the ancestral population. He then predicted that as new genetic variation accumulated in the population, new coadapted gene complexes developed and speciation could occur.

Carson (1968, 1971; see also Carson and Templeton, 1984) proposed a modified version of Mayr's founder principle which he called the "founder–flush" model. In this model, Carson suggested that although drift serves to perturb the coadapted genetic system of the founding population, the founder event itself is followed by rapid population growth due to relaxed selection pressures in the new environment. This is referred to as the "flush" phase of the model. Hence, reduction in variability is minimized as a result of the increase in population size immediately following the founder event. Furthermore, the relaxed selective conditions allow to persist in the population new recombinants that would normally have been selected against in the ancestral population. Then, as the population approaches the carrying capacity of the new habitat, natural selection once again exerts its pressures in regulating effective population size. In contrast to Mayr's model, because of the increased levels of genetic variability permitted in Carson's model, the population is more able to respond to the selective pressures present in changing environmental conditions.

The "genetic transilience" model proposed by Templeton (1980, and Carson and Templeton, 1984) is similar to Carson's founder–flush model in that it also permits high levels of genetic variation in the founder population. However, the genetic transilience model stresses the importance of pleiotropic effects of the genetic environment while deemphasizing the significance of the external environment. That is, Templeton's model predicts that small inbreeding populations that are associated with founder events may lead to stochastic alterations in the relationships of a few major alleles. Such chance alterations can result in drastic changes in the pleiotropic effects that are associated with these major loci. Again, as with Mayr's and Carson's models, such conditions as predicted by the transilience model may lead to shifts toward new adaptive peaks which may then lead to speciation.

SEXUAL SELECTION THEORY

Until recently, it has been generally accepted by most evolutionary biologists that natural selection is the most dominant force in the evolutionary history of the species. To be sure, the ubiquity of this view is well understood when one accepts the notion that populations could ill-afford the incapacity to adjust to the effects of their outer environment. Even Darwin (1871), in his discussion of sexual selection, stated that "sexual selection will also be dominated by natural selection tending towards the general welfare of the species."

Within the past decade, evolutionary biologists have been investigating the significance of sexual selection (see West-Eberhard, 1983, for an excellent review) and its influence on "the mutual adjustment of the sexes to what may be called the intraspecific sexual environments" (Carson, 1978). Of course, it was Darwin (1871) who, more than a century ago, proposed the concept of sexual selection to explain the sexually dimorphic characters he observed in a wide range of organisms. Fisher (1930) examined the genetic consequences of Darwin's idea and presented a model which inferred that female choice and male character would coevolve very rapidly ("runaway selection") in an interbreeding population. O'Donald (1977, 1980) applied a mathematical treatment to Fisher's interpretations and confirmed that the evolution of a male-limited character could result in a correlated response in the female's preference for that character. Lande (1981, 1982) developed models involving polygenic characters to confirm the runaway features of Fisher's original ideas.

In all of the models of sexual selection developed thus far, from Darwin to Lande (and others), it is assumed that two factors act to counterbalance the runaway process of sexual selection. On the one hand, female preference for a certain male character tends to select for extreme forms of that character. On the other hand, natural selection exerts its forces to maintain the optimum male phenotype to survive in its environment. Thus, an essential component of these models of sexual selection is the role of natural selection in checking the runaway process that results from the genetic coupling of male attributes and female preference. Mayr (1972) stated that "natural selection will surely come into play as soon as sexual selection leads to the production of excesses that significantly lower the fitness of the species. . . ."

Kirkpatrick (1982) examined the evolution of sexual selection, using a two-locus model in which one locus codes for a male trait and the second codes for female preference for that trait. His conclusions, for the most part, agree with Lande's polygenic models. However, Kirkpatrick extended his investigation by considering the consequences of introducing a second set of female preference for a different male trait into the population. He showed that at equilibrium there is "no inherent tendency for one mating preference to replace another," and stated that "this refutes that intuitive notion that selection will necessarily favor mating preferences for male genotypes that are superior under natural selection. . . ." Kirkpatrick concluded by writing: "The notion that evolution will necessarily extricate a species from the maladaptive tendencies of sexual selection is unfounded."

MALE MATING ABILITY AND FEMALE CHOICE

In most sexually reproducing species, the mating process can be divided into two distinct phases. The first phase involves the bringing together of the two sexes into close proximity of each other. The second phase, namely courtship, if displayed properly (i.e., in the right sequence and frequency), may result in copulation. Among some animals, the importance of the first phase is minimized because of the social organization within the species (e.g., species in which males control a harem of females) or because of permanent bonds which form during breeding seasons (monogamous species). But for other animals, males form only temporary associations during mating and make little or no investment beyond that of providing gametes. In these cases, it is generally believed that female choice plays an important role in mating, although there seems to be little empirical evidence for female choice (Thornhill and Alcock, 1983). It has been suggested that "coyness," or reluctance on the part of females to accept the courtship overtures of any male, is evidence for female choice in animal reproduction (Mayr, 1972). Female choice is also inferred from the observation that certain males are more successful in copulating with females. That is, it seems that females are able to discriminate among males and "choose" to mate with those possessing particular attributes. The idea that there is a range of male mating types segregating within an interbreeding population is certainly well accepted. In order for sexual selection to operate, there has to be variability among males from which females will "select" the best "fit" male. Thus, any heritable character that gives an individual male an advantage over other males in attracting mates will be under strong sexual selection and will increase in frequency in succeeding generations.

There has been a great deal of controversy surrounding Darwin's theory of sexual selection, due principally to confusion in the definition of fitness. It is not within the scope of this paper to resolve such controversies, but at the same time it is important to identify the constraints of the sexual selection theory as it applies to this paper. I will assume sexual selection to be the differential mating ability of males regardless of their contribution to the overall genetic fitness of the species. That is, male mating success is a manifestation of his courtship ability which is a heritable character.

Evidence for female choice is limited, and laboratory studies provide only ambiguous support for mate discrimination via female choice (see review in Thornhill and Alcock, 1983). Thus, the notion that there is a range of mating types segregating in the females as well is not as readily recognized by behaviorists The ability of females to discriminate among males is not well understood, even though it is generally accepted that there is variation in male mating ability. However, these two phenotypes are not mutually exclusive, and if indeed there is variation among males in mating ability, there must also be selection for female preference (or discrimination) for males with special features. Furthermore, there appears to be a high genetic correlation between homologous characters of the two sexes, such as body size, which may be due to pleiotropy (Lande, 1981). Similarly, there may be a high genetic correlation between female discrimination and male mating ability such that highly discriminant females will accept only those males that perform a superior courtship pattern. Although female mating preference may not be mutually pleiotropic with male mating ability, assortative mating for behaviorally superior males inevitably results in a positive genetic correlation between these two characters.

ASYMMETRICAL MATE PREFERENCE AND DIRECTION OF EVOLUTION

In a recent series of papers (Kaneshiro, 1976, 1980; Arita and Kaneshiro, 1979; Watanabe and Kawanishi, 1979; Ahearn, 1980; Wasserman and Koepfer, 1980; Heth and Nevo, 1981; Markow, 1981; and others), there has been much discussion on the usefulness of studies of mating behavior for inferring "direction of evolution" among closely related species (see Kaneshiro, 1983, and Giddings and Templeton, 1983, for reviews of these studies). In all of these studies, the investigators have reported that a frequent outcome of mate preference experiments between a pair of closely related species is asymmetrical sexual isolation. That is, females of one population (A) may accept the courtship displays of males from another population (B), resulting in successful matings; but in the reciprocal combination, females of the B population refuse matings with males of the A population. Kaneshiro (1976) first suggested that such shifts in behavior are a result of the severe drift conditions and the genetic revolution that accompany founder events. Subsequently, Kaneshiro (1980, 1983) hypothesized that the courtship requirements of females in derived populations are simplified during the early stages of founder events when the population size is reduced. A relaxation of sexual selection may accompany founder events because successful establishment of the new population is likely to be enhanced if the behavioral pattern is simplified. More specifically, during periods of small populations size, females that are less discriminant will be most likely to mate and leave progeny. On the other hand, a highly discriminant female may never encounter a suitable male who is capable of satisfying her courtship requirement. Thus, there may be a shift in the distribution of female mating types toward females that are less discriminating than others. During the genetic revolution that accompanies founder events, such a shift in the mating patterns may become fixed in the new population.

In this chapter, I present an intuitive model as a possible explanation for the role sexual selection plays in species formation. I attempt to make a case for the view that sexual selection is the most important factor in the initial stages of species formation, and that the forces of natural selection only supplement the genetic changes stimulated by sexual selection rather than acting as a counterbalancing factor. Also discussed is the dynamics of the sexual selection process and its role during small population size due to founder effects. Although there are little genetic data to support some of these ideas, some empirical data are presented which seem to indicate the validity of these concepts.

SOME EVIDENCE FROM BEHAVIORAL STUDIES OF HAWAIIAN *DROSOPHILA*

The complex courtship patterns observed in many of the Hawaiian drosophilids (Spieth, 1966, 1968, 1974) offer evolutionists an opportunity to test some of the classical concepts of sexual selection theory. Ahearn (1980), Ahearn et al. (1974), Arita and Kaneshiro (1979), Kaneshiro (1976, 1980), Kaneshiro and Kurihara (1981), Ohta (1978), and Spiess and Carson (1981) have reported on mate preference studies on various species of Hawaiian *Drosophila*. Except for Spiess and Carson's (1981) study, all of the others reported on experiments that involved mating of a single male placed in a chamber with two females (classically referred to as "male choice" experiments), one of which was homogamic with the male. The results of these studies indicated that females from ancestral populations were more discriminant in mate choice than were females from derived populations. At the same time, males from the derived populations were less successful in mating with females of either populations than were males from the ancestral population.

Barton and Charlesworth (1984) criticized the "male choice" method of mate preference experiments, suggesting that the observed asymmetries are merely the result of differential mating propensities of the females. In certain instances where these tests showed a significant excess of homogamic matings in one combination and an excess of heterogamic matings in the reciprocal combination, this argument of differences in the mating propensity of the females rather than in female choice may be a valid one. However, it is still interesting to note that even in these cases there appears to be a sequential increase in "mating propensity" from the putative ancestral females (based on geological inferences) to the most derived females but a corresponding decrease in male mating ability in the derived population. On the other hand, evidence for female choice in these experiments does exist in instances where although there might be a significant excess of homogamic matings in one combination, there is approximately equal mating propensity between the two females in the other reciprocal. This suggests that one of the females is discriminating against the heterogamic male whereas the other female is accepting the courtship overtures of both males. The combination in which both females accept the courtship of the male in approximately equal numbers serves as a control for the question of unequal mating propensity. Unless one argues that in certain combinations female mating propensity overrides the effects of female choice, whereas in others female choice is a real phenomenon, it would be less than convincing to argue that the asymmetries we have been observing in these experiments are the result of differential mating propensity in the females.

Ahearn (1980) and Arita and Kaneshiro (1979) provide some empirical evidence that the direction of asymmetry observed in these studies might be related to bottleneck effects during population crashes and the effects of founder events. Arita and Kaneshiro reported asymmetrical mate preference between two strains of *D. adiastola*, a picture-winged Hawaiian *Drosophila*. One strain had been maintained in the laboratory over a six-year period and had been subjected to a number of unintentional population crashes. The other strain had been in the laboratory for about a year and had been maintained in a large and healthy condition throughout. When mate preference experiments were conducted using these two strains, it was found that females of the older laboratory strain which had gone through several bottlenecks, had a lowered receptivity threshold and that it accepted the courtship of males from either strain. On the other hand, females from the newer strain chose to mate with males from its own strain more frequently than with males from the older strain. Thus, the direction of asymmetry is exactly as would be predicted by the asymmetry model. Ahearn (1980) showed similar results in comparisons between two strains of *D. silvestris*. Barton and Charlesworth (1984) criticized Ahearn's methodology in that she had "marked one of the two types of females she was studying with paint" and that "the pattern of asymmetry she describes is consistent with lowered mating success for the marked females. . . ." However, Kaneshiro and Kurihara (1981) reported that although it was evident that the marking method described by Ohta (1978) and subsequently by Ahearn (1980) had no effect on mating preference of the females, the validity of this method was further tested by alternating the population of females to be marked throughout their study. The asymmetries observed in Kaneshiro and Kurihara's study and hence all previous studies could not have been biased by the marking method used.

EVIDENCE FROM OTHER STUDIES

Giddings and Templeton (1983) and Kaneshiro (1983) reviewed a number of mate preference studies on organisms other than Hawaiian *Drosophila* in which similar asymmetries have been observed. Sperlich's (1964) study of *D. subobscura* in Europe, Bicudo's (1978) work on the *D. prosaltans* group, McPhail's (1969) data on behavioral isolation in the stickleback fish *Gasterosteus*, and Powell's (1978) study of *D. pseudoobscura* all appear to lend support to the asymmetry model. Except for Powell's (1978) work, these will not be elaborated further. Powell's study together with a follow-up investigation by his student D. Dodd (Dodd and Powell, 1986; and Powell, Chapter 11, this volume) deserve further comment since theirs are the first laboratory experiments designed to test the founder effect. Powell demonstrated rapid evolution of premating isolation in populations of *D. pseudoobscura* that had been subjected to a series of founder–flush cycles, although he found no evidence of postmating barriers. Unfortunately, Powell calculated a joint isolation index for both reciprocal crosses in each pairwise combination which often masked the effects of the flush–crash cycles. Indeed, Barton and Charlesworth (1984) were misled by Powell's data and noted that the strains "that were passed through bottlenecks did not show much isolation from the ancestral stock, although they were isolated from each other to some extent." However, by recalculating Powell's data, Kaneshiro (1983) reported that striking asymmetry was found in three of the eight strains when tested against the ancestral strain. In each case, the direction of asymmetry was exactly as predicted by the asymmetry model. Subsequent studies by D. Dodd (Dodd and Powell, 1986) showed that the asymmetries observed in Powell's earlier studies were repeatable in the same strains which had been maintained in Powell's laboratory over the past several years. Their study showed that the direction of asymmetry between the derived strains and the ancestral strain was still consistent with the asymmetry model and that the Kaneshiro hypothesis was upheld in all cases.

RESULTS OF SELECTION FOR HIGH AND LOW DISCRIMINATION IN FEMALES

I summarize here the results of some selection experiments (details of which will be published elsewhere) with Hawaiian *Drosophila* involving high and low mate-discrimination in females. Briefly, the selection protocol involved 10 males of *D. silvestris* which were placed separately into 10 sequentially numbered mating chambers into which single virgin females were introduced. During the peak sexual activity period (between 0730 hours and 1030 hours) for this species, matings among the 10 pairs were observed and recorded. Following this period, each female was removed from the mating chamber and placed into a separate vial overnight.

The next morning, females that had mated the previous day were replaced with virgin females, and each of the 10 females were placed in the next higher-numbered chamber. That is, female from chamber 1 was now placed into chamber 2, that from chamber 2 into chamber 3, and so on, with the female in chamber 10 now being placed into chamber 1. Females were rotated in this manner in each successive mating period over a five-day period so that each male would have had a chance to court and mate with five different females. Out of 10 replicates of 10 males (i.e., total of 100 males), 150 matings were observed (Table 13.1). Thirty-four males failed to mate at all after seeing five different females each; 39 males mated with one or two females (54 matings or 36% of the matings); whereas 27 males mated with three, four, or five females (96 matings or 64% of the matings). Carson (personal communication), in an independent set of experiments, observed nearly identical proportions of males mating with zero, one or two, and three to five females. Out of each replicate, there were also a small number of females that appeared to be highly discriminant, and did not accept the courtship of any of the five males she had encountered. These females were then placed in a chamber with those males that had mated with the most females during the five-day period. It was postulated that progeny from these kinds of matings would contain daughters that were also highly discriminant and sons that had superior mating qualities. When these strains were tested against the original strain in mate choice experiments, significant asymmetries were observed in exactly the direction predicted by the asymmetry model (Table 13.2). That is, females from the selected "high" line were highly discriminant against the courtship overtures of males from the original strain, whereas males from the selected line were successful in satisfying the courtship requirements of females from both strains.

It was also possible to select for the opposite extreme—low discriminant females. This was accomplished by taking the 34% of the males that were not able to satisfy the courtship requirements of the five females placed in their chambers on successive days, and each day placing a virgin female into each of their chambers. When a female finally accepted the courtship ritual of one of these males, it was hypothesized that this female was one of very low discrimination and that progeny produced from these kinds of mating would contain daughters that were also non-discriminant and sons that had poor mating ability. Again, when these strains were tested against the original strain in mate preference experiments, significant asymmetries were observed in exactly the direction predicted by the asymmetry model but in the exact opposite of that observed in the tests with the "high" line (Table 13.2). That is, females from the "low" line accepted males from its own line as well as those from the original strain, but females from the original strain discriminated against males from the selected low line.

The data from the selection experiments clearly indicate that there is indeed a range of mating types segregating in both sexes and that there is

Table 13.1. Mating frequencies of 100 *D. silvestris* males

| | Number of females mated with | | | | | | |
	0	1	2	3	4	5	Totals
Number of males	34	24	15	16	7	4	100
Number of matings	0	24	30	48	28	20	150
Frequency of matings	0	0.16	0.20	0.32	0.19	0.13	1.0
Percentage of males mating with 0, 1–2, or 3–5 females (percentage of females mated with)	34%(0%)	39%(36%)		27%(64%)			

Table 13.2. Mate preference tests among strains of *D. silvestris* selected for high and low female discrimination

Male	Female		Total matings	Frequency, homogamic	Frequency, heterogamic	I	C[a]
U34B4 orig	U34, orig	23d low	31	0.39	0.61	−0.32	−1.23
23d5D low	U34, orig	23d low	37	0.68	0.32	+0.36	+2.19
U34B4 orig	U34, orig	8k high	33	0.73	0.27	+0.46	+2.64
8k10K high	U34, orig	8k high	31	0.29	0.71	−0.42	−2.34

Symbols: U34B4, original male; U34, original female; 23d5D, low line male; 23d, low line female; 8k10K, high line male; 8k, high line female.

[a]At the 5% level, the null hypothesis that mating is random is accepted when $-1.96 < C < +1.96$.

a strong correlation between female discriminatory ability and male mating ability in this species. By selecting for both of these characters simultaneously, it was possible to obtain selected lines at both ends of the distribution in as few as two generations of selection.

AN ALTERNATIVE EXPLANATION FOR SOME CLASSICAL CONCEPTS OF SEXUAL SELECTION THEORY

In light of the results of the studies discussed above, I am proposing an alternative explanation for some of the classical ideas about the sexual selection theory. First, it is a widely accepted view that the Fisherian runaway process due to sexual selection is counterbalanced by the forces of natural selection when such sexually selected characters develop into exaggerated features that reduce the survivability of the individuals that inherit them. This had led to the notion that sexual selection based on female choice may be a stabilizing agent in maintaining an optimal sexually selected phenotype (Carson and Teramoto, 1984). Carson and Bryant (1979) reported on the discovery of an incipient morphotype in populations of *D. silvestris* and described the evolution of a new character which they believed to be important in the courtship ritual of the males. Males from populations on the southwestern (Kona) side of the Island of Hawaii had only two rows of cilia on the antero- and posterodorsal surface of the front tibia, whereas males from the northeastern (Hilo) side had an extra row of cilia on the dorsal surface in between the two rows. The extra row of cilia in populations on the Hilo side of the island were deemed to be a derived character since the related ancestral species on the adjacent older islands had only two rows of cilia on the foretibia. In some selection experiments involving high and low cilia number in a laboratory strain of the extra row population, Carson and Teramoto found that "selection in either direction away from the population mean produces dysgenic lines that cannot be maintained in the laboratory. This suggests that, in the natural population of *silvestris*, some form of stabilizing selection is operating to maintain an optimal number of cilia." Because the cilia on the foretibia is known to be important in the courtship repertoire of the males, Carson and Teramoto suggested that sexual selection is the stabilizing force. However, Kaneshiro and Kurihara (1981) showed that in mate preference tests, males from the Kona side populations were more successful in mating with females from the Hilo side than were males from their own population. These data seem to indicate that the extra row of cilia may not be under strong sexual selection. Rather, it is suggested that the extra row of cilia is maintained pleiotropically, owing to selection on some other phenotypic (behavioral) trait.

I propose the following ideas. If we assume that there is a range of mating types segregating within a single interbreeding population with highly successful males and unsuccessful males (Fig. 13.1A), and highly discriminant females and nondiscriminant females (Fig. 13.1B), one might imagine the following scenario. In this population, highly successful males would obviously be selected for, since these types of males are able to satisfy the courtship requirements of most of the females in the population. On the other hand, highly discriminant females would be selected against, since the frequency of highly successful males are low, and therefore, males that would be able to satisfy the courtship requirements of these females would be encountered only infrequently. Thus, there may be a differential selection for mating types in the two sexes; that is, selection *for* highly successful males but selection *against* unsuccessful males; selection *for* less discriminant females but selection *against* highly discriminant females. Because of the genetic correlation between these two phenotypes (i.e., male mating ability and female discrimination), the differential selection between the two sexes acts as the stabilizing agent for the sexual selection process (Fig. 13.2).

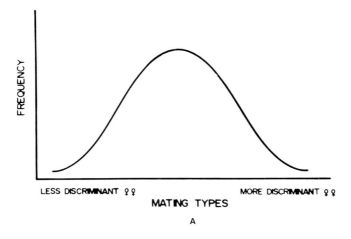

LESS DISCRIMINANT ♀♀   MORE DISCRIMINANT ♀♀

MATING TYPES

A

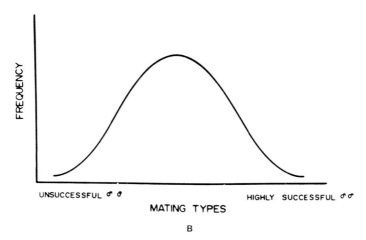

UNSUCCESSFUL ♂♂   HIGHLY SUCCESSFUL ♂♂

MATING TYPES

B

Figure 13.1. (**A**) Distribution of female mating types segregating in a single inter-breeding population. (**B**) Distribution of male mating types segregating in a single interbreeding population.

Now let us suppose that the extra row of cilia in *silvestris* is a pleio-tropic effect of male mating ability but not necessarily a character that is directly acted upon by sexual selection. If, indeed, males with superior mating ability have a higher cilia number than less successful males, and if, indeed, the hypothesis of differential selection between the sexes pro-posed above is the stabilizing force, then one could easily maintain the polymorphism for the cilia number observed by Carson and Bryant (1979) and Carson and Lande (1984). One could also explain the reason why females from the Hilo side would readily accept males from the Kona side which lack the extra row of cilia since, in this case, it is not the extra row of cilia that is under strong sexual selection. Furthermore, under this

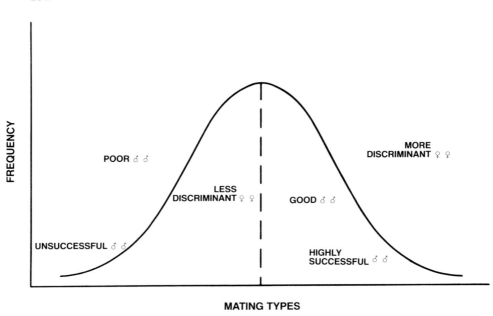

Figure 13.2. Graph depicting "differential selection model": Whereas less discriminant female and behaviorally "good" males are selected *for*, highly discriminant females and behaviorally "poor" males are selected *against*. See text for further discussion.

model, it is not necessary to invoke natural selection to counter the runaway process proposed by Fisher and others.

Additional evidence for the views discussed above derives from data on cilia number in males from the same lines selected for high and low female discrimination (Val, Carson, and Kaneshiro, unpublished data). It was observed that there was a significant increase in mean cilia number in the foretibia of males from the "high" (behaviorally selected) line and a dramatic decrease in mean cilia number in the "low" line (Fig. 13.3). These data would seem to indicate either a genetic correlation or a pleiotropic effect between cilia number and female discrimination and male mating ability. With this system, and given the validity of the differential selection model, it is clear how polymorphism in cilia number can also be maintained in the natural population.

DIFFERENTIAL SELECTION IN THE SEXUAL SELECTION PROCESS AND ITS INFLUENCE ON SPECIATION

Given the possibility of differential selection between the two sexes, the sexual selection process can be viewed as a truly dynamic system. There can be shifts in the frequency of mating types segregating in the population during small population size or during founder events, for example.

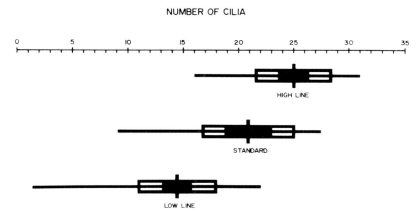

Figure 13.3. Effect of selection for high and low female-discrimination on cilia number on dorsal surface (row 5a) of front tibia of male *D. silvestris*. The diagram represent the ranges, means, standard deviations, and 95% confidence limits (darkened) of cilia number in the high and low lines, compared with the standard laboratory strain.

As discussed above, during founder events, there would be selection against those females that are highly discriminant in mate choice since the probability of encountering males that would satisfy their courtship requirements is much reduced. During the relaxed selection phase of Carson's founder–flush model (Carson, 1968, 1982; Carson and Templeton, 1984), there might be a shift in the frequency toward less discriminant females because these are the females most likely to contribute genetic material to the next generation. As the population grew in size, sexual selection would once again become a dominant force in mate selection and would play an important role in effective population structure.

The flexibility of the mating system in permitting such shifts in the distribution of mating types is crucial especially during the initial stages of colonization of a new habitat. Otherwise, if sexual selection acted merely as a stabilizing force in maintaining an optimal phenotype, the colonization and establishment of a new population following a founder event would be extremely rare. During such shifts, there might be corresponding shifts in the overall genetic environment of the population such that various coadapted genetic systems might be affected. The destabilized genetic environment sets the stage for changes that are conducive to speciation. Note that I do not imply in any sense that the founder event is a macroevolutionary event and that such an event results in an "instant" new species. Rather, I suggest that the phenomenon described above is merely the "entering wedge" of the speciation process and that the incipient population continues to respond to the forces of sexual selection as well as natural selection in completing the speciation process (see also Ringo, 1977); thus small genetic changes accumulate over many generations until irreversible reproductive barriers evolve.

CONCLUDING REMARKS

For nearly a century, the role of sexual selection as an important factor in the speciation process has been largely ignored by evolutionary biologists. Only within the past decade has there been renewed interest in investigating the sexual environment as a major component of genetic adaptations during speciation. The complex balance between male and female behavior may be easily perturbed during small population size such as during founder events, and may result in disorganization of coadapted genetic systems. The dynamics and plasticity of the sexual selection process, as discussed above, permit the population to overcome any effects of such drastic reduction in population size and to recover from the genetic disorganization that may accompany such populational events. In rebuilding its coadapted system, selection may cause a shift to a novel adaptive peak which may become reproductively isolated from the parental population. Thus, sexual selection may be a pivotal feature of the speciation process and may indeed play a prominent role in the origin of new species.

LITERATURE CITED

Ahearn, J. N. 1980. Evolution of behavioral reproductive isolation in a laboratory stock of *Drosophila silvestris*. Experientia 36:63–64.

Ahearn, J. N., H. L. Carson, Th. Dobzhansky, and K. Y. Kaneshiro. 1974. Ethological isolation among three species of the planitibia subgroup of Hawaiian *Drosophila*. Proc. Natl. Acad. Sci. USA 71:901–903.

Arita, L. H., and K. Y. Kaneshiro. 1979. Ethological isolation between two stocks of *Drosophila adiastola* Hardy. Proc. Hawaii. Entomol. Soc. 13:31–34.

Barton, N. H., and B. Charlesworth. 1984. Genetic revolutions, founder effects, and speciation. Annu. Rev. Ecol. Systematics 15:133–164.

Bicudo, H. E. M. C. 1978. Reproductive isolation in *Drosophila prosaltans* (saltans group). Brazil. J. Genet. 1:11–28.

Carson, H. L. 1968. The population flush and its genetic consequences. In R. C. Lewontin (ed.), Population Biology and Evolution, pp. 123–137. Syracuse University Press, Syracuse, NY.

Carson, H. L. 1971. Speciation and the founder principle. Stadler Genet. Symp. 3:51–70.

Carson, H. L. 1978. Speciation and sexual selection in Hawaiian *Drosophila*. In P. F. Brussard (ed.), Ecological Genetics: The Interface, pp. 93–107. Springer-Verlag, New York.

Carson, H. L., and P. Bryant. 1979. Genetic variation in Hawaiian *Drosophila*. VI. Change in a secondary sexual character as evidence of incipient speciation in *Drosophila*. Proc. Natl. Acad. Sci. USA 76:1929–1932.

Carson, H. L., and R. Lande. 1984. Inheritance of a secondary sexual character in *Drosophila silvestris*. Proc. Natl. Acad. Sci USA 81:6904–6907.

Carson, H. L., and A. R. Templeton. 1984. Genetic revolutions in relation to speciation phenomena: the founding of new populations. Annu. Rev. Ecol. Systematics 15:97–131.

Carson, H. L., and L. T. Teramoto. 1984. Artificial selection for a secondary sexual character in males of *Drosophila silvestris* from Hawaii. Proc. Natl. Acad. Sci. USA 81:3915–3917.

Darwin, C. 1859. The Origin of Species. Modern Library, New York.

Darwin, C. 1871. The Descent of Man and Selection in Relation to Sex. Modern Library. New York.

Dobzhansky, Th. 1940. Speciation as a stage in evolutionary divergence. Am. Natur. 74:312–321.

Dobzhansky, Th. 1970. Genetics of the Evolutionary Process. Columbia University Press, New York.

Dodd, D. M. B., and J. R. Powell. 1986. Founder–flush speciation: an update of experimental results with *Drosophila*. Evolution 39(6):1388–1392.

Fisher, R. A. 1930. The Genetical Theory of Natural Selection. Clarendon Press, Oxford.

Futuyma, D. J., and G. C. Mayer. 1980. Non-allopatric speciation in animals. Syst. Zool. 29:254–271.

Giddings, L. V., and A. R. Templeton. 1983. Behavioral phylogenies and the direction of evolution. Science 220:372–377.

Heth, G., and E. Nevo. 1981. Origin and evolution of ethological isolation in subterranean mole rats. Evolution 35:259–274.

Kaneshiro, K. Y. 1976. Ethological isolation and phylogeny in the planitibia subgroup of Hawaiian *Drosophila*. Evolution 30:740–745.

Kaneshiro, K. Y. 1980. Sexual isolation, speciation and the direction of evolution. Evolution 34:437–444.

Kaneshiro, K. Y. 1983. Sexual selection, and direction of evolution in the biosystematics of Hawaiian Drosophilidae. Annu. Rev. Entomol. 28:161–178.

Kaneshiro, K. Y., and J. S. Kurihara. 1981. Sequential differentiation of sexual isolation in populations of *Drosophila silvestris*. Pac. Sci. 35:177–183.

Kirkpatrick, M. 1982. Sexual selection and the evolution of female choice. Evolution 36:1–12.

Lande, R. 1981. Models of speciation by selection on polygenic traits. Proc. Natl. Acad. Sci. USA 78:3721–3725.

Lande, R. 1982. Rapid origin of sexual isolation and character divergence in a cline. Evolution 36:213–223.

Markow, T. A. 1981. Mating preferences are not predictive of the direction of evolution in experimental populations of *Drosophila*. Science 213:1405–1407.

Mayr, E. 1942. Systematics and the Origin of Species. Columbia University Press. New York.

Mayr, E. 1963. Animal Species and Evolution. Harvard University Press, Cambridge, MA.

Mayr, E. 1970. Populations, Species, and Evolution. Belknap Press of Harvard University Press, Cambridge, MA.

Mayr, E. 1972. Sexual selection and natural selection. In B. Campbell (ed.), Sexual Selection and the Descent of Man, 1871–1971, pp. 87–104. Aldine, Chicago.

Mayr, E. 1982. Processes of speciation of animals. In C. Barigozzi (ed.), Mechanisms of Speciation, pp. 1–19. Alan R. Liss, New York.

McPhail, J. D. 1969. Predation and the evolution of a stickleback *(Gasterosteus)*. J. Fish. Res. Bd. Canada 26:3183–3208.

Muller, H. J. 1942. Isolating mechanisms, evolution and temperature. Biol. Symp. 6:71–125.

O'Donald, P. 1977. Theoretical aspects of sexual selection. Theor. Popul. Biol. 12:298–334.

O'Donald, P. 1980. Genetic Models of Sexual Selection. Cambridge University Press, Cambridge.

Ohta, A. T. 1978. Ethological isolation and phylogeny in the grimshawi species complex of Hawaiian *Drosophila*. Evolution 32:485–492.

Paterson, H. E. H. 1982. Perspective on speciation by reinforcement. S. Afr. J. Sci. 78:272–275.

Powell, J. R. 1978. The founder–flush speciation theory: an experimental approach. Evolution 32:465–474.

Ringo, J. M. 1977. Why 300 species of Hawaiian *Drosophila?* The sexual selection hypothesis. Evolution 31:694–696.

Sperlich, D. 1964. Chromosomale Strukturanalyse and Fertilitatsprufung an einer Marginalpopulation von *Drosophila suboobscura*. Z. Verebungsl. 95:73–81.

Spiess, E. B., and H. L. Carson. 1981. Sexual selection in *Drosophila silvestris* of Hawaii. Proc. Natl. Acad. Sci. USA 78:3088–3092.

Spieth, H. T. 1966. Courtship behavior of endemic Hawaiian *Drosophila*. Univ. Texas Publ. 6615:245–313.

Spieth, H. T. 1968. Evolutionary implications of the mating behavior of the species of *Antopocerus* (Drosophilidae) in Hawaii. Univ. Texas Publ. 6818:319–333.

Spieth, H. T. 1974. Courtship behavior in *Drosophila*. Annu. Rev. Entomol. 19:385–405.

Templeton, A. R. 1980. The theory of speciation via the founder principle. Genetics 94:1011–1038.

Thornhill, R., and R. Alcock. 1983. The Evolution of Insect Mating Systems. Harvard University Press, Cambridge, MA.

Wasserman, M., and H. R. Koepfer. 1980. Does asymmetrical mating preference show the direction of evolution? Evolution 34:1116–1124.

Watanabe, T. K., and M. Kawanishi. 1979. Mating preference and the direction of evolution in *Drosophila*. Science 205:906–907.

West-Eberhard, M. J. 1983. Sexual selection, social competition, and speciation. Q. Rev. Biol. 58:155–183.

VII

FOUNDER EFFECTS AND GENETIC CHANGES IN NATURAL POPULATIONS

14

Human Evolution and the Founder–Flush Principle

JAMES V. NEEL

In 1984, for the first time, I publicly took major issue with the advocates of "punctuated equilibria" in evolution, at least as applied to hominid evolution (Neel, 1984). In this chapter, I reiterate and expand that thesis. In a book that has such distinguished zoologists as contributors, I hesitate to stray beyond a consideration of hominids, but perhaps I might be permitted, in my closing remarks, to advance a few cautious generalizations.

THE FACTS OF HOMINID EVOLUTION

At the moment there seems reasonable agreement that the human line of evolution separated from the other primates some 5–8 million years ago. Although I am sure we all would agree on the problems incurred in assigning species status in a continuous time series such as this, most students of human evolution seem comfortable with the position that the degree of morphological change over this interval permits the recognition of four human species. If we also agree that the rate of evolution should be measured in terms of generations rather than absolute time, then it follows that this constitutes relatively rapid evolution: Assuming an average of 15–20 years per generation, this process occurred in approximately 250,000–500,000 generations. A number of the advocates of the concept of punctuated equilibrium in the evolutionary process (Eldredge and Gould, 1972; Gould, 1977; Gould and Eldredge, 1977; Stanley, 1981) have in some of their writings cited the human as an example of their concept. They would argue that there were relatively long periods within these 5–8 million years when little or no evolution was occurring, the speciations being crowded into quite brief periods, during which, given the high average rate of change, the rate of evolution was almost explosive.

The pursuit of their thesis has been relatively unimpeded by a strong fossil record. This situation is changing rapidly (reviewed in Tobias, 1982), and it may soon be possible to address the question from the fossil record alone. In this chapter, I approach the question from a somewhat different standpoint. I suggest that recent developments in our understanding of the population structure of primitive, tribal man, as well as new knowledge concerning human mutation rates, provide an adequate framework for evolution of this rapidity. I suggest, furthermore, that although there may have been some irregularities in the tempo of the hominid evolutionary process, it is extremely difficult to conceive of relatively long periods of stasis, given the population structure I will be describing.

Over the past 20 years, our group has been conducting a many-faceted study of some of the least acculturated Amerindian tribes still to be found in South America, as well as companion studies on certain more acculturated but still genetically intact tribes in both South and Central America. Dozens of colleagues, only a few of whom will be mentioned in this brief presentation, have contributed to whatever success we have enjoyed. Our principal focus of interest has been the Yanomama, of Southern Venezuela and Northern Brazil, a tribe of some 12,000–15,000 persons whose first sustained contacts with non-Indians began in the early 1950s. Although in what follows I draw on our findings from this group as illustrative of various facets of human population structure during the generations of human evolution, I want to stress, very strongly, that we are aware of the many ways the Yanomama must differ from more primitive cultures. In particular, the Yanomama now derive some 70–80% of their total calories from the produce of slash-and-burn agriculture. This fact alone should have major impact on diet, population densities, migrational patterns, etc. However, they are the closest approach to truly primitive man which exists in our hemisphere. Unfortunately, the surviving hunter-and-gatherers to be found elsewhere in the world, who might be expected to more closely approximate the population structure of truly primitive man than do the Yanomama, are probably so modified by their contacts with civilization as to be in many respects even less representative of primitive man than are the Yanomama. I would also scarcely maintain that the differences discussed below between the social structure of tribal humans and a chimpanzee band arose full-blown at the time of the divergence of the two lineages; the social structure we encounter now has its own evolutionary history.

SOME FACTS CONCERNING PRIMITIVE MAN

We turn now to consider some of the essential details of the primitive cultures we have studied.

Mutation Rates

The genetic variation between individuals upon which selection acts is of course the accumulation of mutation pressures over many generations. We are accustomed to thinking of "current" mutations—that is, in the preceding generation—as making a relatively small contribution to the total. This may be so, but it appears that this relatively small contribution may be larger than we previously believed. Since the mid-1970s, our group has been pursuing two different approaches to the question of human mutation rates. Both are geared to the frequency with which mutation alters the electrophoretic behavior of a battery of proteins selected for study only because they may be conveniently and clearly resolved in an electrophoretic system—that is, with no reference to their genetic variability. One approach is direct; the other, indirect.

The indirect approach can be pursued only in relatively undisturbed tribal populations. It necessitates making a number of assumptions, none of which is fully met. One surveys a tribe for the average frequency per protein examined of "rare" variants, whose frequency is presumably maintained by mutation pressure. (For these purposes we define a "rare variant" as one that is found in less than 2% of the population, or, if more frequent, is restricted to a single tribe, or to closely related adjacent tribes.) Thus, for these purposes we exclude from consideration polymorphisms of wide distribution. On the assumption that the tribe is in genetic equilibrium (i.e., constant population size, no in- or outmigrants) and that the variants are neutral in their phenotypic effects and destined for ultimate loss, one can calculate the mutation rate necessary to account for the findings, by at least three different formulae. Elsewhere we have discussed in some detail the effect of departures from the assumptions on the estimates (Neel et al., 1986). Our most recent estimate of this mutation rate, based on some 544,038 locus tests in 12 different tribes, is 1.3×10^{-5}/locus per generation (Neel et al., 1986). It is for many reasons difficult to attach a standard error to that estimate, but it is approximately 0.4×10^{-5}. This mutation rate pertains to nucleotide substitutions that result in an amino acid substitution which alters either the net charge of the protein or its configuration in such a way that electrophoretic mobility is changed. It appears that about one-third of all possible nucleotide substitutions will have that effect. The corrected estimate of the mutation rate therefore becomes approximately 4×10^{-5}/locus per generation. This estimate does not include the frequency of mutation resulting in small deletions, duplications, and insertions.

The direct electrophoretic approach to the study of mutation is much more advantageously pursued in a civilized setting, but limited data can be obtained also for tribal populations. With this approach, one simply examines a series of children for electrophoretic variants not present in either parent. When such a variant is found, one carries out the most detailed studies possible to detect any discrepancy between legal and biological parentage. With current methods, such discrepancies can be detected with about 98% accuracy. Recently, we summarized the data of our group and of others on this point (Neel et al., 1986). The mutation rate for the total sample was 0.3×10^{-5} locus per generation, with 95% confidence limits of 0.1×10^{-5} and 0.8×10^{-5}.

Taken at face value, this rate is about one-quarter the rate estimated from the Amerindian studies. Such are the errors of estimation that the difference is of borderline statistical significance. Currently, I have adopted the position that although mutation rates are possibly somewhat higher in tropical-dwelling tribal populations, it is somewhat surprising how well the two estimates agree, given the many assumptions of the indirect approach.

Let us consider the implications of even the lower of these two estimates for total gametic mutation rates. If we assume that it requires 1000 nucleotides to specify the average protein in our battery, then the nucleotide mutation rate becomes $[(0.3 \times 3) \times 10^{-5}]/1000 \cong 1.0 \times 10^{-8}$. With *at least* 2×10^9 nucleotides in the human haploid genome, the gametic mutation rate becomes at least 20 nucleotide substitutions per gamete. This rate, again, does not include the occurrence of small duplications or deletions, for which the detailed analysis at the DNA level of genetic variants is providing so much evidence. However, although it is clear from the studies of the globin and hypoxanthine phosphoribosyltransferase (HPRT) loci, among others, that such events do occur, there are as yet no good estimates of their relative frequency in the germ line. If they were only half as frequent as nucleotide-type mutations, this would bring the gametic rate up to some 30 mutational events per gamete. This to me is a staggering estimate. How does an animal such as the human, with a low reproductive potential, accommodate such rates? The point to be made in the present context, however, is simply that the amount of variation fed into the system by mutation appears to be sufficient to support a continuous, rapid rate of evolution; we do not need to postulate special bursts of change in the course of human evolution.

"Rogue Cells" in Human Lymphocyte Cultures: Is There Genomic Clustering of Mutational Events?

In the early 1970s, we made chromosomal studies a special focus of one of our expeditions to the Yanomama, examining 100 cultured lymphocytes from each of 49 individuals for evidence of chromosomal damage. The expectation was that, living as far as they did from herbicides, pesticides, automobile exhaust, and chemical dumps, their chromosomes should be remarkably free of the kinds of damage encountered in about 2% of the cells cultured from individuals enjoying the blessings of civilization. To our astonishment, about 1 in each 200 cells exhibited extreme cytogenetic damage (Bloom et al., 1970). Two years later, when we attempted to repeat the observation on individuals from the same village (but not necessarily the same persons), the frequency of these cells had decreased to 1 in 5000 (Bloom et al., 1973). These observations disappeared quietly beneath the great waves of "DNA-ness" breaking over genetics; I am sure that most of our colleagues thought that the genetic damage we had observed had resulted from too much tropical heat and contamination by too many tropical organisms. One of the few references to the finding is in a review by Cowell (1982), who pointed out that what we had referred to as "numerous fragments" were, in fact, identical in appearance with the "double minutes" seen in the cells of some patients treated for malignancies with a chemical agent, most notably methotrexate.

Then, in 1983, Hsu published a photograph of one such cell he had encountered in an otherwise cytologically unremarkable, presumably Caucasian person living in the United States. Fox et al. (1984) then observed, among specimens from 153 commercial and sports divers studied in the United Kingdom, from each of whom 100 cultured lymphocytes were examined, one or more such cells in each of six men. No such cells were observed in 127 controls studied with identical techniques. The subjects were presumably Caucasian. Tawn, Cartmel, and Pyta (1985), in a study designed to collect base line data by scoring 200 cultured lymphocytes from each of 12 young, presumably Caucasian subjects, found such cells in two individuals; when the scoring was extended to 500 cells, there were four such cells in one man and five in another. When these two persons were restudied three months later, the observation could not be repeated, in keeping with our own experience.

Several years ago, on one of my annual trips to Japan in connection with our genetic studies in Hiroshima and Nagasaki, Dr. Awa, in charge of the extensive cytogenetic program there, related that he had been seeing the kinds of cells we had described, unrelated to radiation. These observations have now been published (Awa and Neel, 1986). Figure 14.1 shows two such cells, from a single individual. Altogether, over a span of 18 years, 24 such cells were observed in a total of 102,170 cultured lymphocytes derived from 9818 persons. None of these persons had been exposed to radiation from the atomic bombs. The frequency of these cells is thus about 1 in 4000. Their appearance is not related to the age or sex of the individual, or the year or season of study.

There can now be no doubt that such cells are a fact of lymphocyte life. The mechanism responsible for their production remains completely mysterious. Whether they occur in other somatic cells or in germ cells is completely unknown. Their biological significance could be trivial or enormous. Most of these cells will of course self-destruct at the first cell division in which they engage. Let us assume, however, that a small fraction of the least affected successfully navigate mitosis and, in the germline, even meiosis. Do we then have, in a somatic cell in which the rearrangement involves new juxtapositions for oncogenes, the basis for a

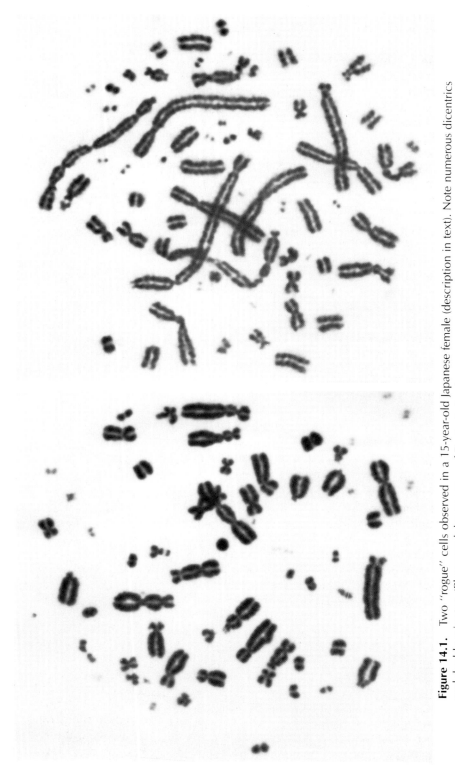

Figure 14.1. Two "rogue" cells observed in a 15-year-old Japanese female (description in text). Note numerous dicentrics and double minutes. (Photograph by courtesy of Dr. A. A. Awa)

clonal malignancy? And, in a gamete, is this the basis for the chromosomal reorganization so widely discussed in evolutionary theory? In this connection, I point out that in view of the very heavy selection against these cells at mitosis, their frequency of occurrence must be very significantly higher than the frequency with which they are observed.

How general this phenomenon is throughout the animal and plant kingdoms has yet to be determined. No other species has been subjected to the amount of karyotyping of presumably normal cells as has *Homo sapiens*. A phenomenon of this rarity could easily have escaped attention in even such genetically well-studied organisms as *Drosophila* and the mouse. Although a burdensome undertaking, it would be of great interest to have comparable data for the mouse. The nature of the mouse karyotype has posed difficulties for classical cytogenetics, but the phenomenon under discussion should be readily recognizable.

There is as yet no clear evidence from humans for a role—in germinal mutation—of transposons, mutator genes, or hybrid dysgenesis. Subjects receive a great deal of attention in an experimental setting, and in the very nature of the situation it may be extremely difficult to identify such phenomena in human populations. On the other hand, it seems unlikely that humans are completely immune to all such factors. I suggest that with or without such phenomena, we are beginning to sense much greater frequencies and potentialities for mutational events than seemed possible only a few years ago.

Differential Fertility

Let us now turn from a consideration of the possibilities for change at the genomic level to the possibilities for change at the population level, the most obvious mechanism for the latter being differential fertility. I would suggest there are certain advantages to studying this subject in human populations, with whose members one can communicate much more readily than with members of most other animal populations, and among whom the bonds of marriage largely determine gene flow. A certain measure of differential fertility is inevitable in any society in which death may intervene during the reproductive period. Human social structure provides for additional differential fertility. As has been observed for most other primitive groups, Amerindian villages have headmen. The institution of headmanship is probably ancient in hominid history, blending into the position of the dominant male in primate troops. The Yanomama are polygynous; Chagnon and collaborators (1979) have demonstrated that the Yanomama headmen, by virtue of the additional wives which are their prerogative, father twice as many children as the average Indian.

This is only the tip of the differential fertility iceberg. To obtain a more complete picture, we developed a Monte Carlo type of simulation program which, incorporating the observed demographic parameters of several Yanomama villages, enabled us to extrapolate where observation and the very inadequate histories supplied by the Yanomama failed (MacCluer et al., 1979). The input to this simulation was the actual composition—by age, sex, and pedigree relationships—of four Yanomama villages. We identified in this population a cohort of 86 boys aged 0–9 years, and asked, how many grandchildren does the simulation program predict? Forty-seven of the males left no grandchildren. At the other extreme, one male left 33. The most common expectation was 1–5. If our simulation results are correct—and there is of course no way to be certain, short of living with the population for two generations—the possibilities for founder effect and genetic change in allele frequencies in this structure surpass what most of us might have imagined. Whereas male strategy is to maximize reproductive differentials, female strategy is to strive for an even spacing of children—a spacing imposed by the need to remain mobile and by the necessity for prolonged breastfeeding of infants. Under these circumstances, selection on given traits may be operating somewhat differently in males versus females.

The differential in reproductive expectation for a cohort of young males we have just described is in part the result of an estimated prereproductive mortality of about 50%, this figure inclusive of infanticide. Although there is undoubtedly a genetic component in who lives and who dies, who reproduces and who does not, our study of the Yanomama has impressed upon me the extent to which survival and reproduction are accidents of time and place. This is to say, there is a strong stochastic element in tribal population structure. Our simulations and mathematical treatments reveal that selective coefficients of as much as 5% actually have a relatively small impact on the survival of a mutation (Neel and Thompson, 1978; Thompson and Neel, 1978). This finding requires us to postulate that many of the mutations or gene combinations that contributed to human evolution conferred rather larger selective advantages than those conventionally visualized in human genetics.

Village Propagation by Fission–Fusion

The demographic characteristics of the Yanomama are such that they are increasing at the rate of about 0.5–1.0% per year (Neel and Weiss, 1975). Although this rate is obviously much higher than the average rate of increase for primitive man, the result permits us to document an extremely important aspect of primitive population structure. As villages expand, they reach a point at which either because of decreasing surrounding resources or because of tensions within the village, they tend to split or fission. One faction stays within the current village, whereas the other moves off to found a new village. These splits are highly structured socially, usually along lineal lines. We have been so fortunate in the course of our fieldwork as to come upon three pairs of villages, each pair of which is the product of a fission event within the preceding 10 years. We could thus reconstitute the original village more or less, and, on the assumption of simple hypergeometric sampling, determine how random the split had been. Available for the quantification of this question were, for each individual, the results of typings for seven codominant, polymorphic genetic systems, from which a genetic distance function, Δ, could be calculated (Smouse et al., 1981).

One of these splits was atypical. A new missionary came into the field and built his home just across a river from an existing missionary home. About half of the village which had located near the established missionary now moved across the river, the motivation being, we believe, to have better access to the trade goods brought in by the missionary. The two villages remained on excellent terms. The other two fissions were more typical, the villages separating by one or two days' walk, and being on less than friendly terms. We developed an expected distribution for the genetic divergence between the two fission products *if there was no social structure guiding the fission,* by generating an empirical expectation based on 10,000 repetitions of the split directed by a random number generator. Figure 14.2 illustrates the results for the two unfriendly fissions. For both of these, the genetic distance between the two products is well outside the expectation if the split had been unstructured.

It seems reasonable to assume that as hominid populations extended throughout the world, they propagated in much this fashion although, in preagricultural days, the units were undoubtedly smaller. To anticipate a point to be made later, I see a clear analogy between a small group detaching itself from the main body of a tribe, to become the nidus of a new

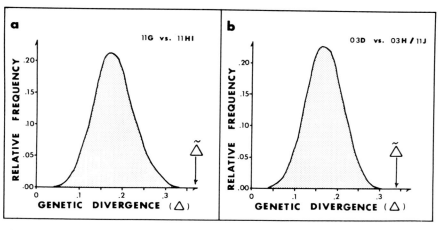

Figure 14.2. The genetic result of two village fissions, each resulting in two "daughter" villages. The curve represents the distribution of Δ (genetic divergence) values expected from simple hypergeometrical sampling of each of the parent villages. The arrow indicates the value actually observed in each instance. Further discussion in text.

tribe, this pattern occurring repeatedly, and the pattern of "island hopping" thought to be so important in the speciation of the Hawaiian *Drosophila*.

Village and Tribal Microdifferentiation

Because of the genetic typings mentioned in the previous section, it is possible to be quite specific concerning the degree of genetic differentiation between the some 50 Yanomama villages that we have studied. Allele frequencies for seven codominant genetic systems studied in these villages have been incorporated into a measure of genetic distance (Δ) between villages (Ward et al., 1975 et seq.). The extent of the intervillage genetic differences is expressed in the statement that the average distance between Yanomama villages is about 70–80% as great as the genetic distance between the Amerindian tribes we have studied. The diversity between tribes, in turn, is expressed in the statement that it is about 50% as great as between the major ethnic groupings of humankind (cf. Smouse et al., 1982).

A question repeatedly raised is, how well does the pattern of diversity revealed by genetic markers agree with the pattern revealed by various morphological traits, since the latter, although having a genetic basis, may also be molded by environmental influences? For a subset of seven of these villages, we used available data on 12 anthropometric measurements and nine dermatoglyphic traits. For these seven villages, distance matrices for the morphological traits were derived by the D^2 measurement of Mahalanobis. We compared and contrasted these various distance matrices by four techniques and tested the significance of the observed correspondences, by three different procedures (Neel et al., 1974). Despite the fact that we regard the data used to test the degree of correspondence as suboptimal, there were highly significant correspondences between the various matrices based on these quite different types of measurements and, in addition, similar significant correspondences with the geographical distances between villages. In general, the correlations of the dermatoglyphic traits with the other two biological traits or geographical distance were less than were the other intermeasurement correlations. Data on dental morphology were also available for seven villages. Employing Pearson's coefficient of racial likeness, we found highly significant correlations in the Yanomama between village dental microdifferentiation (as measured by eight traits) and the distances revealed by genetic polymorphisms or geography (Brewer-Carias et al., 1976). We conclude, then, that the genetic microdifferentiation which geneticists have employed so extensively in recent years to describe population subdivision is relevant to other ways of classifying populations, and is much more convenient.

An unusual and potentially insightful aspect of the genetic microdifferentiation we are discussing is the frequent occurrence in Amerindian tribes of "private polymorphisms"—that is, alleles with a frequency $\geqslant$

1%, which are restricted to a single tribe or to linguistically closely related tribes. Thus far we have encountered eight examples of such polymorphisms (reviewed in Neel, 1980). They presumably result from a mutation which, through drift or selection, has become entrenched within a tribe but because of the reproductive isolation of the tribe, has not spread to adjacent tribes. Some of these private polymorphisms have allele frequencies of 10%, implying considerable antiquity. Failure to spread to other tribes during this time implies a higher degree of tribal isolation than most investigators would have believed was the case. From the frequency with which such polymorphisms have been observed, we can calculate that the probability is about 1 in 40 that any protein that can be analyzed electrophoretically will, in any given tribe, yield a private polymorphism. Partly as a consequence of these private polymorphisms, we find that the profile of locus heterozygosity is quite different in tribal populations than in civilized populations (Neel, 1978). This difference is illustrated in Figure 14.3. In a tribal population, in contrast to a conglomerate civilized population, there are relatively more "monomorphic" loci but also a few unique low-frequency polymorphisms. This profile presents a different "face" to selection than the allele frequencies in a civilized population.

The implications of this microdifferentiation for human evolution appear obvious. Earlier we discussed village fissionings and the relatively large genetic distances between the fission products, distances that arise as a result of the social structuring in the event. In general, the fission products will remain members of the same group—that is, tribe. However, repeatedly in the past, as humans extended their range, there must have been another outcome. A single band or village might wander so far from the other units of the tribe that it became the nidus for what would become a new tribe. The point to be made in the present context is that at the very moment of its inception, both because of this microdifferentiation and the nonrandom fissioning discussed earlier, this "new" tribe would differ genetically in many ways from its predecessor. Many of these breakaway bands probably did not survive, but those that did presented the opportunity for the process to repeat itself, and could provide the basis for what I have termed "stepwise" or 'hippety-hop' evolution (Neel, 1984)—in no way to be confused with "punctuated equilibrium"!

CONCLUSIONS

Proposal 1: These facts and this population structure provide a sufficient framework for rapid human evolution.

The view of the genetic structure of primitive man that has slowly been forced upon me by the evidence is one of populations in a genetic flux that approaches turmoil, at both the genomic and the population level. At the genomic level, point mutations, varying from nucleotide substitutions

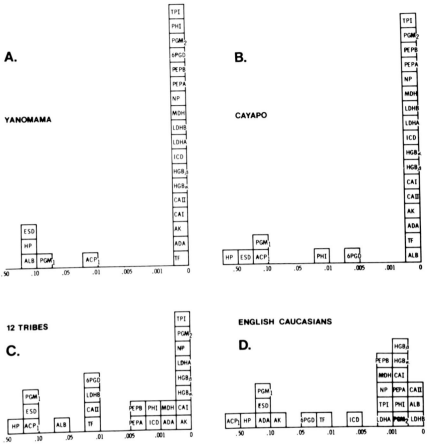

Figure 14.3. The distribution of locus heterozygosities in two Amerindian tribes (**A,B**), in a pool of 12 tribes (**C**), and in Caucasians studied in London by Harris et al. (1974) (**D**). Note the decrease in apparently "monomorphic" loci in Caucasoids, presumably due to the distribution of tribal private polymorphisms across the entire population through tribal intermixture.

to small deletions, duplications, and insertions, are far more common than were visualized a few years ago. To what extent these events are influenced by transposons and similar agents of periodically increased mutation rates is unknown. At the population level, we have the view of a reshuffling of the genetic deck with each generation, a reshuffling that constantly creates new combinations of gene frequencies (and, by inference, genetic attributes), which are in sharp competition with one another and interact with the various factors in their environment.

Until relatively recent times, the primary unit of human organization was the tribe, a loose confederation of villages or bands speaking a common language and sharing a common culture. It is difficult to say when tribal structure first emerged in human evolution, but one can assume that coalitions of bands of humans, at peace with each other but collectively hostile to neighboring bands, are an ancient feature of human social organization. Evolution would thus be characterized by a shifting balance between bands and tribes, of quite different collective genotypes, in the course of which some would disappear even as new ones emerge. No one has ever been present at the birth of a tribe, and now, with the disintegration of tribal structure, no one ever will, but I have indicated in the foregoing my favorite scenario.

We would seem to have, therefore, a sufficient basis for rapid human evolution. Are long periods of evolutionary stasis within the 5–8 million years of human evolution a viable concept? I think not. Once our ancestors embarked on bipedalism and a life-style that depended more on outsmarting than outmuscling or outspecializing the competition, evolution could scarcely stand still. On the other hand, I do not mean to imply that in the stepwise process I envision, the steps were equally graduated and of equal significance. They may have been quite irregular, but they all could have occurred within the framework of the mutational and Mendelian processes as understood today; there is no need to postulate mysterious punctuations and even more mysterious periods of stasis.

Proposal 2: The framework for human evolution that I have described bears a striking resemblance to the "founder–flush" hypothesis invoked to account for the rapid speciation of Hawaiian Drosophila.

In closing, let me draw attention to a point that by now has become apparent to the reader of this chapter. There is a remarkable parallelism between the views we have been led to adopt concerning the basis for rapid human evolution, and the views concerning the basis for the rapid evolution of Hawaiian *Drosophila* subsumed under the term "founder–flush" (Carson, 1968, 1975). Space limitations do not permit a point-by-point comparison, and I would not wish to carry the analogy to the extreme. The basic similarity lies in the postulate of a mechanism for the emergence of relatively small gene pools with marked differences from their predecessor gene pools—pools which then expand dramatically, with the creation of many intercompeting demes, some of which will survive, some of which will not. On general principles—given the social nature of humans—I would not expect the "founder–flush" effect to be as dramatic in human evolution as in the evolution of species where a single impregnated female reaching an unoccupied ecological niche can found a new population. On the other hand, the small bands in which early humans aggregated surely differed no less genetically than the villages of an Indian tribe, and the rapid expansion of one of those bands under favorable conditions (a flush) must have been an important component of the human evolutionary process. The accumulating evidence at the DNA level for "bottlenecks" in human evolution (reviewed in Jones and Rouhani, 1986) would seem to favor this point of view. As one who began his professional life as a *Drosophila* geneticist, and over the years has followed with great interest the work of Hamp and his associates, I have found this convergence of viewpoints an exciting example of the unusual intellectual unity of genetics.

ACKNOWLEDGMENTS

The financial support of the Department of Energy and the National Science Foundation is gratefully acknowledged.

LITERATURE CITED

Awa, A. A., and J. V. Neel. 1986. Cytogenetic "rogue" cells: what is their frequency, origin, and evolutionary significance? Proc. Natl. Acad. Sci USA 83:1021–1025.

Bloom, A. D., J. V. Neel, K. W. Choi, S. Iida, and N. Chagnon. 1970. Chromosome aberrations among the Yanomama Indians. Proc. Natl. Acad. Sci. USA 66:920–927.

Bloom, A. D., J. V. Neel, T. Tsuchimoto, and K. Meilinger. 1973. Chromosomal breakage in leucocytes of South American Indians. Cytogenet. Cell Genet. 12:175–186.

Brewer-Carias, C., S. leBlanc, and J. V. Neel. 1976. Genetic structure of a tribal population, the Yanomama Indians. XIII. Dental microdifferentiation. Am. J. Phys. Anthropol. 44:5–14.

Carson, H. L. 1968. The population flush and its genetic consequences. In R. C. Lewontin (ed.), Population Biology and Evolution, pp. 123–137. Syracuse University Press, Syracuse, NY.

Carson, H. L. 1975. The genetics of speciation at the diploid level. Am. Natur. 109:83–92.

Chagnon, N. A., M. V. Flinn, and T. F. Melacon. 1979. Sex-ratio variation among the Yanomama Indians. In N. A. Chagnon and W. Irons (eds.), Evolutionary Biology and Human Social Behavior: An Anthropological Perspective, pp. 290–320. Duxbury Press, North Scituate, MA.

Cowell, J. K. 1982. Double minutes and homogeneously staining regions: gene amplification in mammalian cells. Annu. Rev. Genet. 16:21–59.

Eldredge, N., and S. J. Gould. 1972. Punctuated equilibria: an alternative to phyletic gradualism. In T. J. M. Schopt and J. M. Thomas (eds.), Models in Paleobiology, pp. 82–115. W. H. Freeman, San Francisco.

Fox, D. P., F. W. Robertson, T. Brown, A. R. Whitehead, and J. D. M. Douglas. 1984. Chromosome aberrations in divers. Undersea Biomed. Res. 11:193–204.

Gould, S. J. 1977. The return of hopeful monsters. Nat. Hist. Mag. June–July:22–30.

Gould, S. J., and N. Eldredge. 1977. Punctuated equilibria: the tempo and mode of evolution reconsidered. Paleobiology 3:115–151.

Harris, H., D. A. Hopkinson, and E. B. Robson. 1974. The incidence of rare alleles determining electrophoretic variants: data on 43 enzyme loci in man. Ann. Hum. Genet. 37:237–253.

Hsu, T. C. 1983. Genetic instability in the human population: a working hypothesis. Hereditas 98:1–9.

Jones, J. S., and Rouhani, S. 1986. How small was the bottleneck? Nature 319:449–450.

MacCluer, J. W., J. V. Neel, and N. A. Chagnon. 1971. Demographic structure of a primitive population: a simulation. Am. J. Phys. Anthropol. 35:193–207.

Neel, J. V. 1978. Rare variants, private polymorphisms, and locus heterozygosity in Amerindian populations. Am. J. Hum. Genet. 30:465–490.

Neel, J. V. 1980. Isolates and private polymorphisms. In A. Eriksson (ed.), Population Structure and Genetic Disorders, pp. 173–193. Academic Press, London.

Neel, J. V. 1984. Human evolution: many small steps but not punctuated equilibria. Perspect. Biol. Med. 28:75–103.

Neel, J. V., and E. A. Thompson. 1978. Founder effect and the number of private polymorphisms observed in Amerindian tribes. Proc. Natl. Acad. Sci. USA 75:1904–1908.

Neel, J. V., and K. M. Weiss. 1975. The genetic structure of a tribal population, the Yanomama Indians. XII. Biodemographic studies. Am. J. Phys. Anthropol. 42:25–51.

Neel, J. V., C. Satoh, K. Goriki, M. Fujita, N. Takahashi, J. Asakawa, and R. Hazama. 1986. The rate with which spontaneous mutation alters the electrophoretic mobility of polypeptides. Proc. Natl. Acad. Sci. USA 83:389–393.

Neel, J. V., F. Rothhammer, and J. C. Lingoes. 1974. The genetic structure of a tribal population, the Yanomama Indians. X. Agreement between representations of village distances based on different sets of characteristics. Am. J. Hum. Genet. 26:281–303.

Neel, J. V., H. W. Mohrenweiser, E. D. Rothman, and J. M. Naidu. 1986. A revised indirect estimate of mutation rates in Amerindians. Am. J. Hum. Genet. 38:649–666.

Smouse, P. E., V. J. Vitzthum, and J. V. Neel. 1981. The impact of random and lineal fission on the genetic divergence of small human groups: a case study among the Yanomama. Genetics 98:179–197.

Smouse, P. E., R. S. Spielman, and M. H. Park. 1982. Multiple-locus allocation of individuals to groups as a function of the genetic variation within and differences among human populations. Am. Natur. 119:445–463.

Stanley, S. M. 1981. The New Evolutionary Timetable. Basic Books, New York.

Tawn, E. J., C. L. Cartmel, and E. M. T. Pyta. 1985. Cells with multiple chromosome aberrations in control individuals. Mutat. Res. 144:247–250.

Thompson, E. A., and J. V. Neel. 1978. The probability of founder effect in a tribal population. Proc. Natl. Acad. Sci. USA 75:1442–1445.

Tobias, P. V. 1982. The antiquity of man: human evolution. In B. Bonné-Tamir (ed.), Human Genetics, Part A: The Unfolding Genome, pp. 195–214. Alan R. Liss, New York.

Ward, R. H., H. Gershowitz, M. Layrisse, and J. V. Neel. 1975. The genetic structure of a tribal population, the Yanomama Indians. XI. Gene frequencies for 10 blood groups and the ABH-Le secretor traits in the Yanomama and their neighbors: the uniqueness of the tribe. Am. J. Hum. Genet. 27:1–30.

15

Coadaptive Changes in Speciation via the Founder Principle in the Grimshawi Species Complex of Hawaiian *Drosophila*

ALAN T. OHTA

Coadaptive gene complexes (the results of coadaptation) have been proposed as the major reorganizing unit during speciation (Carson, 1982) and have a primary role in models of speciation via the founder principle. Dobzhansky (1948, 1950) first defined the concept of coadaptation, but Wallace (1968) provided a broader, more meaningful definition: "Genes are said to be coadapted if high fitness depends upon specific interactions between them." Prakash and Lewontin (1968, 1971) demonstrated the coadaptation of alleles into tightly linked systems, whereas Allard et al. (1972) and Clegg et al. (1972) showed that coadaptation occurred in response to environmental parameters. Hartl (1977), Kuhn (1971), and Templeton et al. (1976) have shown that differing genetic environments can also lead to coadaptive changes. Most studies have shown associations between chromosomal or protein variants and given environmental, geographical, or genetic conditions, but few studies have shown associations between coadapted complexes and speciation, the crux of speciation being the development of isolating mechanisms between populations.

The importance of island populations in the study of evolution has been noted since the formulation of evolutionary theory. Indeed, both Darwin and Wallace solidified their theories of evolution after observing the differentiation present among insular populations because oceanic islands may reproduce evolutionary events occurring on the continents except on a smaller, simpler scale. As such, the islands of the Hawaiian archipelago form a superb natural laboratory in which to study the dynamics of the evolutionary processes and, in particular, speciation via the founder principle.

The high islands of Hawaii are a relatively new group of islands (the oldest island, Kauai, is approximately 4–5 million years old) and was formed as the Pacific Plate passed over a "hot spot" in the Earth's mantle (Dalrymple et al., 1973), creating a chain of islands which are successively younger to the southeast. These islands are among the most isolated in the world, and consequently few colonizers were successful in migrating here; as a result, the native terrestrial biota were very fragmented (Zimmerman, 1970). The topography of the high islands has produced a diversity of habitats, from high alpine regions to wet rain forests and dry deserts within a few kilometers of one another. The diversity of habitats, the isolation of the archipelago, and the isolation within and between islands have led to rapid speciation in many of the colonizers of this island group by founder events.

Probably the most extensively studied group of the endemic Hawaiian biota has been the Hawaiian Drosophilidae and, in particular, the picture-winged species group of this family. Of the more than 100 picture-winged species of Hawaiian *Drosophila*, all but three are single-island endemics (the Maui complex of islands—Maui, Molokai, and Lanai—is considered a single biological unit as these islands were joined at least once during the Pleistocene). *D. grimshawi* is one of these exceptional species and inhabits all of the major high islands except the island of Hawaii, where a very close relative, *D. pullipes*, is found. Studies carried out on various populations of *D. grimshawi* have shown them to have differences indicative of allopatric populations in the dynamic stages of speciation.

Montgomery (1975) studied the breeding site ecology of the picture-winged species and found an interesting ecological differentiation among populations of *D. grimshawi*. Montgomery was able to rear Maui complex (Lanai, Molokai, and Maui) *D. grimshawi* from field-collected substrates representing 12 different plant families, whereas Oahu and Kauai *D. grimshawi* and *D. pullipes* were reared only from the rotting bark of one genus of plant, *Wikstroemia*. Thus, Maui complex *D. grimshawi* are considered to be ecological generalists, whereas Kauai and Oahu *D. grimshawi* and *D. pullipes* are considered to be ecological specialists in their breeding site ecology. Studies were conducted to further investigate differentiation among these populations. In this chapter, I review some of the studies investigating populational differentiation in coadapted systems of *D. grimshawi*, and present data from more recent investigations. Although these "new" data are not completely analyzed here, they clearly show differentiation among populations of *D. grimshawi*. Finally, I discuss the mechanisms by which species formation is proceeding and its relation to speciation via the founder principle, and propose a possible scenerio for the sequential colonization of the island chain.

OVIPOSITIONAL BEHAVIOR

The specificity observed in the larval feeding sites are, in many holometabolous insects, controlled by the adult ovipositional behavior (Brues, 1920; Schoonhoven, 1968). Therefore, the differentiation observed in the breeding site ecology of insular populations of *D. grimshawi* and in *D.*

pullipes appears to be due to a differentiation in the ovipositional behavior of the adult female which dictates the breeding environment of the larvae.

A number of different factors have been shown to influence the ovipositional behavior of insects (see Craighead, 1921; Thorsteinson, 1960; Hershberger and Smith, 1967; Schoonhoven, 1968; Singer, 1971; Golini and Davis, 1975; Mitchell, 1975; LeBerre and Launois-Luong, 1976; Deseo, 1976; Saringer, 1976; Szentesi, 1976; Rausher, 1978; Jaenike, 1982; Takamura, 1984), indicating the highly coadapted nature of this behavior. The influence of genetic factors involved in ovipositional behavior has also been investigated with conflicting results. Some studies have shown ovipositional behavior to be polygenically controlled (Del Solar, 1968; Leslie and Dingle, 1983), whereas others have shown this behavior to be controlled by a few major genes and modifiers (Huettel and Bush, 1972). Carson and Ohta (1981) observed the segregation of ovipositional specificity in F_1, F_2, and backcross Oahu–Maui and Oahu–Molokai (specialist–generalist) *D. grimshawi* hybrids and reported this "ecobehavioral" phenotype to be controlled by as little as a single-gene, two-allele system in which generalism is dominant to specialism.

Current studies of the segregation of this trait are being carried out on Kauai–Lanai (specialist–generalist) hybrids from stocks representing these island populations. These stocks were used because the Kauai and Lanai populations are the apparent ancestral specialist and generalist *D. grimshawi* populations, respectively (Ohta, 1978). The methods used in this study are similar to those described in Carson and Ohta (1981). Some of the data obtained are reported in Table 15.1. These studies indicate that the Kauai–Lanai hybrids possess more specialist tendencies than either the Oahu–Maui or Oahu–Molokai hybrids, since F_1 specialist segregants were observed in the Kauai–Lanai hybrids whereas all F_1 hybrids of both the Oahu–Maui and Oahu–Molokai crosses were generalists. In an effort to determine whether the deviation, from our earlier study (Carson and Ohta, 1981), was due to the Kauai or to the Lanai stock,

Table 15.1. Ovipositional specificity in controls and interisland hybrids of Kauai and Lanai *Drosophila grimshawi*

Cross female × male	Female progeny with specialist tendencies		Female progeny with generalist tendencies	
	No. observed	%	No. observed	%
Kauai × Kauai	31	86.1	5	13.9
Lanai × Lanai	0	0.0	38	100.0
Lanai × Kauai	38	27.9	98	72.9
F_1 × F_1	13	43.3	17	56.7
F_1 × Lanai	13	24.1	41	75.9
F_1 × Kauai	29	96.7	1	3.3
Lanai × F_1	5	13.9	31	86.1
Kauai × F_1	9	81.8	2	18.2

Kauai–Maui hybrids were obtained and the segregation ratios of these hybrids for ovipositional behavior were observed (Table 15.2). The segregation ratios of the Kauai–Maui hybrids appear to be consistent with those of the earlier, Carson and Ohta, study in which Maui and Molokai generalists were hybridized with the Oahu specialist. These data indicate that the Lanai population is responsible for the F_1 specialist segregants in the Kauai–Lanai hybrids. Thus, it appears that the Lanai population is maintaining more of the specialist tendencies of the ancestral Kauai population than either the Maui or Molokai populations. This finding is consistent with the phylogeny obtained by Ohta (1978) using the Kaneshiro theory of asymmetrical isolation, as these data show the Lanai population to be the most ancestral of the Maui complex populations of *D. grimshawi*.

The current segregation data, furthermore, indicate that the regulation of ovipositional behavior in these flies may be more complex than was initially proposed. Unfortunately, the nature of the phenotype and the classification method used, require that this behavior be treated as a threshold trait, making it difficult to estimate with any accuracy the number of genes involved in the regulation of this behavior. More detailed familial studies are currently being completed, however, which will allow a more complete analysis of this trait.

To aid in the determination of the number of genetic factors involved in ovipositional behavior, selection for generalist ovipositional behavior in the Kauai specialist stock and for specialist ovipositional behavior in the Lanai generalist stock was attempted. Although the selection for a Kauai generalist failed to produce any response (the selected lines failed to produce progeny after five generations of selection, and no significant response was observed), the selection for a Lanai specialist produced a very strong response after a severe bottleneck in the fourth generation of selection (Table 15.3). As mentioned above, because ovipositional behavior must be treated as a threshold trait and because of the severe bottleneck in the fourth generation, a calculation of the heritability of this trait

Table 15.2. Ovipositional specificity in controls and interisland hybrids of Kauai and Maui *Drosophila grimshawi*

Cross female × male	Female progeny with specialist tendencies		Female progeny with generalist tendencies	
	No. observed	%	No. observed	%
Kauai × Kauai	31	86.1	5	13.9
Maui × Maui	0	0.0	30	100.0
Maui × Kauai	0	0.0	42	100.0
F_1 × F_1	15	31.3	33	68.7
F_1 × Maui	3	5.3	53	94.7
F_1 × Kauai	17	41.5	24	58.5
Maui × F_1	3	3.7	79	96.3
Kauai × F_1	—	—	—	—

Table 15.3.   Selection for specialism in ovipositional behavior in Lanai *Drosophila grimshawi*

Generation (pair no.)	No. pair-matings	No. ovipositing	No. females with specialist tendencies	No. females with generalist tendencies
First	52	50	6	44
Second(4)	6	6	0	6
(11)	28	27	3	24
(19)	19	14	1	13
(24)	34	32	4	28
(27)	7	6	0	6
Total	94	85	8	77
Third (11,18)	15	14	1	13
(11,27)	14	13	0	13
(11,29)	6	6	0	6
(24, 7)	14	10	0	10
(24,16)	10	5	0	5
(24,18)	19	17	0	17
(24,19)	1	1	0	1
Total	79	66	1	65
Fourth (11,18,13)	16	6	3	3
Fifth (11,18,13, 4)	23	8	6	1
(11,18,13, 6)	21	7	7	0
(11,18,13, 7)	2	2	2	0
Total	46	17	15	1

may be of little value. However, the speed of the response to selection, especially in light of the fact that selecction was possible only in the female (since ovipositional behavior is strictly a female trait), indicates that only a few genes may be involved in the regulation of this behavior. Furthermore, the lack of response to selection in the second, third, and fourth generations followed by a very strong response in the fifth generation seems to indicate that this coadapted system may possess highly additive genetic variation. These data also confirm the finding above, that the Lanai population possesses some specialist behavioral traits.

HYBRID STERILITY

Intraspecific hybridizations of interisland stocks of *D. grimshawi* were completed by Ohta (1980). Hybridizations between Oahu specialist and Maui and Molokai generalist stocks produced fully fertile F₁ hybrid progeny, although the F_2 and backcross male progeny of these crosses showed varying degrees of sterility. The pattern of sterility (i.e., highest sterility in the F_2 males, with less sterility in male progeny of the backcross using F_1 females, and finally with the lowest sterility in the male progeny of the backcross using F_1 males) led to the conclusion that, with the assumption of no recombination in *D. grimshawi* males, ecologically distinct populations of *D. grimshawi* have differing coadapted gene complexes. It also appears that no such differentiation in coadapted gene complexes has taken place between ecologically similar populations, since interisland hybrids of Maui–Molokai and Oahu–Kauai stocks of *D. grimshawi* showed no detectable reductions in fertilty in F_1, F_2, and backcross male progeny.

Two additional hybridizations have recently been completed between Kauai and Maui and Kauai and Lanai stocks. Both reciprocal crosses were attempted, but only crosses between generalist females (Lanai and Maui) and specialist (Kauai) males were successful. The failure of the crosses between Kauai females and Lanai and Maui males was probably due to the asymmetrical isolation observed among these populations by Ohta (1978), who showed, in male-choice experiments, that Kauai females are very reluctant to mate with males from any other *D. grimshawi* population. The data obtained from these crosses, presented in Tables 15.4 and 15.5, are consistent with each other and with the data obtained by Ohta (1980). It appears, therefore, that some degree of hybrid sterility exists whenever ecologically distinct (specialist and generalist) populations are hybridized, whereas hybridizations between ecologically similar populations are fully fertile.

SEXUAL BEHAVIOR

Two other highly coadapted systems, male aggression and courtship behavior, have been studied in *D. grimshawi* populations. Both of these are important sexual behaviors as these flies display lek (Spieth, 1968) and very elaborate courtship (Spieth, 1966, 1984), behaviors that are crucial to their reproductive success. Ringo (1976) and Ringo and Hodosh (1978) studied qualitative and quantitative characters of male aggressive behavior among the Maui and Molokai generalist and Oahu specialist populations of *D. grimshawi* and *D. pullipes*. They found that generalist populations (Maui and Molokai) appeared to have little population differentiation in male aggressive behavior, whereas a considerable difference was observed between these populations and the Oahu specialist. Furthermore, although differences were observed between *D. pullipes* and Oahu *D. grimshawi*, these two populations appeared to be closer to each other than either was to Maui or Molokai *D. grimshawi* populations. Thus, it appears that differentiation in male aggressive behavior was

Table 15.4. Sperm condition in the male progeny of crosses between Kauai and Lanai *Drosophila grimshawi*

Cross female × male	Male progeny with motile sperm		
	No. observed	%	*n*
Kauai × Kauai	60	95.2	63
Lanai × Lanai	52	100.0	52
Lanai × Kauai	171	100.0	171
F_1 × F_1	17	51.5	33
F_1 × Lanai	34	53.1	64
F_1 × Kauai	8	26.7	30
Lanai × F_1	47	100.0	47
Kauai × F_1	18	41.9	33

Table 15.5. Sperm condition in the male progeny of crosses between Kauai and Maui *Drosophila grimshawi*

Cross female × male	Male progeny with motile sperm		
	No. observed	%	*n*
Kauai × Kauai	60	95.2	63
Maui × Maui	57	98.3	58
Maui × Kauai	35	100.0	35
F_1 × F_1	16	44.4	36
F_1 × Maui	16	61.5	26
F_1 × Kauai	13	54.2	24
Maui × F_1	65	100.0	65
Kauai × Maui	—	—	—

greater between ecologically distinct conspecific populations than between ecologically similar populations of two different species.

Although courtship behavior was not studied directly, Ohta (1978) conducted male-choice experiments among populations of *D. grimshawi* and *D. pullipes* which provide a gross indication of differences in courtship behavior. The failure of one population to accept the courtship overtures of another was an indication that some differentiation in courtship behavior existed between those populations tested. Ohta observed varying degrees of asymmetrical mating preferences among the *D. grimshawi* populations and *D. pullipes* and, utilizing the Kaneshiro hypothesis (Kaneshiro 1976, 1980, 1983), was able to propose a sequence of island colonizations consistent with the age of the islands. It appears that the island colonizations occurred in stepwise fashion down the island chain, with the most ancestral population being the Kauai *D. grimshawi*.

CYTOLOGY

Insular populations of *D. grimshawi* are highly conserved both cytologically (Carson and Stalker, 1968; Carson and Sato, 1969; Carson et al., 1970) and morphologically, although some chromosomal differentiation has occurred between specialist and generalist populations. The generalist populations possess a polymorphic 4a inversion sequence, whereas all specialist populations (including *D. pullipes*) are monomorphic for the standard polytene chromosomal sequence. Another polytene chromosomal difference has been observed in the third chromosome of generalist and specialist populations (Stuart et al., 1981). The 3–13 band of this chromosome appears as a dark thick band in the specialists (including *D. pullipes*), whereas in the generalists there is a narrow band at this position. Furthermore, this band has been characterized as the region of 18/28S rRNA production in *D. heteroneura* and *D. silvestris*. The function of, and reason for these chromosomal changes are still unclear but are certainly worth further investigation.

DISCUSSION

Three models of speciation via the founder principle have been proposed by Mayr (1942, 1954, 1982), Carson (1968, 1971, 1975, 1978), and Templeton (1980, 1981, 1982). All three models (Mayr's genetic revolution, Carson's founder–flush, and Templeton's genetic transilience) of speciation rely on changes in the coadapted systems, through drift, to initiate differentiation between progenitor and ancestral populations. However, the type and amount of genetic changes that are required for speciation differ among these models. (For a detailed review of the critical differences, see Carson and Templeton, 1984.)

Mayr emphasizes "peripatric speciation" in which the isolated (founder) population experiences a considerable loss of genetic variability due to drift. This increase in homozygosity in turn affects the existing coadapted systems of the founder population and can lead to a reorganization of these coadapted gene complexes, eventually resulting in a "genetic revolution." Thus, although it is not explicitly stated, Mayr implies that a considerable portion of the genome is affected during the speciational process.

Carson's founder–flush model, a modification of Mayr's model, proposes that drift, associated with the founder event, disrupts ancestral coadapted systems. In addition, as a result of a change in environmental conditions in the founder population, a relaxation of selection triggers a population flush in which genetic variants previously selected against are now allowed to survive. A disorganization of the "closed" coadapted gene complexes results, which, as the population reaches its carrying capacity and as selective forces are reestablished, rapidly reorganizes, owing to high levels of additive genetic variation present in these polygenic systems. Thus, in Carson's model, speciational changes occur in response to external selective forces acting primarily on the "closed" coadapted systems with high additive genetic variance.

Templeton's genetic transilience model is similar to Carson's model in that Templeton also proposes that high levels of genetic variability are maintained during the founder event. However, in the genetic transilience model, changes in the frequencies of major loci in coadapted systems are brought about by drift. These changes perturb the genetic environment and lead to alterations in the pleiotropic balances of the founder population. A reweighting of the pleiotropic effects associated with these major genes results, and large amounts of variation at a number of modifier loci allow a rapid response to these altered pleiotropic balances. Thus, in Templeton's view, the significant changes leading to speciation occur in response to the internal environment of the organisms and in only a few major genes and their modifiers.

It is apparent that differentiation of several tightly coadapted systems has occurred among the insular populations studied; differentiation in the sexual, male aggressive, and ovipositional behaviors, as well as differences in hybrid sterility and cytology, have been observed. These data indicate that a period of relaxed selection may have accompanied the founder event, as described by Carson, in order for these "closed" genetic systems to be disrupted. As selection was reestablished, in the Maui complex populations of *D. grimshawi*, a reorganization of the coadapted gene complex regulating ovipositional behavior rapidly occurred, owing to the high levels of additive genetic variance and relatively few genes involved in this behavior. Moreover, this change may have occurred in response to a change in the external selective forces. The islands of Kauai and Oahu are inhabited by another closely related generalist species, *D. crucigera*, which is not found on either the Maui complex or the island of Hawaii, and the relaxation of competition with *D. crucigera*, on the Maui complex, may have aided in the ecological change observed in these populations. Interestingly, among the more than 100 species of picture-winged Hawaiian *Drosophila*, only two truly polyphagous species are known to exist: *D. crucigera* and Maui complex *D. grimshawi*.

The change in ovipositional behavior then, appears to have occurred as described by Carson's founder–flush model of speciation. This change, however, may have been only the entering wedge of species differentiation and was probably followed by other genetic changes. The alteration of ovipositional behavior may have altered the pleiotropic balance of the founder population and produced the concomitant changes we observed in male aggressive behavior, hybrid sterility, and the chromosomes. This agrees with Templeton's model of genetic transilience, although the initial populational change here is an adaptive one. Courtship behavior, on the other hand, appears to have evolved independently of ovipositional behavior and is due primarily to drift via the founder event. This is consistent with the Kaneshiro hypothesis (1976) which is based on stochastic changes in sexual behavior during the founding of a new population. Indeed, the data seem to show that a genetic revolution (i.e., a change in a large portion of the genome) may have occurred during the founding of these populations since some degree of differentiation can be observed in all of the coadapted systems studied.

Thus far our studies of the insular populations of *D. grimshawi* and *D. pullipes* suggest the following scenerio of island colonizations: The asymmetrical mating preference observed among these populations indicate, by the Kaneshiro hypothesis, that the Kauai specialist population was the progenitor of all *D. grimshawi* populations. This is also consistent with the geological history of the islands, since Kauai is known to be the oldest of the high islands (McDougall, 1979). Again, utilizing the Kaneshiro hypothesis to infer the direction of evolutionary events, we find that Kauai appears to have then given rise to both the Oahu and Lanai populations. The Oahu population seems to have ended its colonizations, whereas the Lanai population then founded both *D. pullipes,* on the island of Hawaii, and either the Maui or Molokai population of *D. grimshawi* (for the sake of discussion, let us assume that the Lanai population founded that of Maui). It is thought that the generalists evolved during the founding of the Maui population from Lanai, because a specialist *(D. pullipes)* inhabits the younger (approximately 400,000 years old) island of Hawaii. Thus, the generalism event may have taken place on Maui before the founding of the Molokai population which was subsequently formed from Maui. Lanai was then recently recolonized by founders from the Maui–Molokai generalists (possibly during the Pleistocene, when the islands of the Maui complex were joined, and after the founding of *D. pullipes*) and has retained some of its earlier specialist tendencies possibly because of the 4a chromosomal inversion which may have arisen to tie up coadapted gene complexes, some of which may be involved in hybrid sterility (Ohta, 1980).

Speciational changes in populations of *D. grimshawi* appear to be occurring in response to an adaptive shift in the breeding site ecology of these flies. Differentiation in several, seemingly unrelated coadapted systems appears concomitant with the change in ovipositional behavior and thus may have some linkage and/or pleiotropic relationship with this behavior. Indeed, hybrid breakdown among populations of *D. grimshawi* is observed only between ecologically distinct populations; ecologically similar populations show no breakdown in the hybrids. Our evidence shows, then, that the change in an adaptive behavior may be a major factor in speciational changes in *D. grimshawi* populations, whereas the differentiation observed in some of the other coadapted systems may have occurred as a result of linkage and/or pleiotropic effects.

ACKNOWLEDGMENTS

I wish to thank my colleagues Kenneth Kaneshiro and Kelvin Kanegawa for their help in field collections and my student assistant Dee Gushiken for her aid in the laboratory. I also wish to thank John Ringo, an anonymous reviewer, and Kenneth Kaneshiro for their comments on this paper. The work of the author was supported in part by the National Science Foundation Grant BSR 81-17985 and the Hawaiian Evolutionary Program and the Office of Research Administration at the University of Hawaii, Manoa.

LITERATURE CITED

Allard, R. W., G. R. Babbel, M. T. Clegg, and A. L. Kahler. 1972. Evidence for coadaptation in *Avena barbata*. Proc. Natl. Acad. Sci. USA 69:3043–3048.

Brues, C. T. 1920. The selection of food plants by insects, with special reference to Lepidopterous larvae. Am. Natur. 54:313–332.

Carson, H. L. 1968. The population flush and its genetic consequences. In R. C. Lewontin (ed.), Population Biology and Evolution, pp. 123–127. Syracuse University Press, Syracuse, NY.

Carson, H. L. 1971. Speciation and the founder principle. Stadler Symp. 3:51–70.

Carson, H. L. 1975. The genetics of speciation at the diploid level. Am. Natur. 109:83–92.

Carson, H. L. 1978. Speciation and sexual selection in Hawaiian *Drosophila*. In P. F. Brussard (ed.), Ecological Genetics: The Interface, pp. 93–107. Springer-Verlag, New York.

Carson, H. L. 1982. Speciation as a major reorganization of polygenic balances. In C. Barigozzi (ed.), Mechanisms of Speciation, pp. 411–433. Alan R. Liss, New York.

Carson, H. L. and A. T. Ohta. 1981. Origin of the genetic basis of colonizing ability. In G. G. E. Scudder and J. L. Reveal (eds.), Evolution Today, Proc. 2nd Int. Cong. Syst. and Evol. Biol., pp. 365–370. Hunt Institute for Biological Documentation, Carnegie-Mellon University, Pittsburg, PA.

Carson, H. L., and J. E. Sato. 1969. Microevolution within three species of Hawaiian *Drosophila*. Evolution 23:493–501.

Carson, H. L., and H. D. Stalker. 1968. Polytene chromosome relationships in Hawaiian species of *Drosophila*. I. The *D. grimshawi* subgroup. Univ. Texas Publ. 6818:335–354.

Carson, H. L., and A. R. Templeton. 1984. Genetic revolutions in relation to speciation phenomena: the founding of new populations. Annu. Rev. Ecol. Systematics 15:97–131.

Carson, H. L., D. E. Hardy, H. T. Spieth, and W. S. Stone. 1970. The evolutionary biology of Hawaiian Drosophilidae. In M. K. Hecht and W. S. Steere (eds.), Essays in Evolution and Genetics in Honor of Theodosius Dobzhansky, pp. 437–543. Appleton-Century-Crofts, New York.

Clegg, M. T., R. W. Allard, and A. L. Kahler, 1972. Is the gene the unit of selection? Evidence from two experimental plant populations. Proc. Natl. Acad. Sci. USA 69:2474–2478.

Craighead, F. C. 1921. Hopkin's host-selection principle as related to certain cerambycid beetles. J. Agric. Res. 22:189–220.

Dalrymple, G. B., E. A. Silver, and E. D. Jackson. 1973. Origin of the Hawaiian Islands. Am. Sci. 61:294–308.

Del Solar, E. 1968. Selection for and against gregariousness in the choice of oviposition sites by Drosophila pseudoobscura. Genetics 58:275–282.

Deseo, K. V. 1976. The oviposition of the Indian meal moth (Plodia interpunctella Hbn., Lep. Phyticidae) influenced by olfactory stimuli and antennectomy. In T. Jermy (ed.), The Host-Plant in Relation to Insect Behavior and Reproduction, pp. 61–65. Plenum, New York.

Dobzhansky, Th. 1948. Genetics of natural populations. XVII. Experiments on chromosomes of Drosophila pseudoobscura from different geographical regions. Genetics 33:588–602.

Dobzhansky, Th. 1950. Genetics of natural populations. XIX. Origin of heterosis through natural selection in populations of Drosophila pseudoobscura. Genetics 35:288–302.

Golini, V. I., and D. M. Davis. 1975. Relative response to colored substrates by ovipositing blackflies (Diptera: Simulidae). I. Oviposition by Simulium (Simulium) verecundum Stone and Jamenback. Can. J. Zool. 53:521–535.

Hartl, D. L. 1977. Mechanism of a case of genetic coadaptations in populations of Drosophila melanogaster. Proc. Natl. Acad. Sci. USA 74:324–328.

Hershberger, W. A., and M. P. Smith. 1967. Conditioning in Drosophila melanogaster. Anim. Behav. 15:259–262.

Huettel, M. D., and G. L. Bush. 1972. The genetics of host selection and its bearing on sympatric speciation in Procecidochanes (Diptera: Tephritidae). Entomol. Exp. Appl. 15:465–480.

Jaenike, J. 1982. Environmental modification of oviposition behavior in Drosophila. Am. Natur. 119:784–802.

Kaneshiro, K. Y. 1976. Ethological isolation and phylogeny in the planitibia subgroup of Hawaiian Drosophila. Evolution 30:740–745.

Kaneshiro, K. Y. 1980. Sexual selection, speciation, and the direction of evolution. Evolution 34:437–444.

Kaneshiro, K. Y. 1983. Sexual selection and direction of evolution in the biosystematics of Hawaiian Drosophilidae. Annu. Rev. Entomol. 28:161–178.

Kuhn. D. T. 1971. Coadaptation of the Payne inversion with a previously unrelated genetic background in Drosophila melanogaster. Evolution 25:207–213.

LeBerre, J. R., and H. Launois-Luong. 1976. Finding of feeding and egg-laying sites by the migratory locust, Locusta migratoria L. In T. Jermy (ed.), The Host-plant in Relation to Insect Behavior and Reproduction, p. 137. Plenum, New York.

Leslie, J. F., and H. Dingle. 1983. A genetic basis of oviposition preference in the large milkweed bug, Oncopeltus fasciatus. Entomol. Exp. Appl. 34:215–220.

Mayr, E. 1942. Systematics and the Origin of Species. Columbia University Press, New York.

Mayr, E. 1954. Change of genetic environment and evolution. In J. Huxley (ed.), Evolution as a Process, pp. 157–180. Allen & Unwin, London.

Mayr, E. 1982. Processes of speciation in animals. In C. Barigozzi (ed.), Mechanisms of Speciation, pp. 1–19. Alan R. Liss, New York.

McDougall, I. 1979. Age of shield-building volcanism of Kauai and linear migration of volcanism in the Hawaiian Island chain. Earth Planet. Sci. Lett. 46:31–42.

Mitchell, R. 1975. The evolution of oviposition tactics in the bean weevil, *Callosobruchus maculatus* (F.). Ecology 56:696–702.

Montgomery, S. L. 1975. Comparative breeding site ecology and the adaptive radiation of picture-winged *Drosophila* (Diptera: Drosophilidae) in Hawaii. Proc. Haw. Entomol. Soc. 22:65–103.

Ohta, A. T. 1978. Ethological isolation and phylogeny in the grimshawi species complex of Hawaiian *Drosophila*. Evolution. 32:485–492.

Ohta, A. T. 1980. Coadaptive gene complexes in incipient species of Hawaiian *Drosophila*. Am. Natur. 115:121–131.

Prakash, S., and R. C. Lewontin. 1968. A molecular approach to the study of genic heterozygosity in natural populations. III. Direct evidence of coadaptation in gene arrangements of *Drosophila*. Proc. Natl. Acad. Sci. USA 59:398–405.

Prakash, S., and R. C. Lewontin. 1971. A molecular approach to the study of genic heterozygosity in natural populations. V. Further direct evidence for coadaptation in inversions of *Drosophila*. Genetics 69:405–408.

Rausher, M. D. 1978. Search image for leaf shape in a butterfly. Science 200:1071–1073.

Ringo, J. M. 1976. A communal display in Hawaiian *Drosophila* (Diptera: Drosophilidae). Ann. Entomol. Soc. Am. 69:209–214.

Ringo, J. M., and R. J. Hodosh. 1978. A multivariate analysis of behavioral divergence among closely related species of endemic Hawaiian *Drosophila*. Evolution 32:389–397.

Schoonhoven, L. M. 1968. Chemosensory basis of host plant selection. Annu. Rev. Entomol. 13:115–136.

Singer, M. C. 1971. Evolution of food-plant preference in the butterfly *Euphydryas edita*. Evolution 25:383–389.

Spieth, H. T. 1966. Courtship behavior of endemic Hawaiian *Drosophila*. Univ. Texas Publ. 6615:245–313.

Spieth, H. T. 1968. The evolutionary implications of sexual behavior in *Drosophila*. Evol. Biol. 2:157–191.

Spieth, H. T. 1984. Courtship behaviors of the Hawaiian picture-winged *Drosophila*. Univ. Calif. Publ. Entomol. No. 103.

Stuart, W. D., J. G. Bishop, H. L. Carson, and M. B. Frank. 1981. Location of the 18/28*S* ribosomal RNA genes in two Hawaiian *Drosophila* species by monoclonal immunological identification of RNA–DNA hybrids in situ. Proc. Natl. Acad. Sci. USA 78:3751–3754.

Szentesi, A. 1976. The effect of the amputation of head appendages on oviposition of the bean weevil, *Acanthoscelides obtectus* Say (Coleoptera: Bruchidae). In T. Jermy (ed.), The Host-Plant in Relation to Insect Behavior and Reproduction, pp. 275–281. Plenum, New York.

Takamura, T. 1984. Behavior genetics of choice of oviposition sites in *Drosophila melanogaster*. IV. Differentiation of oviposition force in the *melanogaster* species sub-group. Jpn. J. Genet. 59:71–81.

Templeton, A. R. 1980. The theory of speciation via the founder principle. Genetics 94:1011–1038.

Templeton, A. R. 1981. Mechanisms of speciation—a population genetic approach. Annu. Rev. Ecol. Systematics 12:23–48.

Templeton, A. R. 1982. Genetic architectures of speciation. In C. Barigozzi (ed.), Mechanisms of Speciation, pp. 105–121. Alan R. Liss, New York.

Templeton, A. R., C. F. Sing, and B. Brokaw. 1976. The unit of selection in *Drosophila mercatorum* I. The interaction of selection and meiosis in parthenogenetic strains. Genetics 82:349–376.

Thorsteinson, A. J. 1960. Host selection in phytophagous insects. Annu. Rev. Entomol. 5:193–218.

Wallace, B. 1968. Topics in Population Genetics. Norton, New York.

Zimmerman, E. C. 1970. Adaptive radiation in Hawaii with special reference to insects. Biotropica 2:32–38.

16

Founder Effects and the Evolution of Reproductive Isolation

ALAN R. TEMPLETON

As evidenced by this symposium, many evolutionary biologists believe that founder events can serve as a critical trigger to the process of speciation. However, this proposition is not without controversy (e.g., see Carson and Templeton, 1984, and Barton and Charlesworth, 1984). One way of testing the idea that founder events can trigger speciation is to examine the problem experimentally. One difficulty with the experimental approach is the amount of time that must be invested. Although founder event speciation is regarded as "rapid" by its proponents, Carson and Templeton (1984) emphasized that the founder event is merely the trigger, and that the genetic events leading to actual speciation occur over many generations following the initial founder event. Hence, to a human experimenter, empiricially studying founder effects implies a commitment of several years of work, even for such rapidly reproducing organisms as *Drosophila*. One such long-term experiment is that reported by Powell (1978; Chapter 11, this volume), which was initiated in 1972 and which has been supportive of the idea that founder events can indeed facilitate speciation.

An alternative experimental design was offered by Templeton (1979a). This design utilized the capacity for parthenogenetic reproduction that is present in natural, sexually reproducing populations of *Drosophila mercatorum* to simulate the most severe founder effect possible—a single haploid genome. The rationale for this design was not to represent parthenogenesis as a realistic or commonplace type of founder event in nature (although it certainly does occur in nature; see Templeton, 1982), but rather to deliberately push the essence of the founder–speciation models to their absolute extremes in the hope of speeding up the evolutionary process to the point where empirical study would be more feasible. If no evidence for speciation were obtained under these extreme conditions, it would cast serious doubt on the validity of founder-induced speciation. If evidence for speciation were obtained, it would not prove that less drastic founder events induce speciation, but it would certainly make that proposition more plausible. In any event, these experiments provide a unique opportunity to study empirically the genetic basis of rapid speciation events.

Templeton (1979a) reported the initial results of these experiments. Starting with a collection of 34 isofemale lines established from the natural, sexually reproducing population of *D. mercatorum* in the vicinity of Kamuela, Hawaii (collected under the guidance of Hampton L. Carson), two self-sustaining parthenogenetic lines (K23-0-Im and K28-0-Im) were established that reproduce by a parthenogenetic mechanism known as gamete or pronuclear duplication (Carson et al., 1969; Templeton, 1983). Under this mode of reproduction, meiosis is normal and produces an haploid egg nucleus. The haploid egg nucleus then undergoes spontaneous mitotic cleavage divisions, and two of the cleavage nuclei fuse to form a diploid nucleus that then undergoes normal development. Hence, the parthenogenetic strains are diploid with normal meiosis, but totally homozygous. Each parthenogenetic strain traces its origin to a single haploid egg nucleus. Templeton (1979a) showed that K23-0-Im and K28-0-Im rapidly evolved coadapted complexes that were incompatible with both their sexual ancestors and with each other. Subsequent work with K28-0-Im has documented that this strain underwent drastic evolutionary changes in life history, morphology, and behavior (Templeton, 1982, 1983). Moreover, these changes are not found in other parthenogenetic isolates, and thus are not attributable to parthenogenesis *per se*. Hence, there is no doubt that the extreme fonder event leading to the establishment of K28-0-Im has caused rapid evolution in a broad array of phenotypes.

Speciation in many sexually reproducing diploid animals is commonly defined in terms of the evolution of isolating mechanisms that prevent reproduction with ancestral or closely related species. Such isolating barriers can also be critical in the successful establishment of a diploid parthenogenetic species from a sexually reproducing ancestor (Templeton, 1982, 1983). In terms of parthenogenetic speciation, any factor that diminishes the chances for sexual reproduction contributes to the establishment of a parthenogenetic species. The purpose of this chapter is to examine the existence and genetic basis of some isolating mechanisms and factors that make sexual reproduction unlikely that have evolved in K28-0-Im. Although K28-0-Im is a parthenogenetic stock, these studies include both pre- and postmating barriers. This can easily be accomplished because gamete duplication per se imposes no direct barriers to sexual reproduction since the resulting parthenogenetic females are diploid with normal meiosis. Indeed, the vast majority of parthenogenetic isolates will mate with males when given the opportunity. Moreover, the evolution of premating barriers in K28-0-Im was documented shortly after the stock was established (Carson et al., 1977); thus the premating barrier to be studied in this report is not attributable to a gradual decline in mating ability under parthenogenetic reproduction (Carson et al., 1982).

MATERIALS AND METHODS

Stocks

The focus of these experiments is the parthenogenetic strain K28-0-Im. The origins of this stock are briefly described above, and more details are available in Templeton (1979a,b). The parthoenogenetic strain K23-0-Im, described above in the same papers as K28-0-Im, was also used in some experiments. In addition to these strains, two sexual stocks were used. The first is K74-Bi. This stock was founded by pooling the progeny of the Kamuela isofemale lines that were established during the isolation of K23-0-Im and K28-0-Im. It has been kept in large population sizes (at least 200 flies per generation) since then. The other sexual stock is S-sl v pm vl-Br14. This stock bears visible mutations on all the major chromosomes of the *mercatorum* genome: *spotless (sl)*, an X-linked recessive characterized by lacking a spot at the base of the middle supraorbital bristle; *vermillion (v)*, a recessive red-eye color mutation on the metacentric chromosome; *plum (pm)*, a recessive plum-eye color mutation on the acrocentric I chromosome; and *veinless (vl)*, a recessive wing mutation on the acrocentric II chromosome (further details are in Templeton, 1983). These visible mutations were introduced by 14 rounds of backcrossing onto the parthenogenetic stock S-1-Im to produce the stock S-sl v pm vl-Br14 (Templeton et al., 1976). For convenience of reference, S-sl v pm vl-Br14 will sometimes be abbreviated to "S" and K28-0-Im to "K."

Mating Experiments

To test female sexual receptivity, a single newly eclosed female of the strain to be tested was placed in a shell vial containing a freshly yeasted cornmeal–molasses–agar *Drosophila* food. Two sexually mature males (at least 48 hours after eclosion) of the S strain were then placed in the shell vial with the female. After a week, all three flies were transferred to a second shell vial, and after a second week all flies were discarded. A female was scored as having mated within the first week if sexually produced progeny emerged from the first shell vial, and she was scored as having mated during the second week if sexual progeny emerged only from the second vial. If no sexual progeny emerged from either vial, she was scored as not having successfully mated.

Male Sterility

Males were pair mated to S females in a fresh shell vial, then transferred to a second vial after one week. At the end of the second week, all males that had failed to produce sexual progeny were dissected as follows. The males were knocked out with CO_2 and were placed next to a drop of 0.7% saline solution. By use of two sharp dissecting needles, the entire male reproductive tract was pulled into the saline solution. The testes were then detached and placed in a fresh drop of saline solution on a microscope slide. A cover slip was dropped over the testes and gently tapped. The testes were then examined with a compound microscope to check for the presence of motile sperm. Before a set of presumably sterile males was scored, a known fertile male was scored to ensure that the solutions were not inhibiting sperm motility.

Isozyme Scoring

Certain flies were scored for their esterase A phenotypes, which are determined by a metacentric locus. Esterase A was scored by use of the starch gel system described in Templeton et al. (1976).

RESULTS

Premating Sexual Isolation

Carson et al. (1977) and Templeton (1983) had previously documented that K28-0-Im females are extremely reluctant to mate with males of several strains, including their immediate sexual ancestors. However, of all the strains tested, the sexual isolation is most extreme against S-sl v pm vl-Br14 males (Templeton, 1983). The isolation against this strain is so extreme that it can be scored as a yes/no discrete character. For this reason, all mating tests were performed with S-sl v pm vl-Br14 males.

Table 16.1 presents the mating propensities of various females with S males. As is readily evident, K28-0-Im females show by far the least propensity to mate with S males. This lack of mating success cannot be attributed to geographical isolation because the K74-Bi females mate very well with S males, even thought they come from a sexual stock isolated from the very same females caught at Kamuela that gave rise to K28-0-Im. The premating isolation cannot be attributed to parthenogenesis because the K23-0-Im females mate very well with S males, despite the fact that this parthenogenetic strain was isolated at the same time and from the same ancestral population as K28-0-Im. Finally, the data of Table 16-1 indicate that this premating isolation depends upon recessive genetic factors, for it totally disappears in K $\times$ S F_1 hybrid females.

To examine the genetic basis of this premating isolation, F_1 males from the few successful matings between K and S were backcrossed to K females. Twenty-eight percent of the 43 crosses set up reproduced sexually; hence the premating isolation of the K females is not as strong with hybrid males as with S males.

The K $\times$ S F_1 males have a K-type X chromosome, and are S/K het-

Table 16.1. Percentage of females mating to S-sl v pm vl-Br14 males

Female strain	Percentage mating in two weeks	Sample size
K28-0-IM	4	204
S-sl v pm vl-Br14	69	71
K × S hybrids	92	26
K23-0-Im	87	15
K74-Bi	97	33

erozygotes for all autosomes. Because there is no crossing-over in male *mercatorum*, the resulting female backcross progeny will be homozygous for the K-type X chromosome, and either K homozygous or K/S heterozygous independently at each autosome with an expected 1 : 1 ratio. Because the visible markers from the S stock are all recessive, all females will have a wild-type phenotype regardless of their autosomal genotype. The backcross females were then mated with S males by use of the standard mating design. Because a cross with S males also constitutes a test-cross, the genotype of the backcross females that mated with S males could be inferred from the phenotypes of their progeny.

As can be seen from Table 16.2, of the eight possible autosomal genotypes, only four were present within the females that mated with S males during the first week. All four of these genotypes are K/S heterozygous for the acrocentric II autosome, with all other autosomal genotypes at the metacentric and acrocentric I autosomes being about equally likely. Table 16.2 also indicates that 105 backcross females did not mate at all. Even though these females did not mate, some information is obtainable concerning their genotypes. First, the K and S strains differ at the esterase A locus, which is lined to v on the metacentric autosome (Templeton, 1979a). Hence, the metacentric chromosomal genotype can be determined by scoring for esterase. Unfortunately, there are no isozyme markers available for the two acrocentric autosomes. However, the backcross females can reproduce parthenogenetically. Because meiosis is normal and the resulting parthenogenetic progeny are homozygous, a heterozygous female has a 50 : 50 chance of producing a visibly marked parthenogen. Hence, when the phenotypes of the parthenogenetic progeny are scored, the most likely genotypes of the unmated females can be calculated by the use of Bayes's theorem.

The most likely genotypes of the unmated females are given in Table 16.2. The genotypes that are missing in the mated females are present in highest frequency in the unmated females. These results are very clear-cut; the sexual isolation displayed by K females toward S males maps exclusively to the K-type acrocentric II autosome on a genetic background fixed for the K-type X.

To investigate the role of the X chromosome, K × S F_1 males were

Table 16.2. Numbers of backcross females mating with S males, as a function of female genotype

Backcross genotype	Mated in one week	Mated in two weeks	Unmated (most likely[a])
$\dfrac{+\ \ v\ \ pm\ \ vl}{+\ \ +\ \ +\ \ +}$	29	0	0
$\dfrac{+\ \ v\ \ pm\ \ +}{+\ \ +\ \ +\ \ +}$	0	1	15
$\dfrac{+\ \ v\ \ +\ \ vl}{+\ \ +\ \ +\ \ +}$	36	0	1
$\dfrac{+\ \ +\ \ pm\ \ vl}{+\ \ +\ \ +\ \ +}$	26	0	1
$\dfrac{+\ \ v\ \ +\ \ +}{+\ \ +\ \ +\ \ +}$	0	3	19
$\dfrac{+\ \ +\ \ pm\ \ +}{+\ \ +\ \ +\ \ +}$	0	0	13
$\dfrac{+\ \ +\ \ +\ \ vl}{+\ \ +\ \ +\ \ +}$	25	2	0
$\dfrac{+\ \ +\ \ +\ \ +}{+\ \ +\ \ +\ \ +}$	0	2	56

[a]The probabilities of all possible genotypes from an unmated female were determined by scoring an isozyme locus on the metacentric autosome and by using Bayes's theorem on the observed phenotypes of the unmated female's parthenogenetic progeny. The unmated female's genotype was determined as the most probable one on the basis of the isozyme results and Bayes's theorem.

backcrossed to S females. From the resulting backcross progeny, males were chosen that were wild-type for all autosome markers. These males would be K/S heterozygotes for all autosomes, just like the original F_1 males, but they would be hemizygous S (and hence spotless) at the X chromosome. These spotless males were then crossed to K females, and the resulting female progeny were then tested for premating isolation with S males, as shown in Table 16.3.

In great contrast to Table 16.2, all eight genotypes mated during the first week. Hence, the extreme degree of mating isolation displayed by K females depends upon simultaneous homozygosity for the K-type X and acrocentric II chromsomes. However, the acrocentric II chromosome still has a strong effect on mating isolation in flies that are K/S heterozygotes for the X. Only 50% of the flies that were K/K homozygous for the acrocentric II autosome mated, as opposed to 89% for the F/S acrocentric II heterozygoes (contingency chi-square = 44.21 with 1 d.f.; < .0001).

To gain some insight into the number of loci and/or linkage relationships to visible markers, the F_1 females produced by the K × S crosses were allowed to reproduce parthenogenetically. (Recall that meiosis is normal

Table 16.3. Numbers of backcross females heterozygous at the X chromosome that mate with S males, as a function of their autosomal genotype

sl/+ backcross genotype	Mated in one week	Mated in two weeks	Unmated (most likely[a])
$\dfrac{sl\ \ v\ \ pm\ \ vl}{+\ \ +\ \ +\ \ +}$	35	0	6
$\dfrac{sl\ \ v\ \ pm\ \ +}{+\ \ +\ \ +\ \ +}$	14	0	7
$\dfrac{sl\ \ v\ \ +\ \ vl}{+\ \ +\ \ +\ \ +}$	22	0	3
$\dfrac{sl\ \ +\ \ pm\ \ vl}{+\ \ +\ \ +\ \ +}$	30	0	3
$\dfrac{sl\ \ v\ \ +\ \ +}{+\ \ +\ \ +\ \ +}$	12	0	6
$\dfrac{sl\ \ +\ \ pm\ \ +}{+\ \ +\ \ +\ \ +}$	17	0	15
$\dfrac{sl\ \ +\ \ +\ \ vl}{+\ \ +\ \ +\ \ +}$	33	0	3
$\dfrac{sl\ \ +\ \ +\ \ +}{+\ \ +\ \ +\ \ +}$	13	0	27

[a] The unmated female's genotype was determined as the most probable one on the basis of the isozyme results and Bayes's theorem.

under parthenogenesis.) The genotype at the visible markers of the F_2 parthenogenetic progeny (F_2-Im) can be immediately scored from their phenotypes since all flies are totally homozygous. The mating propensity of these F_2-Im females was then tested against S males as a function of their genotype for the X-linked marker sl and the acrocentric II marker v. The results are shown in Table 16.4. There is no detectable association of failure to mate with either of the visible markers. Hence, either the loci responsible for mating behavior are located far away from the visible markers, or the mating behavior is due to several loci scattered over the chromosome, leading to a generalized recombinational breakdown that obscures linkage relationships to known markers.

A standard F_2 generation was also produced by crossing K × S F_1 females with F_1 males. The resulting F_2 sexual females were then test-mated to S males, with the results given in Table 16.5. Unfortunately, in this case no information about the females' X-linked genotype is available from the females' phenotypes because all females receive a sl^+ allele from their F_1 male father. In compensation, it is easier to obtain larger sample sizes with sexual reproduction, and with these larger sample sizes there is a significant association of failure to mate with the vl acrocentric II marker. Hence, there is evidence for detectable, but loose, linkage.

Table 16.4. The numbers of parthenogenetic F_2 females mating and not mating with S males as a function of their genotypes for the X-linked marker sl and the acrocentric II autosomal marker vl

A. Two-locus analysis			
F_2-Im genotype			
sl/vl	$+/vl$	$sl/+$	$+/+$
Mated	17	10	8	10
Unmated	33	25	39	34

Contingency chi-square $= 4.02$ (3 $d.f.$)

B. Single-locus analyses			
F_2-Im genotype		F_2-Im genotype	
vl	$+$	sl	$+$
Mated	27	18	25	20
Unmated	58	73	72	59
Chi-squares (1 $d.f.$)	3.32		.005	

Table 16.5. The number of sexually produced F_2 females that mated and failed to mate with S males as a function of their veinless phenotype

	F_2 female phenotype	
	vl	$+$
Mated	46	153
Unmated	8	101
Chi-square (1 $d.f.$)		12.12

Male Sterility

Given that mating has occurred, various postmating barriers can now play a role in reproductive isolation. One such postmating barrier is male sterility. Table 16.6 shows the results when K × S F_1 males are backcrossed to K females. The various backcross males will all be hemizygous for the K-type X chromosome, and should have the eight different autosomal combinations shown in Table 16.6 with equal probability. These backcross males were then mated to S females, and their genotypes inferred from the phenotypes of their sexual progeny. As shown in Table 16.6, 99 out of 230 males failed to reproduce, and hence their genotypes could not

Table 16.6. The number of fertile and sterile backcross males as a function of autosomal genotype

Backcross genotype	Observed fertile	Expected[a] fertile
$\dfrac{+\ \ v\ \ pm\ \ vl}{Y\ \ +\ \ +\ \ +}$	36	28.75
$\dfrac{+\ \ v\ \ pm\ \ +}{Y\ \ +\ \ +\ \ +}$	0	0
$\dfrac{+\ \ v\ \ +\ \ vl}{Y\ \ +\ \ +\ \ +}$	36	28.75
$\dfrac{+\ \ +\ \ pm\ \ vl}{Y\ \ +\ \ +\ \ +}$	29	28.75
$\dfrac{+\ \ v\ \ +\ \ +}{Y\ \ +\ \ +\ \ +}$	0	0
$\dfrac{+\ \ +\ \ pm\ \ +}{Y\ \ +\ \ +\ \ +}$	0	0
$\dfrac{+\ \ +\ \ +\ \ vl}{Y\ \ +\ \ +\ \ +}$	30	28.75
$\dfrac{+\ \ +\ \ +\ \ +}{Y\ \ +\ \ +\ \ +}$	0	0
Sterile	99	115

[a]Expected values were calculated under the hypothesis that sterility is caused by homozygosity for the acrocentric II autosome from K28-0-1m. Resulting chi-square = 5.94 (4 d.f.).

be inferred. These males were observed courting and copulating with the females; therefore, an examination of their testes was performed after the test period. The testes of these males were not as tightly coiled nor as pigmented as the testes of fertile males. Sperm were present in the testes of these males, but none were motile.

As can be seen from Table 16.6, the fertile males are found only in the four genotypic categories characterized by heterozygosity for the acrocentric II autosome. It is therefore likely that K/K homozygosity on this X chromosomal background causes male sterility. This hypothesis was tested by calculating the chi-square from the data in Table 16.6 under the hypothesis of normal Mendelian segregation and assortment and of sterility being due only to homozygosity for the K-type acrocentric II autosome. The resulting chi-square is 5.94 with 4 degrees of freedom, which is not significant at the 5% level.

The genetic background of these backcross males is fixed for the K-type X chromosome. To investigate the possible role of the X chromosome in male sterility, the crosses given in Figure 16.1 were undertaken. These crosses produce an acrocentric II autosome that with increasing probability is of the K type, against an otherwise S genetic background. Note that the recombination that occurs to introduce the K-type acrocentric II autosome onto the S genetic background is accomplished through parthenogenetic reproduction. Hence, there is no bias for fertile males.

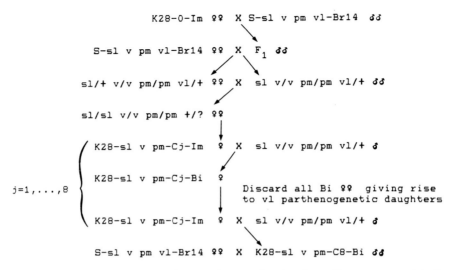

Figure 16.1. The crosses used to produce males that have K-type acrocentric II autosomes and an S-type X chromosome. Slanting arrows and the letters "Bi" (bisexual) indicate sexually produced progeny, whereas vertical arrows and the letters "Im" (impaternate) indicate parthenogenetically produced progeny. Further details are given in the text.

All males from these crosses were completely fertile. In addition, 100 males from the last cross were testcrossed to S females, as indicated in Figure 16.1. All of these males were fertile. Hence, the male sterility requires simultaneous hemizygosity for the K-type X chromosome and homozygosity for the K-type acrocentric II autosome.

Possible associations between the visible markers sl and vl were investigated by crossing the F_2 males with S females. The results are shown in Table 16.7. Only the genotype of the fertile F_2 males could be inferred, and, as can be seen from Table 16.7, there is no significant association between the visible markers and male fertility. Hence, there are either multiple loci per chromosome, or very loose linkage with the visibles. Table 16.7 provides some support for more than one locus per chromosome, as 37 F_2 males were sterile and 87 were fertile. The expected numbers under the hypothesis that sterility is caused by hemizygosity for a single X-linked locus and by homozygosity for a single autosomal locus are 15.5 and 108.5, respectively. The resulting chi-square is 34.08 with 1 degree of freedom, which implies a very strong rejection of the two-locus hypothesis.

Linkage of Female Mating Isolation with Male Sterility

Both female mating isolation and male sterility depend upon interactions between the K-type X chromosome with the K-type acrocentric II autosome. There are two possible explanations for this pattern. First, both

Table 16.7. The genotypes of the fertile F_2 males at the X-linked sl locus and the autosomal vl locus and the total numbers of fertile and sterile F_2 males

	Fertile F_2 male genotype				F_2 male	
	vl/vl	$+/vl$	$+/+$		Obs.	Exp.[a]
sl:	10	29	8	fertile:	87	108.5
$+$:	9	24	7	sterile:	37	15.5
chi-square $= 5.94$ (5 $d.f.$)				Chi-square $= 34.08$ (1 $d.f.$)		

[a]Expected values under the hypothesis that male sterility requires hemizygosity for one locus on the X chromosome and homozygosity for one locus on the acrocentric II autosome.

phenotypes are merely pleiotropic expressions of a single genetic syndrome; second, two different genetic complexes are responsible that happen to be on the same set of chromosomes. If the second explanation is true, the syndromes should be separable through genetic recombination. To determine whether this is possible and to produce a linkage map of the acrocentric II autosome, the crosses shown in Figure 16.2 were performed.

First, K × S F_1 males were backcrossed to K females, and the resulting backcross females were crossed to S males. The only females that mated were $+/vl$ heterozygotes at the acrocentric II autosome. Consequently, recombination between the K- and S-type acrocentric II autosome can occur in these females. The male offspring of the mated backcross females were retained and sorted into two phenotypically distinct categories: vl/vl and $+/vl$. In these males, one vl allele is brought in from the S female parent, and the other, potentially recombinant, acrocentric II autosome is derived from the $+/vl$ mother. Because these males have at least one intact S-type acrocentric II autosome, all are fertile. These males were then crossed with K females. The resulting progeny would inherit one intact K-type acrocentric II autosome from their K mother, and the other autosome would either be an intact S-type or the potentially recombinant autosome inherited from the $+/vl$ backcross females (recall that there is no recombination in *mercatorum* males). If the male parent passed on the intact S-type autosome, the male progeny should be fertile and the female progeny sexually receptive. However, if the male parent passed on the chromosome derived from the $+/vl$ backcross female, we would have a standard testcross for mating and sterility. Because segregation in the male parents cannot be controlled, a minimum of 15 male and 15 female progeny from a single male parent were characterized for sterility and mating behavior, respectively, in order to ensure that true segregation was occurring, as opposed to just a sporadically sterile male or unreceptive female. Because this mapping experiment is so laborious and time-intensive, only 26 chromosomes were so characterized. The results are shown in Table 16.8.

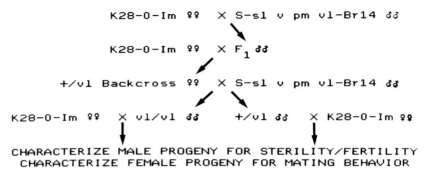

Figure 16.2.   The crosses used to map male fertility and female mating behavior to the visible marker *vl* on the acrocentric II autosome. Further details are given in the text.

Although the sample size is small, the data are consistent with a simple 3-point testcross. Hence, there is no evidence that more than one autosomal locus affects either female mating behavior or male sterility. A second important conclusion is that female mating behavior and male sterility are genetically separable, and indeed they seem to be about 27 map units apart. Consequently, male sterility and female sexual isolation are, at least in part, distinct genetic systems that just happen to be linked, although not closely. Finally, neither male fertility nor female mating isolation is closely linked to *vl*, although mating isolation is the closer of the two. This result is consistent with the previously discussed chi-squares, which show a loose association between *vl* and mating isolation but no detectable association between *vl* and male fertility.

DISCUSSION

Besides the premating barrier of female sexual isolation and the postmating barrier of male sterility, another barrier that isolates K from S is an F_2 breakdown in viability (Templeton, 1979a). This F_2 breakdown is also apparent in the appearance of flies that have morphologically abnormal abdomens *(aa)*. Templeton et al. (1985) have worked out the genetic basis of *aa* and have discovered that it depends upon two closely linked X-linked loci that interact epistatically with each other and with genes on all the major autosomes and on the Y chromsome. The molecular basis of the X- and Y-linked factors has also been revealed (Desalle and Templeton, 1986; Desalle et al., 1986) with one of the X-linked elements corresponding to insertions that go into the coding region of many of the X-linked 28*S* rRNA genes, and the second X-linked element controlling the preferential replication (or lack thereof) of noninserted 28*S* genes over inserted 28*S* genes during the formation of polytene tissue. The Y-linked element corresponds to a deletion of the rDNA cluster on the Y chro-

Table 16.8. Numbers of the various recombinant chromosome types and the resulting acrocentric II linkage map from the mapping experiments involving the visible marker veinless (*vl*), female mating behavior, and male sterility

Inferred genetic state of characterized chromosomes				Number of chromosomes
vl	mated	fertile		4
+	unmated	sterile		11
vl	unmated	sterile		3
+	mated	fertile		1
vl	mated	sterile		0
+	unmated	fertile		6
vl	unmated	fertile		1
+	mated	sterile		0
			Total	26

Inferred genetic map

vl	mating		fertility
	19	27	

mosome. No molecular basis has yet been worked out for the autosomal elements. Interestingly, the major autosomal influence on *aa* maps to the acrocentric II autosome. However, *aa* is recombinationally separable from both female sexual isolation and male sterility, so that although all three of these isolating barriers depend upon epistatic interactions between the X chromosome and the acrocentric II autosome, they are separate genetic systems.

This similarity of linkage relationships between three apparently unrelated isolating barriers could have been due to chance alone, but there may have been some nonrandom factors operating as well. The theory of genetic transilience predicts that founder effects will preferentially involve X-linked elements (Templeton, 1986), and this has certainly been the case in these experimental results. However, this does not explain why the major autosomal components of these three barriers reside on the acrocentric II chromosome, which constitutes only about 25% of the autosomal genome. One reason might be hitchhiking effects. As shown by Templeton (1982, 1983), the transition from sexual reproduction to parthenogenesis selects for females with high fecundity. This same type of selection might occur in natural founder–flush events involving sexual populations as well, because the founding females will find themselves in an unexploited environment that allows for rapid population expansion—conditions that should favor those females that put more energy into reproduction. One of the phenotypic manifestations of certain combinations of the *aa* genes is to increase female fecundity, and there is indeed evidence for direct selection on *aa* during the formation of new parthenogenetic stocks (Templeton, 1983). This would induce particularly strong selection on the acrocentric II autosome, as it is the major autosomal modifier of *aa*. Hence, hitchhiking effects should be more probable with this autosome, and this might explain the similarity of the linkage patterns observed here.

The fact that strong premating and postmating isolating mechanisms arose in this parthenogenetic stock certainly indicates that this extreme type of founder event can trigger the evolution of events that underlie speciation. Indeed, these isolating barriers are so effective that K28-0-Im females continue to reproduce parthenogenetically even in the presence of fertile males, and sexual analogues of K28-0-Im cannot be maintained (Templeton, 1983). Hence, K28-0-Im is a truly parthenogenetic strain that cannot revert to sexual reproduction. This one-way leap from sexual to parthenogenetic reproduction is an excellent example of transilient evolution.

These results do not prove that the less extreme founder events associated with sexual populations in nature could likewise trigger speciation, but the founder-event theory of speciation has passed this test of extremes. What is now needed experimentally is to determine how much one can relax these extreme conditions and still observe the evolution of reproductive isolation. The experimental results obtained to date by Powell (1978), Ahearn (1980), and Arita and Kaneshiro (1979) all support the idea that less extreme founder events can trigger the evolution of reproductive isolation, although in these cases the degree of reproductive isolation was less extreme than that reported here.

Regardless of whether or not these experimental results mimic, albeit in extreme form, naturally occurring founder events, the results obtained with K28-0-Im clearly indicate that this is a successful laboratory model of speciation that is amenable to genetic analysis. Consequently, these studies yield certain insights concerning the genetic basis of newly evolved reproductive isolating mechanisms. The first implication concerns the reality of coadapted gene complexes. All three isolating barriers discussed in this chapter depend upon strong epistatic interactions between loci. In no case could the isolating barrier be attributed to a single locus. Likewise, the isolating barrier could not be attributed to a polygenic complex consisting of many minor genes of additive effect. The fact that the phenotypes map to only a pair of chromosomes, and the fact that only one (female reproductive isolation and male sterility) or two *(aa)* segregating units are observable upon the chromosomes that have been mapped indicate that very few genes are involved. Consequently, this experimental speciation system strongly supports the idea that coadapted gene complexes can provide the genetic basis for speciation, but at the same time it demonstrates that these coadapted gene complexes can consist of only a handful of loci with major phenotypic effect and do not have to involve large numbers of loci scattered throughout the genome.

ACKNOWLEDGMENTS

This chapter is dedicated to Dr. Hampton L. Carson in honor of his 70th birthday. The work reported in this paper traces to Hamp in many ways. I was first introduced to *Drosophila mercatorum* as an experimental organism by Hamp in an undergraduate laboratory course in 1969. While I was a graduate student at Michigan, Hamp kindly provided me with stocks of *mercatorum* for my research, and during a postdoctoral stay with Hamp in 1974, the stocks and parthenogenetic strains discussed in this chapter were isolated with his aid and advice. In addition to the experimental system, Hamp's exciting ideas concerning founder events and speciation provided the intellectual motivation for this study.

This work was supported by National Institutes of Health Grant R01 AG02246. I would also like to thank Brian Charlesworth and an anonymous reviewer for their comments on an earlier draft of this paper.

LITERATURE CITED

Ahearn, J. N. 1980. Evolution of behavioral reproductive isolation in a laboratory stock of *Drosophila silvestris*. Experientia 36:63–64.

Arita, L. H., and K. Y. Kaneshiro. 1979. Ethological isolation between two stocks of *Drosophila adiastola* Hardy. Proc. Hawaiian Entomol. Soc. 13:31–34.

Barton, N. H., and B. Charlesworth. 1984. Genetic revolutions, founder effects, and speciation. Annu. Rev. Ecol. Systematics 15:133–164.

Carson, H. L., and A. R. Templeton, 1984. Genetic revolutions in relation to speciation phenomena: the founding of new populations. Annu. Rev. Ecol. Syst. 15:97–131.

Carson, H. L., I. Y. Wei, and J. A. Niederkorn, Jr. 1969. Isogenicity in parthenogenetic strains of *Drosophila mercatorum*. Genetics 63:619–628.

Carson, H. L., L. T. Teramoto, and A. R. Templeton. 1977. Behavioral differences between isogenic strains of *Drosophila mercatorum*. Behav. Genet. 7:189–197.

Carson, H. L., L. S. Chang, and T. W. Lyttle. 1982. Decay of female sexual behavior under parthenogenesis. Science 218:68–70.

Desalle, R., and A. R. Templeton. 1986. The molecular through ecological genetics of abnormal abdomen. III. Tissue specific differential replication of ribosomal genes modulates the abnormal abdomen phenotype in *Drosophila mercatorum*. Genetics 112:877–886.

Desalle, R., J. Slightom, and E. Zimmer. 1986. The molecular through ecological genetics of abnormal abdomen. II. Ribosomal DNA polymorphism is associated with the abnormal abdomen syndrome in *Drosophila mercatorum*. Genetics 112:861–875.

Powell, J. R. 1978. The founder–flush speciation theory: an experimental approach. Evolution 32:465–474.

Templeton, A. R. 1979a. The unit of selection in *Drosophila mercatorum*. II. Genetic revolutions and the origin of coadapted genomes in parthenogenetic strains. Genetics 92:1256–1282.

Templeton, A. R. 1979b. The parthenogenetic capacities and genetic structures of sympatric populations of *Drosophila mercatorum* and *Drosophila hydei*. Genetics 92:1283–1293.

Templeton, A. R. 1982. The prophecies of parthenogenesis. In H. Dingle and

J. P. Hegmann, (eds.), Evolution and Genetics of Life Histories, pp. 75–101. Springer-Verlag, New York.

Templeton, A. R. 1983. Natural and experimental parthenogenesis. In M. Ashburner, H. L. Carson, and J. N. Thompson (eds.), The Genetics and Biology of *Drosophila,* Vol. 3c, pp. 343–398. Academic Press, London.

Templeton, A. R. 1986. Genetic systems and evolutionary rates. In K. S. W. Campbell (ed.), Rates of Evolution, pp. 218–234. Australian Academy of Sciences Press, Canberra.

Templeton, A. R., C. F. Sing, and B. Brokaw. 1976. The unit of selection in *Drosophila mercatorum.* I. The interaction of selection and meiosis in parthenogenetic strains. Genetics 82:349–376.

Templeton, A. R., T. Crease, and F. Shah. 1985. The molecular through ecological genetics of abnormal abdomen in *Drosophila mercatorum.* I. Basic genetics. Genetics 111:805–818.

Genetic Imbalance, Realigned Selection, and the Origin of Species

HAMPTON L. CARSON

EVOLUTIONARY GENETICS, A CHANGING SUBJECT

In these final decades of the twentieth century, evolutionary geneticists find their world greatly altered by the landmark discoveries of molecular genetics. One of the most fundamental findings is the fact that coding sequences make up only a small part of an individual's DNA. According to Britten (1986), the single-copy gene coding regions, those that are especially subject to the operation of natural selection, amount to only about 5–10% of the total. Reciprocally, the great bulk of the hereditary material thus appears to be very largely a neutral and randomly drifting sink made up of pseudogenes, silent redundancies, introns, repetitive sequences, retrovirus DNA, and transposable elements.

To what extent this DNA functions in the life of the genome harboring is unknown. Igorance of function, however, is not grounds for dismissing all this DNA as unimportant junk; the 5–10% will surely be revised upward as new information about the genome is added, but a large drifting sink will certainly remain. Meanwhile, the sink has turned out to be very useful for phylogenetic evolutionary study. It has spawned a new and fascinating kind of molecular evolutionary genetics that views noncoding DNA as a powerful monitor of the rate of genetic divergence between species with geological time. This new science is making a strong bid to stand alongside paleontology in its capacity to provide insight into certain aspects of the biological past.

THE ROLE OF THE GENETICAL NATURALIST

The population geneticist or ecologist is not concerned with those elements that are genetically neutral. On the contrary, he or she attempts to concentrate on the origin of genetic attributes that fuel the biologically meaningful properties of the different species and the adaptations that serve fitness. Since Darwin's day, speciation and adaptation have been amenable to microevolutionary deductions made from phenomena seen in keenly observed contemporary populations. About 50 years ago, Ford (1931) and Dobzhansky (1937) insisted on the application of genetic techniques to the study of variation within natural species. Their efforts signaled the birth at that time of a new science, the genetics of natural populations (see Lewontin et al., 1981). My own researches, as well as those of many of the illuminating essays in this book, have attempted to build on this Neo-Darwinian foundation.

What does the future hold for this type of evolutionary genetics? With the aid of some of the new molecular techniques, it should become easier to recognize and, accordingly, to concentrate study on that part of the total DNA that is particularly relevant to the important adaptive properties that preoccupy the genetical naturalist. The task is a very large one, and we may find, for example, that the molecular biologist will remain aloof from it as he or she becomes almost exclusively preoccupied with the exquisite genetic details. At a different time and integrative level, this same sort of obstacle to the development of evolutionary genetics appeared shortly after the rediscovery of Mendelism. Accordingly, in the flush of the recent molecular discoveries and techniques, the broader picture of evolution is likely to get relatively little attention until most of the new facts about the structure and function of the genome are in place. These tools now make it possible to dissect and reveal previously hidden details of the genome. For the first time in the history of genetics, crucial techniques for a real understanding the hereditary material are available.

THE ROLE OF QUANTITATIVE GENETICS

Some older ideas about adaptative evolution, however, can already be laid to rest as being oversimplified and naive in view of what has already been learned about the organization of the genome. For example, adaptation in populations is not accomplished simply by selection for, and fixation of, alleles of single genes of individually large effect. All important characters that have been genetically analyzed turn out to have a complex multigenic basis. Although a lead gene may sometimes be identified, the character itself is likely to be strongly molded by many genes having modifying effects. The evolutionist must become increasingly involved with the genetics of quantitative characters, a view very cogently argued, for example, by Lande (1980).

Deemphasis of single genes of large effect, of course, does not apply to the interests of the medical geneticist, who frequently finds that major genes and chromosome aberrations are important etiologic agents in human genetic disease. The effects of such genes, however, are usually strongly affected by the genes of the rest of the total genotype. Here lies an important difference between evolutionary and medical genetics. Natural selection differentially multiplies only those genetic variants that act positively on the fitness of the individual. "Normalizing" natural selection militates against those genes that are deleterious, either removing them completely from the population or driving them down to a rarity close to the frequency of the mutation rate.

THE ONE GENE, ONE ADAPTATION CONCEPT CAN BE REJECTED

In the same vein, the love affair of evolutionary genetics with electrophoresis of soluble proteins has disappointed those who were seduced by the notion that natural selection could accomplish genetic adjustment by simply adding up the effects of a few loci encoding biochemical variants. This is not to say that all biochemical polymorphisms are neutral, but 20 years of data and controversy in the literature suggest that most of the electrophoretically observed allelic variation within the species is, in fact, a part of the neutral sink (Kimura, 1983). Although alleles at these loci are affected by random walk, and although species appear to drift apart genetically with time, there is a startling lack of evidence that relates such loci to the signal adaptive differences so strikingly manifest between species.

Nuclear cytologists also have had their affairs with conspicuous types of variation at the chromosomal level that do not appear to be proximate to important selective differences between populations of closely related species. For example, the heterochromatin of the cytologist consists of blocks of highly repetitive DNA. This material has attracted a lot of attention because it forms conspicuous patterns of variability both within and between species. The functions of this material, as well as that of the various somewhat less repetitive categories of DNA, such as transposable elements and retroviruses, remain unknown.

Perhaps the whole notion of seeking a function for this kind of DNA in the biological economy of the species whose genome bears it is erroneous. Much of this DNA may have been originally extraneous and may have become horizontally associated with that of the "host" species some time during its past history. Following the insinuating event, it could have been perpetuated vertically through one or, indeed, many generations. Such DNA may have its own intrinsic variability systems that may lead the evolutionist to assume that observed genetic variability in these is part of the genetic system of the host. I use the term "host" here with some misgivings since the word implies that such associated DNA is of a defined parasitic nature. There may be some support for such a view, but more facts are needed. A perfect molecular commensal would be one that rides along on the chromosome without harming its host. In this case, the ultimate in efficiency by an associated molecule would appear to be a capacity to integrate into a replicon system such as that of the very chromosomal system itself (see Orgel and Crick, 1980, and Dolittle and Sap-ienza, 1980, for discussion of molecular "selfishness"). These DNA sys-tems, however, are under no obligation to obey the classical postulates of Robert Koch (see King, 1952) for the etiology of a host–parasite relation-ship. For example, there is no necessity for foreign associated DNA to be, at this time level, capable of culture in an existence separate from that of its "host." Mitochondrial DNA might serve as an example. I will leave it to shrewder minds to work out the terminology in the event that new data force some sort of "host–parasite" view upon us.

SELECTION AND BALANCE

A quite different fate, however, has befallen those favorites of the popu-lation cytologist, the chromosome aberrations. Inversions and transloca-tions, for example, do not relate to or underlie biologically important phenotypic characters in a one-to-one relationship. In this regard, they resemble electrophoretic variants and major deleterious genes. On the other hand, they clearly reveal a basic property of the diploid genome: the phenomenon of selective balance. Polymorphic aberrations, such as inversions, are commonly observed to persist in populations, and the evi-dence strongly indicates that, although other explanations are possible, they are held there by Fisher's (1930) phenomenon of heterozygote su-periority of fitness.

The existence of heterozygous balance underlying fitness in natural populations is not confined to systems wherein conspicuously segregating chromosome aberrations occur. Inversions and some heterozygous het-erochromatic blocks may serve to mark an underlying balance rather than themselves be the direct cause of balance. For example, there is strong evidence that the condition of heterozygous balance is the main reason for the persistence in populations of many genes with major deleterious effects, including some lethals, semilethals, subvitals, and sterility genes and gene groups. It is possible that many apparently balanced electropho-retic polymorphisms display this condition because they are included within a block of balanced genes. It is important to stress, however, that the heterosis that appears to maintain single, easily observed genetic vari-ants does not necessarily result from the effects provided by a single locus with an overdominant effect. Rather, the variant under observation is likely to be part of a larger block of genes held in a coadaptive manner by balancing selection.

The theoretical framework just described is essentially that which has formed the principal working hypothesis of modern quantitative genetics. The animal and plant breeder, applying directional selection to members of a finite and effectively small population, is able to proceed by fixing some additive genetic variance. Much of the variation he or she sees, however, is difficult to fix. This suggests that epistasis is exceedingly important and predicts that the systems of balance will spawn dysgenic recombinations as a price for the high fitness of some individuals that are heterozygous for blocks of genes.

The foregoing paragraphs attempt to arrive at the upshot of a very large literature that was keynoted, with somewhat different emphases, by Mather (1949) and Dobzhansky (1955). Almost all important characters of the organism have a complex and polygenic basis (see especially the discussion of Wright, 1982). This view may have been temporarily eclipsed by the remarkable developments of the electrophoresis period. The "electrophoretic view" of the genome may have led many into acceptance of the "classical" theory of population structure, as Dobzhansky (1955) called it. On the other hand, evidence continues to mount that balance is the far more acceptable theory. The individual allelic variants are likely to "hitchhike" with, or participate in, a large heterotic balanced block of genes rather than be balanced as individual loci (see Hedrick, 1980). As Mather proposed in the 1940s, relational and internal balance may well be the major response of the complex genetic architecture to both natural and artificial selection.

THE GENETIC CONTEXT OF SPECIATION STUDY

Virtually all of the above discussion has dealt with the apparent manner in which genetic adjustment within a single interbreeding population is accomplished. Essentially, natural selection is viewed as a process that continually maximizes Darwinian fitness, using as its raw materials new mutations and especially recombinational variants that are continually being generated in a sexual population. A set of processes are involved that are very similar to those employed artificially by the animal and plant breeder. Thus, the genetical characteristics of captive or field-plot populations are manipulated with artificial selection in a manner directly comparable to what happens in nature under natural selection.

But what of the origin of species? Can a new species be produced by manipulation of populations by the animal and plant breeder or the experimental evolutionist? How good are the empirical models that attempt to mimic speciation? In attempting to answer such questions, much depends on the definition of a species that is adopted. In turn, this depends on the perspective of the worker making the definition. I admit to a strong genetic bias. A number of genetically inclined biologists have advanced definitions that take into late account the generation of genetic variability in populations and are thus relevant to the evolutionary point of view. These genetical concepts are usually not useful to taxonomists engaged in cataloging each specimen that is collected.

Before proceeding further, I would like to review very briefly two of the classical genetically oriented definitions of the species. I favor a simple definition that can be put together, in slightly altered form, from one of Dobzhansky's discussions (1950):

A species is "the largest and most inclusive reproductive community of sexual and cross-fertilizing individuals that share in a common gene pool."

In this statement, I find a welcome stress on the species as a field for genetic recombination since the ability of the organism to adjust to environmental changes depends on this gene pool. Later in the same 1950 paper, however, Dobzhansky insisted on including another criterion, namely the reproductive isolation of such an "inclusive community." This latter point served earlier as the focus of Mayr's (1942) well-known formulation:

"Species are groups of actually or potentially interbreeding natural populations, which are reproductively isolated from other such groups."

Species definitions invoking reproductive isolation have made their way into most textbooks and have been widely adopted by a whole generation of evolutionists. For reasons to be developed later in this chapter, I feel that the stress on reproductive isolation is misplaced and is neither necessary nor desirable in a genetic–evolutionary definition of the species.

The Dobzhansky definition cited above, which omits mention of isolation, has much to recommend it. Some workers have criticized it since the term "gene pool" has an aura of jargon about it. On the contrary, I contend that most observant naturalists are capable of, and routinely do, recognize the gene pool of the species. Indeed, this is the basis for our ability to recognize the species as an objective natural entity, unlike the higher levels of taxonomic distinction. What a naturalist and species taxonomist sees, in effect, are the normal boundaries of the interbreeding and individually variable group, as gauged by observed morphological, geographical, and ecological criteria.

Perhaps the reader might permit a partially anecdotal exploration of the above idea. When in the Bahamas collecting a species of land crab, *Gecarcinus ruricola* (locally known as "black crab"), I was fortunate to have the assistance of Mr. Pedro Romer, a legendary and highly accomplished West Indian naturalist, employed at the Lerner Laboratory on Bimini. Romer had never had formal training in biology. With his help, I collected about 50 specimens of this species at one locality on a small islet. The phenotypic variation in this sample of specimens was as prodigious as anything I had ever seen before or have since. There was polymorphic color variation from purple to yellow, to gray, to orange, and to red. Many were patterned with white or colored penciled markings etching the background color. Some had brightly colored chelipeds contrasting with the rest of the body. Relatively few were black with small red markings. Sexual dimorphism and size variation was superimposed on all of this. I raised the question with Romer as to whether all these were really the species I was after. He looked at me with the pleasantly disdainful incredulity of an expert familiar with his subject.

"Look, mon" he said emphatically, "these are all black crab." The gene pool of the species was no problem for him.

I had much the same experience on a field trip to the Okefenokee Swamp in Georgia, where I was guided by several members of a family of local naturalists—Mr. Wade Hampton Chesser, his son, Tom, and his grandson, Wade. All three generations had grown up in the swamp and had no formal education in biology. Yet there was not one life-history stage of any species, plant, or animal, that I could point to, even including the puzzling differences between the seedlings of the different pine tree species, that they had not already recognized and for which they had a vernacular name. Like Romer, their species taxonomy was impeccable; they were a living monograph of the biology of the swamp. Nevertheless, when I remarked on this, Tom was apologetic: "We really don't know what these things are; these are just family names that we use."

In view of these personal experiences I was particularly interested in Diamond's (1966) more formal account of the taxonomic acuity of the Fore people of highland New Guinea. For just the vertebrate species alone, Diamond recorded 182 local names. He found an almost perfect one-to-one correspondence with the species as recognized by European taxonomists, including recognition of some particularly confusing ones.

What is being observed, by both amateur and professional naturalists alike, are the variable populations of interbreeding organisms. Technical, detailed examination, of course, will reveal that some of these species are a little ragged or "leaky" at the edges in that a certain amount of hybridization occurs. This has been recorded very widely in nature in both animals and plants, including some of the birds of paradise in New Guinea. These cases do not fool a Fore hunter or other good naturalists, but what is more important, they rarely result in the disintegration of the gene pool of the species. Indeed, hybrids, as plant biologists have long recognized, are the exceptions that serve to prove the rule of the reality of species gene pools in nature. I believe that our genetic definitions of a species do not have to be crafted in a complicated manner in order to help some taxonomist decide the biologically trivial question as into which museum drawer a few hybrid specimens should be put. To the questions posed earlier, I may suggest that the formation of species, unlike phyletic adaptive responses, has not yet been successfully approached artificially in the laboratory. This is not to say that it can never be done, but up to the present time empirical models have failed to convince anyone, and especially the theoretical models based on the origin of reproductive isolation likewise leave much to be desired.

THE ORGANIZATION THEORY OF SPECIATION

I view species formation as a set of population processes involving a stochastic phase that changes population structure followed by a powerful realigned selective phase. As the latter is phyletic, and as we know something about it, I believe most experimental attention must be given to the stochastic phase that sets the stage for selection. The stochastic phase appears to begin in a population that has been divided off from a parent population by a process of spatial separation; that is, a daughter population is formed that has somehow become allopatrically separated. The rules that may apply to such population sundering are expected to be very different from what occurs during intrapopulational phyletic evolution. Curiously, relatively little attention has been paid to the question, what genetic events may transpire during the formation of populations that appear to have the capacity of later becoming incipient species? As the title of the present chapter indicates, I center the discussion here on a theory of speciation that calls for the new population to undergo drastic changes in the organization of its patterns of quantitative inheritance.

The earlier discussion of the organization of the genome into a series of multigenic fitness monitors based on balanced polymorphism is crucial to what I have termed the organizational theory of the origin of species. In several recent theoretical papers (Carson, 1982, 1986a), I have recast some of my earlier ideas about processes that lead to the formation of species. Some theorists (e.g., White, 1978) have listed very many ways in which species might possibly arise. In contrast, my approach has been to identify those populational events that appear to be common to the process in its various expressions. To this end, strong emphasis has been placed on empirical data rather than on theoretical constructions. In particular, an attempt has been made to examine data from both animal and plant examples, so as to promote the unification of theory in a wide biological context (Carson, 1985a).

The general conclusion that I have reached can be stated rather simply. The population of an incipient species arises from a subpopulation of an existing species. This subpopulation is rendered unusual in that its gene pool has been forced to undergo successively, two population phases that have the result of altering its genetic composition rather drastically. The initial phase is a stochastic *disorganization* of a number of the balanced polymorphisms that characterize the gene pool of the parent species. This effect is largely imposed by the interaction of chance environmental factors with random meiotic recombination. This phase is followed by a *reorganization* directed by selection. This results in the building of polymorphisms differing from those of the parent species. These result not only from further recombination but also from novel avenues of selection. This latter process serves to maximize Darwinian fitness around new adaptive modes and in the process spawns a new and unique gene pool. Following the second phase, the population may emerge as a phenotypically recognizable new species with new characters that serve fitness in a novel adaptive manner.

The disorganization phase basically consists of a series of destructive recombinational events, and it is proposed that this occurs in a deme following a founder event, a reduction of a parent population to a vestige, or a series of recombinational episodes conditioned by interspecific hybridization. If, for some reason, a population is too tightly organized to be able to undergo such a cycle, then it may be judged as incompetent for speciation. Many contemporary species appear to be in this category.

The reorganization phase can be considered as an evolutionary response to one of two major types of environmental challenge. Two modes may be recognized: mode A, the construction of an altered system closely proximate to sexual reproduction, or mode B, a novel adaptive system that relates to some facet of the ambient (i.e., nonsexual) environment. In particular, I have recently suggested (Carson, 1985a) that the new population will respond to one or the other of these challenges but not both simultaneously. The characters developing from the initial selective mode, whether it be A or B will be diagnostic of the incipient species.

Restating the theory, it holds that the speciation process consists of two phases: an initial stochastic disruptive phase, followed by a directional phase characterized by selection impinging on a variable and initially somewhat disrupted gene pool. With regard to the actual phenotypic syndrome that comes to be recognized as a new species, the directional phase is of crucial importance. The discussion in this chapter will emphasize this latter phase. Particular attention will be given to the identification of the mode of selection that comes to be applied naturally to the genetically disorganized deme. Before entering this discussion, however, I will list a number of constraints that I believe these considerations place on certain classical theories of species formation. These alterations in my thinking have been outlined and the arguments for them given in detail in my 1982, 1985a, and 1986a papers. As a background to the present discussion of selective modes, I merely restate these points below without further detailed discussion.

1. The species is a field for genetic recombination and change.

As discussed earlier, the species concept adopted here is dynamic and evolutionary in orientation. It stresses the importance of viewing the species as a set of partially connected sexually reproducing, cross-fertilizing populations. The most important of these recombinational systems are diploid. Specifically excluded are apomictic and agamic complexes, self-fertilizing species, and, in considerable measure, nondiploidized polyploids. These latter genetic systems are omitted from consideration on the grounds that the generation of a field of genetic recombinational variability in these cases is impaired to a considerable degree, such that they are not capable of proliferation of new species that can continue the speciation cycle.

This species concept will not serve the systematicist who wishes to retain the word "species" as a cataloging device for organic diversity and who adheres to a basically morphological concept of the species.

2. Evidence indicates that most speciation is allopatric.

Despite the tempting plausability of many of the schemes of nonallopatric speciation, evidence that they play an important role in nature is meager (Futuyma and Mayer, 1980; Paterson, 1981). The existence of multiple morphs in a natural population is more a reflection of elaboration of balanced polymorphism within a population rather than speciation, a process that requires separation and independent evolution of populations.

3. If allopatric speciation is the rule, then reproductive isolation is a byproduct of differential adaptation rather than a major cause of speciation.

Because most allopatric populations are out of contact with one another, the origin of reproductive isolation by natural selection (reinforcement) is unlikely to play a key role in the origin of species. Reproductive isolation may be long delayed in some instances. Thus, although many related species may show reproductive isolation, this does not mean that the currently observed isolation between them was important at the crucial incipient stages. Many authors (e.g., Nei et al., 1983) appear to oversimplify by reducing speciation process simply to the origin of reproductive isolation.

4. Reproductive isolation is an unsatisfactory criterion for recognizing incipient species.

The "isolation concept" of the species (usually referred to as "the biological species concept") should yield to a broader "adaptive concept" that recognizes that a new species comes to center about a novel and, indeed, unique adaptive mode. This adaptive mode is usually, but not exclusively, reflected in morphological change. Behavioral, chromosomal, or physiological changes that are morphologically cryptic appear very widely in newly formed species. The novel adaptive mode may be wholly concerned with behavioral characters related to sexual reproduction—for example, those that result from sexual selection. These characters, arising intrapopulationally, may have the incidental effect of isolating but are not themselves the result of "selective reinforcement."

5. *Positive selection against interspecific hybridization is probably very rare in nature.*

The widespread use of the word "mechanisms" with regard to characters that have the effect of biological isolation (i.e., "isolating mechanisms") implies that there has been positive selection for "devices" that "protect" the species and its adaptations from disintegration by hybridization. There is, however, very little evidence that hybridization is a serious destructive force in nature. In other words, a species that follows by selection its own unique adaptive norm can support substantial hybridization without loss of integrity. Indeed, introgression can be a source of increased genetic variance, some elements of which can serve the sharpening of adaptation.

6. *The gene pool of a species is complexly organized and balanced.*

Summarizing 20 years of detailed empirical investigations of the genetics of natural populations, Dobzhansky (1955) wrote that "the adaptive norm is an array of genotypes heterozygous for more or less numerous gene alleles, gene complexes and chromosomal structures." The extensive genetic variability discovered later with electrophoretic techniques, and even more recently with genic analyses at the DNA level, bears out Dobzhansky's view. As emphasized earlier in this chapter, balanced polymorphism must be accorded a central position. In normal outbred populations, homozygotes for the heterotic genes and gene complexes involved occur only in a minority of individuals and render these individuals more or less inferior to the norm in fitness. It is this basic polygenic balanced organization, the segregation of gene blocks, that gives coherence to the gene pool of a species over the long term. With time, these balances may become a virtually obligatory closed genetic system such that recombination from within these genomic sections produces gene combinations of lowered fitness (see Carson, 1975). Natural or artificial selective adjustments within the species are possible without perturbing the basic balanced system. Thus, genetically distinct intraspecific ecotypes, minor local adaptations, subspecies, clines or artificially altered selection lines or subpopulations are possible without speciation. In the present view, these populations could become species only after a major disorganization event.

THE MAXIMIZATION OF FITNESS

As Darwin so strongly emphasized, the greatest impact of selection is with regard to the slight reproductive edge one individual organism has over another. The concept of Darwinian fitness later became a ruling theorem of population genetics. Adaptations, expressed in individual gen-

otypes, emerge as palpable entities only gradually, and are built generation by generation. To understand the origin of an adaptation, one must
first understand the fitness concept that underlies it.

Various arguments, exposed prominently in the literature, tend to
obfuscate the overriding importance of the genetically unique, sexually
reproduced individual in the selection process. A number of paleontolo-
gists (e.g., Gould, 1980; Vrba and Eldredge, 1984) have adopted a macroevolutionary stance that describes natural selection as a hierarchical
system. In addition to individual selection involving the genetically
unique individual, they variously recognize selection among families,
groups, demes, and even species (see Stanley, 1975). It is not necessary
to deny that some degree of selection can occur at these higher levels in
order to aver that there is only one level in nature at which selection is
really given a prominent field of variability on which to act. This is made
possible only by the generation of large numbers of individual genotypes
in a sexual population. Selection at any other level is impotent to effect
evolutionary change relative to what can happen within a single sexual
population. Sewall Wright made this point eloquently in 1932, and it is no
less true more than a half-century later.

QUANTUM-PHASE SELECTION

Selection as a maximizing process for individual fitness does not condemn
the evolutionary process as it occurs in sexual populations to be utterly
slow and imperceptible. On the contrary, there are periods in the life of a
geographical population when genetic change may be exceedingly rapid
over a relatively small number of generations. The proximate causes of
such a rapid phase can be either environmental or genetic or both. They
are easily modeled empirically in experimental populations (see Carson,
1958, for rapid genetic changes induced in a population; and Crow,
1954, for the genetic consequences of rapid, directional environmental
changes).

In both natural and artificial populations, however, such phases of
quantum selective change are probably very rare indeed. Recognition of
a transient but active selective phase becomes an important and difficult
matter, especially in natural populations where these events may be
masked by the way the data are handled (see Samollow and Soulé, 1983).

The net result is that a population tends to run along, generation by
generation at an equilibrium state, with little phenotypic or genotypic
change. The genetic variability, however, is maintained by balanced heterotic systems wherein directional change is essentially lacking. Another
curious result is that individual fitness differences among members of the
population are temporarily rendered insignificant as they are absorbed
into the equilibrium system. In the case discussed by Samollow and

Soulé, it was the imposition of a novel environmental stress that set directional genetic change in motion.

To the population geneticist interested in significant evolutionary change to the point of species formation, it becomes supremely important to seek out and study the events that transpire during quantum phases of change. Unfortunately, many studies of selection in natural populations, including many of my own on inversion polymorphisms (e.g., Carson and Stalker, 1949), deal with populations in a steady, balanced, and basically unchanging state. Although there were some minor frequency changes, the basic species-wide, multiple-inversion condition of inversion polymorphism in *Drosophila pseudoobscura*, which was first recorded in Dobzhansky in 1944, had not changed three decades later (Anderson et al., 1975).

DOES PUNCTUATED EQUILIBRIUM HAVE AN ANALOGUE IN POPULATION GENETICS?

The notion of punctuated equilibrium as an explanation of certain phenomena displayed in the fossil record has been eloquently argued by Gould and Eldredge (1977 and later). It will be obvious to the reader that what has been described above as balance and directionality in natural populations appears to be at least analogous to the idea of Gould and Eldredge, with the balanced state (equilibrium) interrupted by brief periods of directional change (punctuation). This is certainly true, but many cautions are necessary. At the level of population genetics, no macromutations or mutations leading to abrupt alterations affecting key developmental processes need be assumed or invoked. Rather, shifts in extant polygenic variability and balance over as few as 10–20 generations are wholly adequate to explain the events of the quantum phase (for modeling experiments, see Carson, 1958, and Cannon, 1963). Rapid change is explicable within the context of shifts in Darwinian fitness in sexual populations. Although the change in phenotype may be large over the relatively small number of generations required, the change is wholly within the purview of Darwinian microevolution. Darwinian constructs and basic population genetic principles deduced from contemporary populations are alive and well in evolutionary thinking. Novel macroevolutionary principles and developmental changes adduced as arguments for punctuated equilibrium have taken the form of unfortunate hyperboles. These and the use of a man of straw called "Darwinian gradualism" have clouded the issues and have been responsible for the very poor quality of the debate over punctuated equilibrium. Despite the wholly different data bases, there may yet be grounds for reconciliation of paleontological and population genetic views in this matter (for further discussion, see Carson, 1986b).

GENETIC REORGANIZATION AFTER DISORGANIZATION: THE ROLE OF QUANTUM SELECTION PHASES

Carson and Templeton (1984) have recently considered the founder effect in detail as an event leading to the disorganization of the normal balanced system characteristic of steady-state populations. Accordingly, these arguments will not be repeated here except to emphasize that founder events are surely exceedingly rare in most biological situations. Recognition and identification of the time at which these events occurred may be difficult in species that dwell in modern complex continental ecosystems, especially those of the tropics. The implication is that many speciation events that occurred long ago in past geological epochs were equilibrated genetically at an early time and have remained so since. Thus the majority of continental species may not be undergoing speciational changes at all at the modern time level; they may merely represent an equilibrium or long-lasting steady-state stage. In view of such a possibility, the manifest neoevolutionary events apparent on such geologically recent archipelagos as Hawaii acquire special importance.

THE SEQUENCE OF SELECTIVE MODES IN THE REORGANIZATION PHASE IN NEOSPECIES

When the genetic structure of a population is perturbed, the opportunity arises for a quantum selection phase to subsequently take the population in a new direction. In my 1985a paper, I suggested that the quantum phase that leads in the direction of a new species is likely to affect one of two quite separate selective modes, or syndromes of characters. These are what I called characters leading to an alteration of the details of the sexual process itself (mode A), or characters leading to alterations in the reaction of the organism to the ambient environment (mode B).

The evidence indicates that selective reorganization may proceed by mode A or B but not by both simultaneously. It appears that selection follows the most crucial immediate path toward maximizing individual fitness, so that either reproductive or ambient characters tend to be added sequentially, not synchronically. This means that the incipient stages of speciation in any one case tend to be characterized by only one of the two syndromes of characters. Once a new mode is established, the population will be free to add characters from the other mode by the well-known processes of intrapopulational phyletic evolution. The older well-established species, therefore, ultimately comes to have a mosaic of both kinds of characters. For this reason, it is usually not possible in an older species to interpret which set of characters was established first.

To examine the question of mode sequence, it is useful to identify groups of closely related species that give evidence of being newly evolved (neospecies). In the 1985a paper, I have listed a series of examples of which only a few will be mentioned here. One important implication of this discussion is the fact that mode A (characters specifically pertaining to reproduction) often is in the initial sequential position in animal speciation. This appears to result from the large superstructure of behavioral, morphological, and physiological characters that become associated with sexual reproduction in animals. Mate choice and sexual selection are prominent features of animal systems. Such shifts obviously easily lead to incompatibility between the older and the newer mating system.

Conversely, plants, especially diploid flowering plants, often appear to differentiate initially by mode B. Changes in adaptation to the ambient environment do not result in mating incompatibilities or complexities comparable to those mentioned above for animals. Thus, in such groups of organisms (animal as well as plant) hybridization, both natural and artificial, is a common result.

INITIAL SEXUAL ADAPTATION: AN EXAMPLE

In some organisms—for example, in certain phylads of Hawaiian *Drosophila* species—the quantum phase of selection following the founder event appears to bring about a shift in sexual selection. Adaptation to food, light, temperature, oviposition site, etc. appears to be unaffected in the early quantum selection stage. Lineages of closely related picture-winged species, such as the planitibia species cluster (*differens*, Molokai; *planitibia*, Maui; *heteroneura* and *silvestris*, Hawaii) are differentiated almost wholly by morphological and behavioral secondary sexual characters. In certain genetically analyzed cases (e.g., *silvestris*), these characters show genetic variation both within and between populations. Further, they give evidence of subserving a powerful sexual selection that has as its basis some kind of balanced polymorphism (see especially Carson, 1985b).

Accordingly, it appears that, in this series of speciations, the successive postulated founder events have tended each time to disorganize one, and only one, set of genetically based characters, namely those having to do with sexual selection. Each time a new syndrome of sexual selection is built up in a species, it seems rather more vulnerable, compared with the older balanced systems, to be reorganized following the founder event. Thus this phylad of species and possibly many other Hawaiian *Drosophila* show a sort of evolutionary momentum that is manifested in continual speciational change involving the sexual system.

MOMENTUM IN SPECIATION

"Momentum" is a dangerous term to use, as it may seem to imply the operation of some vague inner orthogenetic force as the basis of a phylogenetic trend. Nothing could be further from my intention. I contend that the maximization of fitness in these populations, periodically disrupted by founder events, is served by the continual and resharpening of high-fitness mate choice via continuing sexual selection.

As I have argued in detail (Carson, 1985a), in some phylads, especially in plants, a comparable set of founder events appears to be followed by a tracking of the ambient environment rather than involving shifts in the sexual reproduction system. Thus the exuberant speciation of the plant genus *Epilobium* on the geologically new soils, slopes, and volcanic exposures in New Zealand produces species with syndromes of characters related to the ambient environment (Raven and Raven, 1976). The sexual reproduction system remains largely unaltered, and hybridization occurs both widely in nature and is easily accomplished in the laboratory.

FUTURE EMPIRICAL WORK IN EXPERIMENTAL EVOLUTION

In conclusion, I would like to point out what may be a redeeming feature of this rather speculative view of the speciation process. Much of the argument depends on stochastic disorganization followed by directional quantum-phase reorganization. Both of these processes can be studied in contemporary fast-breeding organisms that can be handled experimentally in laboratory animals or the plant field-plot. Most experiments in artificial selection have been carried out on organisms or populations chosen for study for various practical reasons such as ease of culture or economic importance. Furthermore, few attempts have been made to select carefully a competent species and to disorganize purposely its gene pool and then apply novel directional selection to the disorganized elements. Moreover, a conscious choice of selective mode as outlined here has almost never been done. The concepts developed in this chapter can indeed be empirically tested, a point that might help resolve uncertainties in the above arguments. Discoveries of molecular genetics alluded to in the introduction may provide a basis for our ability to recognize when it has been possible truly to disorganize the genome by experimental means. Thus, molecular precision could be added to the breeding methods employed. In this and the next century, evolution, the great pervading law of life, can and should be studied intensively in contemporary populations.

LITERATURE CITED

Anderson, W., Th. Dobzhansky, O. Pavlovsky, J. Powell, and D. Yardley. 1975. Genetics of natural populations XLII. Three decades of genetic change in *Drosophila pseudoobscura*. Evolution 29:24–36.

Britten, R. J. 1986. Rates of DNA sequence evolution differ between taxonomic groups. Science 231:1393–1398.

Cannon, G. B. 1963. The effects of heterozygosity and recombination on the relative fitness of experimental populations of *Drosophila melanogaster.* Genetics 48:919–942.

Carson, H. L. 1958. Increase in fitness in experimental populations resulting from heterosis. Proc. Natl. Acad. Sci. USA 44:1136–1141.

Carson, H. L. 1975. The genetics of speciation at the diploid level. Am. Natur. 109:83–92.

Carson, H. L. 1982. Speciation as a major reorganization of polygenic balances. *Mechanisms of Speciation,* pp. 411–433. C. Barigozzi. Alan R. Liss, New York.

Carson, H. L. 1985a. Unification of speciation theory in plants and animals. Syst. Bot. 10:380–390.

Carson, H. L. 1985b. Genetic variation in a courtship-related male character in *Drosophila silvestris* from a single Hawaiian locality. Evolution 39:678–686.

Carson, H. L. 1986a. Sexual selection and speciation. In S. Karlin and E. Nevo (eds.), Evolutionary Processes and Theory, pp. 391–409. Academic Press, London.

Carson, H. L. 1986b. Population genetics, evolutionary rates and Neo-Darwinism In K. S. W. Campbell and M. F. Day (eds.), Rates of Evolution, pp. 209–217. Allen & Unwin, London.

Carson, H. L., and H. D. Stalker. 1949. Seasonal variation in gene arrangement frequencies over a three-year period in *Drosophila robusta* Sturtevant. Evolution 3:322–329.

Carson, H. L., and A. R. Templeton. 1984. Genetic revolution in relation to speciation phenomena: the founding of new populations. Annu. Rev. Ecol. Systematics 15:97–131.

Crow, J. F. 1954. Analysis of a DDT-resistant strain of *Drosophila.* J. Econ. Entomol. 47:993–998.

Diamond, J. 1966. Zoological classification system of a primitive people. Science 151:1102–1104.

Dobzhansky, Th. 1937. Genetics and the Origin of Species. Columbia University Press, New York.

Dobzhansky, Th. 1944. Chromosomal races in *Drosophila pseudoobscura* and *Drosophila persimilis.* Carnegie Inst. Wash. Publ. 554:47–144.

Dobzhansky, Th. 1950. Mendelian populations and their evolution. Am. Natur. 84:401–418.

Dobzhansky, Th. 1955. A review of some fundamental concepts and problems of population genetics. Cold Spring Harbor Symp. Quant. Biol. 20:1–15.

Dolittle, W. F., and C. Sapienza. 1980. Selfish genes: the phenotype paradigm and genome evolution. Nature 284:601–603.

Fisher, R. A. 1930. The Genetical Theory of Natural Selection. Clarendon Press, Oxford.

Ford, E. B. 1931. Mendelism and Evolution. Methuen, London.

Futuyma, D. J., and G. C. Mayer. 1980. Non-allopatric speciation in animals. Syst. Zool. 29:254–271.

Gould, S. J. 1980. Is a new and general theory of evolution emerging? Paleobiology 6:119–130.

Gould, S. J., and N. Eldredge. 1977. Punctuated equilibria: the tempo and mode of evolution reconsidered. Paleontology 3:115–151.

Hedrick, P. 1980. Hitchhiking: a comparison of linkage and partial selfing. Genetics 94:791–808.

Kimura, M. 1983. The Neutral Theory of Molecular Evolution. Cambridge University Press, Cambridge.

King, L. S. 1952. Dr. Koch's Postulates. J. Hist. Med. 7:330–361.

Lande, R. 1980. Genetic variation and phenotypic evolution during allopatric speciation. Am. Natur. 116:463–479.

Lewontin, R. C., J. A. Moore, W. B. Provine, and B. Wallace (eds.). 1981. Dobzhansky's Genetics of Natural Populations, I–XLIII. Columbia University Press, New York.

Mather, K. 1949. Biometrical Genetics. Methuen, London.

Mayr, E. 1942. Systematics and the Origin of Species. Columbia University Press, New York.

Nei, M., T. Maruyama, and C-I. Wu. 1983. Models of evolution of reproductive isolation. Genetics 103:557–579.

Orgel, L. E., and F. H. C. Crick. 1980. Selfish DNA, the ultimate parasite. Nature 284:604–607.

Paterson, H. E. H. 1981. The continuing search for the unknown and the unknowable: a critique of contemporary ideas on speciation. South Afr. J. Sci. 77:113–119.

Raven, P. H., and T. E. Raven. 1976. The genus *Epilobium* (Onagraceae) in Australasia: a systematic and evolutionary study. N. Z. Dept. Sci. Ind. Res. Bull. 216:1–321. Christchurch, Bascands, Ltd.

Samollow, P. B., and M. E. Soulé. 1983. A case of stress related heterozygote superiority in nature. Evolution 37:646–649.

Stanley, S. M. 1975. A theory of evolution above the species level. Proc. Natl. Acad. Sci. USA 72:646–650.

Vrba, E. S., and N. Eldredge. 1984. Individuals, hierarchies and processes: towards a more complete evolutionary theory. Paleobiology 10:146–171.

White, M. J. D. 1978. Modes of Speciation. W. H. Freeman, San Francisco.

Wright, S. 1932. The roles of mutation, inbreeding, crossbreeding and selection in evolution. Proc. VIth Int. Cong. Genet. 1:356–366.

Wright, S. 1982. Character change, speciation and the higher taxa. Evolution 36:427–443.

Index

selective advantage conferred on
humans by, 306

Nair, P. S., 16
National Drosophila Species Resource
Center (Bowling Green, Ohio), 260,
278
National Science Foundation (NSF),
funding Hawaiian *Drosophila*
research by, 24–25, 26
Natural Selection. *See* Selection
Nebraska, 16
Neo-Darwinism, 50, 227, 228–29, 346
Neospecies, 358–59
Network gene, 11–12
Nevo, E., 216
New Guinea, 16, 351
New Zealand, 360
North Africa, 210
Novelties, evolutionary, 67

O'Donald, P., 282
Ohno, S., 229
Ohta, A. T., 285, 286, 317, 318, 319, 321
Oncogenes, 303
Oregon, 124
Ornduff, R., 119
Ovipositional behavior, research on
Grimshawi species, complex, 316–19

Paigen, K., 186
Panaxia, 64
Parsons, P. A., 265
Parthenogenesis, 329, 332, 334–35, 341
Parthenogenetic isolates, 329–42
Partula, 103
Paterson, H. E. H., 280
Peak shifts, 44, 60, 61–62, 63, 218, 221.
See also Shifting balance theory
Pennsylvania, University of, 4, 6, 7, 13
Peripatric speciation, 205–30, 322. *See
also* Speciation
Peripheral populations. *See also*
Populations
founder origin and, 49–50
genetic revolution and, 47
genetic study of inversions among, 16–
17
modes of speciation among, 205–30
Phosphoglucoisomerase (PGI), 90, 92
Phosphoglucomutase (PGM), 90
Phylogenetics, gene regulation and, 193–
98

Phylogenic Inference Package
(Felsenstein), 196
Phylogeny
founder effect and, 176–78
of Hawaiian Drosophilidae, 131–33,
132, 134
Physiology, among species of Hawaiian
Madiinae, 88–90
Plants, founder events, role in speciation
of, 79–95, 99–111, 113–24
Pleiotropic balance, 323
Polyacetylenes, 105–6
Polymorphisms, 133, 186, 195, 357
climatic factors' correlation with
isolates/semi-isolates, 206, 210, 212–
13, 230
level in isolates/semi-isolates, 223–24,
230
private, 308–9
regulatory diversity and, 184, 186, 190
Population
crashes, 286
differential fertility's effect on levels of,
305–6
fluctuation, 54–55
founders' origin from area of species,
49–50, 65
genetics, 10, 22, 182, 190, 357
structure conducive to speciation, 62–
63
Population bottlenecks. *See* Bottlenecks,
population
Population flush. *See* Founder-flush
theory
Populations. *See also individual
population types*
genetic cohesion of natural (parent), 63
genetic distances and genic
differentiation between continuous/
isolated/semi-isolated, 213–16
island, 315–24
laboratory *Drosophila*, 239–49
mutation process in small isolated, 45–
46
peripheral, 15, 16–17, 45, 49–50, 205–30
regulatory diversity in natural, 185–90
speciation and isolating mechanisms
between, 315–24
Postmating barriers
in *D. mercatorum* experiments, 330,
336–38, 340–42
male sterility as, 336–38
premating isolation in the absence of,
239–40, 241–43
Potentilla, 118, 119, 120